Palgrave Studies in Emerging Risk Management and Sustainable Finance

Series Editors
Thomas Walker, Concordia University, Montreal, QC, Canada
Jane McGaughey, Concordia University, Montreal, QC, Canada

This series explores the intricate intersection of artificial intelligence (AI), sustainability, and emerging risk management. It delves into the pressing challenges and opportunities that arise from the rapid advancements in AI technologies, the crucial need for sustainable finance practices, and the complex landscape of emerging technological, environmental, and sociopolitical risks in our ever-changing world. Each book in the series features insights from leading academic researchers, industry experts, and regulators and incorporates real-world case studies that illustrate successful strategies and cautionary tales. The series encompasses topics that are poised to become global concerns ranging from AI, quantum computing, sustainable finance, and emerging risks, whilst addressing contemporary concerns such as the debt crisis, rising inflation, the energy crisis, global warming, and geopolitical tensions. The series embraces an interdisciplinary and holistic approach, bridging the gap between technology, environmental science, finance, and risk management.

The series addresses the growing need for accessible, reliable, and up-to-date information on the intricate relationship between AI, sustainability, and emerging risks. It caters to individuals and organizations seeking to navigate the complexities of these interconnected issues, fostering a deeper understanding of how to harness the potential of AI, promote sustainability, and proactively manage emerging risks for a more secure and sustainable future.

Thomas Walker · Aphrodite Salas ·
Moein Karami · Akram Sadati
Editors

Wiring the Future

Financial Strategies, Challenges, and Opportunities in the Sustainable Energy Transition

palgrave
macmillan

Editors
Thomas Walker
Department of Finance
Concordia University
Montreal, QC, Canada

Aphrodite Salas
Department of Journalism
Concordia University
Montreal, QC, Canada

Moein Karami
Department of Finance
Concordia University
Montreal, QC, Canada

Akram Sadati
Department of Finance
Concordia University
Montreal, QC, Canada

ISSN 3059-2437 ISSN 3059-2445 (electronic)
Palgrave Studies in Emerging Risk Management and Sustainable Finance
ISBN 978-3-032-14716-5 ISBN 978-3-032-14717-2 (eBook)
https://doi.org/10.1007/978-3-032-14717-2

This Palgrave Macmillan imprint is published by the registered company Springer Nature Switzerland AG
The registered company address is: Gewerbestrasse 11, 6330 Cham, Switzerland

Acknowledgements

We acknowledge the financial support provided through the Jacques Ménard—BMO Centre for Capital Markets at Concordia University. In addition, we appreciate the excellent copy-editing and editorial assistance we received from Victoria Kelly, Simone Donders, Hanna Murray, Alison MacLean, Casey Jussaume, and Simon Malépart.

Contents

Notes on Contributors

Petri Ahokangas received his D.Sc. degree (1998) from the University of Vaasa, Finland. Currently, he is Professor of Future Digital Business and Director of the Martti Ahtisaari Institute, Oulu Business School, University of Oulu. His research is in the intersection of entrepreneurship, strategic management, international business, futures research, and action research in various fields of high technology. Specifically, he is interested in business models, strategies, ecosystems, and internationalization within digital, mobile (5G/6G), smart energy, and smart city domains. He has over 300 scientific publications and has published in high-ranking journals such as the IEEE Communications Magazine, IEEE Wireless communications magazine, Technological Forecasting and Social Change, and the Journal of Business Studies.

Anthony Jnr. Bokolo is Senior Researcher at the Department of Applied Data Science, Institute for Energy Technology (IFE), Halden, Norway. He is also an associate professor at the Department of Computer Science and Communication, Østfold University College, Norway. Previously he was a Post-Doctoral Fellow at the Department of Computer Science, Norwegian University of Science and Technology (NTNU), Trondheim, Norway. He currently works on research and development in Energy Informatics, Sustainability, Sustainable Mobility, Urban Mobility, Smart Cities, and Data Ecosystems.

Per Carlborg is an associate professor at Örebro University School of Business where he co-leads the INTERORG Marketing Research Center. He is also guest professor at Linköping University. Per researches sustainability and organizational change, as well as sustainable market shaping, foresighting, and innovation. His work has been published in journals such as *Industrial Marketing Management*, the *Journal of Business Research*, and the *Journal of Cleaner Production*. Per is currently leading an EU-funded project focused on inclusive and just energy transitions. As a teacher, he leads pedagogical projects focusing on hope and engagement in higher education.

Capucine Chapel is a research scientist in Economics at the French National Research Institute for Agriculture, Food and Environment (INRAE), where she works at the UMR Herbivores, Clermont–Auvergne–Rhône-Alpes Center. Her research focuses on sustainable development, with particular attention to energy transitions, ecosystem services associated with green spaces and livestock systems, affordable housing, and the effects of natural disasters. Working at the intersection of environmental, agricultural, and regional economics, she uses applied microeconometrics and spatial analysis to illuminate the mechanisms that can steer societies toward more sustainable and inclusive paths, attentive to both social and territorial dimensions.

Patrícia Pereira da Silva is an associate professor with habilitation at the Faculty of Economics of the University of Coimbra (UC) where she also obtained her Ph.D. in Management. She is a faculty member of the MIT-P Program, where she lectures on Energy Economics and Markets. Since March 2021, she is a member of the National Council of the Environment and Sustainable Development (CNADS). Patrícia is a researcher at the Centre for Business and Economics Research (CeBER, UC), a collaborator at INESC Coimbra, and member of the Scientific Coordination and Management Board of UC's Energy for Sustainability Initiative. She is a founding member of APEEN, a Portuguese Affiliate of the International Association for Energy Economics (IAEE). Patrícia has been a visiting professor at Brazilian, North American, and European universities, acted as principal investigator in several research projects (Portuguese and internationally funded), and participated in many international interdisciplinary projects. Her current research interests include energy economics, markets, and finance.

Rodolfo Damásio de Castro has a degree in physics and a master's degree in Energy Systems Planning from the University of Campinas. He has worked as a consultant for the renewable electricity markets in Brazil and Europe. He is currently a researcher at the INESC Coimbra, completing his Ph.D. in Sustainable Energy Systems at the University of Coimbra. He is an EC2U fellow, and his interests are in quantitative analysis of electricity markets and economics.

Rashid Dehkordi holds a doctoral degree in Economics and Business Administration (2024) and an M.Sc. in Electrical Engineering (2020). He is currently a postdoctoral researcher at the Martti Ahtisaari Institute, Oulu Business School, University of Oulu, Finland. His research, funded by the European Union, explores how new business models gain legitimacy in emerging ecosystems, with a special focus on smart energy and electric vehicles. Rashid has published in leading academic journals, such as *Energy Policy*, *Energy*, and *IEEE Transactions on Engineering Management*, contributing to interdisciplinary research bridging technology, sustainability, and business. His main research interests are business models, business ecosystems, digital platforms, and the legitimation of new technologies.

Nuno Carvalho Figueiredo holds a Ph.D. in Sustainable Energy Systems from the Faculty of Science and Technology of the University of Coimbra with a distinguished career in both the corporative energy market and the academic environment. Professor Nuno also teaches subjects related to energy economics and energy markets at the University of Coimbra and the University of Aveiro.

Jonas Hlynur Hallgrímsson is a senior business analyst at Landsvirkjun, the National Power Company of Iceland. His work includes assessing the competitiveness of Landsvirkjun and its customers in an international context. Jonas has also been involved in demand forecasting, pricing, and securing the firm's energy supply. Prior to joining Landsvirkjun, he worked as an energy economist at the engineering and consulting firm Efla in Iceland. In that role, he worked for many of the key stakeholders in the Icelandic energy sector. Jonas has also worked at the Institute of Economic Studies at the University of Iceland, where he was involved in domestic and international energy projects. He is a lecturer in Energy Economics, Policy and Sustainability at the University of Iceland and has previously lectured in Energy Economics at Reykjavik University.

Nina Hasche is an associate professor at Örebro University School of Business where she co-leads the INTERORG Marketing Research Center. Her research interests include network dynamics as well as change and development processes. Her recent projects have focused on value creation, trust building and resource development in various types of business collaborations. Recently, Nina has also examined transformation processes for a more sustainable society and how different actors collaborate for circular solutions in different sectors. Her work has been published in the *Journal of Business Research, Industrial Marketing Management,* the *Public Management Review,* and the *International Public Management Journal*, among others. Nina has extensive experience from having served as a project leader and work package leader in multiple research projects, which commonly involve collaboration with both public and private organizations.

Quetzalcoatl Hernández-Escobedo is a Mexican researcher who has been dedicated to the study of renewable energies, especially wind energy in the optimization of wind farms. He holds a Mechanical Engineering degree from the Universidad Autónoma Metropolitana, a master's in Finance from the Universidad de Xalapa, and a Ph.D. from the Universidad de Almería. He was a researcher at the National Institute of Electricity and Clean Energies, and a professor at the Faculty of Engineering of the Universidad Veracruzana. He is currently a full-time professor at the Escuela Nacional de Estudios Superiores Unidad Juriquilla of UNAM. He has created research groups, directed national and international projects, has published more than 70 scientific articles in high impact journals, and has garnered more than 1,100 citations to his work. He has supervised more than 25 theses and is ranked Level II in Mexico's National System of Researchers.

Akanksha Jaiswal is an assistant professor in the Department of Management and a dedicated research scholar currently pursuing her Ph.D. She holds dual postgraduate degrees—a Master of Business Administration (MBA) and a Master of Commerce (M.Com.)—and has successfully qualified for the UGC-NET. Her academic and research interests lie primarily in the circular economy, sustainable business practices, and the evolving paradigms of responsible management. With teaching experience across institutions of higher education, Jaiswal has taught a wide range of subjects including General Management, Marketing, Business Communication, and Human Resource Management. In addition to her academic endeavors, Ms. Jaiswal is actively involved in student counselling and career guidance. Her commitment to continuous learning, pedagogical innovation, and academic scholarship mark her as a promising contributor to management education and research.

Moein Karami is a postdoctoral fellow at Concordia University. He holds a B.Sc. in Mechanical Engineering and an MBA in Strategy. In 2021, he graduated with a doctoral degree in Finance from the John Molson School of Business (Concordia University). His Ph.D. research was funded through several prestigious awards including the National Bank Initiative in Entrepreneurship and Family Business Fellowship. He has served as the instructor of numerous finance courses at Concordia University. He has published in journals such as the *Journal of Business Ethics* and *Finance Research Letters*. His main research interests include corporate finance, entrepreneurial finance, fintech, and sustainable finance.

Johan Kask is a professor at the University of Inland Norway's Center for Research on Digitalization and Sustainability (CREDS). His research focuses on the evolutionary dynamics of firms and industries, as well as business models that are innovative, digital, and sustainable. Johan's work appears in journals such as *Industrial Marketing Management Environmental Innovation & Societal Transitions Technological Forecasting & Social Change*, and the *Journal of Business Research*. He serves as Co-Editor of the *Journal of Small Business and Enterprise Development* and sits on the Editorial Review Board of *Industrial Marketing Management*. Currently, Johan is, among other things, working on an EU-funded project focused on inclusive and just energy transitions and is guest-editing a special issue on "Theorizing Organisational Coevolution" for the *International Journal of Management Reviews*.

Vojtěch Klézl is a senior lecturer at Örebro University School of Business with a background in purchasing and supply chain management. His recent research focuses on sustainability and circularity in various contexts, including road freight transport. He has published in journals such as the *Journal of Business Research*, the *Journal of Cleaner Production*, and the *International Journal of Procurement Management*. Currently, Vojtěch leads a scientific project on circular practices in road freight transport and serves as a work package leader in a project on inclusive and just energy transitions. His research involves collaborations with public and private organizations and contributes to policy recommendations. Additionally, Vojtěch has served as a reviewer for academic journals and an evaluator for national funders.

Anith Krishnan is a research scholar in the Department of Electrical Engineering at the National Institute of Technology Calicut. He holds a B.Tech. degree from Cochin University of Science and Technology and an M.Tech. from the University of Kerala. His research focuses on peer-to-peer energy trading, blockchain-based energy markets, and sustainable power systems.

Jakub Kubiczek is an assistant professor at the Department of Economic and Financial Analysis at the University of Economics in Katowice. His research focuses on financial decision-making, green consumer behavior, and sustainability, particularly in the context of how organizations respond to evolving consumer expectations. He has been involved in numerous research projects, including those funded by the National Science Centre (NCN) in Poland, exploring the interplay between consumer behavior, ecological awareness, and economic motivations. In his teaching, Dr. Kubiczek specializes in quantitative methods—such as statistics and econometrics—with a strong emphasis on their practical application to real-world economic and business challenges.

Bino Kuriachan is a research scholar in the Department of Electrical Engineering at the National Institute of Technology Calicut. He holds a B.Tech. degree from Mahatma Gandhi University and an MTech from Cochin University of Science and Technology. His research focuses on distributed battery energy storage systems, power procurement, and tariff rationalization.

Gabriel Linton is an associate professor at the University of Inland Norway and Senior Lecturer at Örebro University. His research focuses on business models, entrepreneurship, and innovation in the context of sustainability and digital transformation. He leads the Smarter Greener Innovation research group. Gabriel has published in journals such as the *Journal of Business Research Industrial Marketing Management R&D Management*, and the *Journal of Small Business & Entrepreneurship*, where he serves on the Editorial Review Board. Besides research, Gabriel has, for example, served as CEO of a university-owned energy startup exploring new business models in the electricity market. He has vast experience with different research projects, including work package leader for several EU Horizon projects. These projects often include collaboration with public and private organizations. He also serves as an expert evaluator for the European Commission.

Nimal Madhu M. is an assistant professor in Electrical Engineering at NIT Calicut. He holds a Ph.D. from AIT Thailand, an MTech from IIT Bombay, and a bachelor's from GEC Thrissur, India. His research spans ML/DL applications in Power Systems, Dynamic System Modeling, Power Markets, and Financial Time Series. He previously worked as a senior research executive at BIIC@AIT and was a faculty member at the Center for Computational Engineering and Networking (CEN) at Amrita Vishwa Vidyapeetham Coimbatore. A recipient of the Bangchak Post-Doctoral Scholarship for three consecutive years, Dr. Nimal has an extensive publication record to his credit.

Jônatas Augusto Manzolli is a postdoctoral researcher at the IMaTS Laboratory at McGill University, Canada. He holds a Ph.D. in Sustainable Energy Systems, an M.Sc. in Energy for Sustainability from the University of Coimbra, Portugal, and a B.Sc. in Electrical Engineering from the University of Campinas, Brazil, which included an exchange at the Technical University of Munich, Germany. His research spans energy transition, optimization, and electric vehicles. In 2020, he received the SYLFF award from the Tokyo Foundation for Young World Leaders. He founded the Power and Energy Society chapter at the University of Coimbra in 2020 and served as vice-chair of the IEEE Student Branch from 2020 to 2021.

Jordan Mitchell is currently a Ph.D. candidate at the University of Iceland with a focus on sustainable finance and sustainable business models. Prior to embarking on his doctoral studies, he collaborated on more than 150 case studies across a wide range of disciplines in over 30 industries on all continents. His cases have been published at Harvard, Ivey, IESE, Darden, and Dalhousie and have appeared in several textbooks. He has also taught business and case study writing seminars in Iceland, Spain, Russia, and China. Aside from his academic pursuits, he is currently Director within Commercial Payments at RBC in Toronto, Canada and previously worked at Sears Canada, Levi Strauss & Co. (Canada) and a number of start-ups in energy, tech, and apparel in Spain and Canada.

Mostafa Mohammadian received a B.Sc. and M.Sc. degree from the Amirkabir University of Technology, Tehran, Iran, in 2014 and 2017, respectively. He is currently working toward a Ph.D. degree with the College of Engineering and Applied Science, at the University of Colorado, Boulder, CO, USA. His research focuses on the intersection of machine learning and power systems, with an emphasis on optimization techniques, and distributed algorithms for solving complex energy system problems such as Optimal Power Flow (OPF).

Vivek Mohan is an assistant professor in the Department of Electrical Engineering at NIT Calicut since January 2023. Previously, he served in the Department of Electrical and Electronics Engineering (EEE) at NIT Trichy (2018–2023) and has 11 years of teaching experience. He holds a B.Tech. in EEE from Amrita Vishwa Vidyapeetham, an MTech in Power Systems from Government Engineering College, Thrissur, and a Doctor of Engineering (Energy FoS) from the Asian Institute of Technology, Thailand. He was a postdoctoral researcher at The Hong Kong Polytechnic University (2017–2018) and led two sponsored projects on P2P Transactive Energy Markets in India from 2019 to 2025. His research interests include power system economics, energy management, optimization, risk and reliability management, and policy regulations.

David Muñoz-Rodríguez is a lecturer in the Department of Applied Physics, Radiology and Physical Medicine at the University of Córdoba. He holds a Ph.D. in Chemistry, complemented by an MBA. He combines teaching with research in renewable energies, energy efficiency, and environmental management. He has supervised four final degree projects and participated in numerous educational innovation projects. He has authored 16 scientific articles in fields such as photovoltaic systems, renewable energies, hydraulic network optimization, and the agri-food sector.

Kofi Nyarko is a professional electrical engineer with over 11 years' experience in teaching, researching, and consultation in the electrical, energy, and environmental fields. He holds a bachelor's degree in Electrical Engineering and a master's degree in Energy & Environmental Engineering. His work career began in 2012 at Koforidua Technical University, Ghana, where he focused on training students on sound electrical and renewable energy engineering principles and practices. While still attached to academia, he performed consultation services for private engineering companies, governments, and international agencies. He has expertise in the implementation of sustainable renewable energy-based power systems and environmental sustainability analysis of energy projects. Currently, he is enrolled in the Ph.D. program at the School of Engineering and Energy, Murdoch University, where he is working to develop a novel framework for implementing sustainable mini-grids in remote communities in developing countries.

Alberto-Jesus Perea-Moreno is Full Professor in the Department of Applied Physics, Radiology and Physical Medicine at the University of Córdoba (Spain). He earned a degree in Agricultural Engineering from the University of Córdoba, where he also completed his Ph.D. His research focuses on renewable energy resources. He has led and participated in numerous national and international projects and has authored more than 80 scientific articles in reputable international journals, with over 1,900 citations and an H-index of 22 (Scopus). Dr. Perea-Moreno was listed in the "World's Top 2% Scientists 2023," ranking released by Stanford University. He currently holds two six-year research periods and one six-year knowledge transfer period. His work plays a key role in advancing sustainable technologies and exploring new energy sources, significantly contributing to the global transition toward cleaner energy systems.

Andre Poyser is a lecturer in the Department of Economics and Finance at the Auckland University of Technology in New Zealand. Andre completed his doctoral studies at the University of Otago. A native of Jamaica, Andre's research has focused on indigenous finance, sustainability and climate finance. Andre holds a Ph.D. in Finance from the University of Otago.

Fabienne Rioux-Gobeil is a Ph.D. student at the Australian National University, focusing on interdisciplinary research within the Research School of Humanities and the Arts. Originally from Montreal, Canada, she holds a master's degree in Environmental Sciences and a Bachelor's degree in Geography from the University of Quebec in Montreal (UQAM). In her Master's thesis, Fabienne examined the territorial governance of energy transition processes and identified a significant gap in Indigenous contexts—an area she intends to further investigate in her doctoral research. Currently, she is conducting a comparative case study of two Indigenous-led renewable energy projects in Australia and Canada.

Fabienne hopes that her analysis will shed light on pathways toward decolonized energy transitions.

Akram Sadati is an M.Sc. graduate of the John Molson School of Business, Concordia University, where she studied Finance. She currently serves as Research Associate at the Jacques Ménard—BMO Center for Capital Markets, Concordia University. She previously worked as a Supervisor of the Data Management team at Maples Group and currently serves as a Senior Business Analyst at Intact Financial Corporation. She has received several scholarships and awards for her academic performance, including the Jacques Ménard—BMO Graduate Scholarship and the New Millennium Student Award. Her research interests include risk management, corporate finance, sustainability, fintech, and machine learning.

Aphrodite Salas is Associate Professor of Journalism at Concordia University in Montreal, Canada, where her research focuses on collaborative journalism with Indigenous communities and the decolonization of journalism education. Her ongoing work is with Inuit research partners in the Arctic village of Inukjuak. Her most recent project, "The Flowing River," is a documentary about the transition from diesel dependency to renewable energy in the polar north. Her previous work, "Arctic Shift to Clean Energy," was screened internationally with the leading non-profit Indigenous Clean Energy.

Sarmistha Sarma is Professor, Consultant, Author, Editor, and Columnist. At present, she is Professor of Marketing (Department of Management) at the Asian Business School NOIDA. She is a Subject Matter Expert to Tata Consultancy Services since 2018. She has authored 5 books in various domains of Marketing and is presently Chief Adviser of the IITM Journal of Business Studies (JBS). Also, she is in the editorial advisory board of Ecoforum Journal in Romania actively promoting research in business management globally. Along with all these roles she is associated with National Council of Educational Research and Training (NCERT) as a teacher and was a member of the team that designed the All India Teacher Training Manual. She is empanelled as a PhD research guide with All India Management Association (AIMA).

Fadi Sayegh is an accomplished researcher and professional engineer specializing in sustainable decision-making, climate change awareness, and proactive adaptation strategies. He has engaged in collaborative research with prestigious institutions, including Universidad Católica de Murcia (UCAM), the Rochester Institute of Technology (RIT), the University of Quebec at Montreal (UQAM), and Sorbonne Paris IV, contributing to advancements in sustainable engineering practices. A recognized leader in his field, he holds professional engineering licenses across multiple Canadian provinces and international jurisdictions. His research has been featured in the *Journal of Environmental Management* and *GeoJournal*. Dr. Sayegh remains dedicated to transforming engineering practices to address global challenges and promote sustainability, earning him a distinguished reputation in both academic and professional circles.

Thorstur Olaf Sigurjónsson is Professor of Business at the University of Iceland whose research explores the intersections of corporate governance, business ethics, and organizational change. His work is grounded in a deep interest in how organizations adapt to changes while maintaining effective governance structures. A key focus of his research lies in the change management processes involved in implementing digitalization and sustainability solutions and in particular, understanding how leadership, strategic alignment, and organizational culture influence the success of these initiatives. In addition to his post at the University of Iceland, Dr. Sigurjónsson previously taught at the Copenhagen School of Business and at Reykjavik University. He has also been a visiting professor at Rotterdam School of Management, IEEM in Uruguay, and UQ Business School in Australia.

Mika Sorvisto received an M.Sc. degree in Electrical Engineering from the University of Oulu, Finland, in 1996. He is currently a senior advisor with Ahola Group, Kokkola, Finland. He has worked in different managerial position within medical technology, consumer electronics, telecommunications, software, e-mobility, logistics, and energy businesses. His research interests include the field of energy ecosystems, business model innovation, and commercialization of new technologies.

Annick Thomassin has over 17 years' experience in the field of social anthropology and political ecology working across a range of interrelated topics including coastal resources (co)-management, small-scale and Indigenous fisheries and environmental stewardship practices, Indigenous economic perspectives, Indigenous-settler state relations, the intersection of Indigenous and Western ontologies, and conceptualizations of sovereignty. Over these years, she has worked closely with Indigenous partners to develop methodologies and methods informed by Indigenous perspectives and privileging two-way knowledge exchanges. Most of her work and studies so far have been focused on the Australian, Canadian, and Vietnamese contexts, with emerging collaborations in Japan (in particular with the Museum of Ethnology (MINPAKU) in Osaka).

Tania Urmee is a distinguished researcher and professor at the School of Engineering and Energy, Murdoch University. Her expertise spans energy planning, policy, modeling, and framework development, with a focus on supporting the Sustainable Development Goals (SDGs). She has developed energy roadmaps for ASEAN and Pacific nations and has worked across diverse regions, including Bangladesh, Fiji, Kenya, Ghana, Indonesia, and Australia. A pioneer in solar home systems through microfinance in Bangladesh, she has authored over 100 publications, including books, journal articles, and conference papers. Her research covers renewable energy integration, energy storage, microgrids, sustainable energy, and climate policy, and social, economic, and political issues of the energy transition. She is committed to fostering collaboration between academia, industry, and policymakers to advance sustainable energy solutions and drive the transition toward a low-carbon future.

Thomas Walker holds an MBA and Ph.D. degree in Finance from Washington State University. Prior to his academic career, he worked for several years in the German consulting and industrial sector. His research interests are in banking, emerging risk management, fintech, and sustainable finance, and he has published over ninety journal articles in these areas. He is the lead editor of twenty books on sustainable financial systems, sustainable real estate, sustainable aviation, environmental policy, emerging risk management, innovations in social finance, and water risk management (among others). In 2021, he became the Director for the Jacques Ménard—BMO Centre for Capital Markets at Concordia University and the Concordia University Research Chair in Emerging Risk Management (Tier 1). Since 2025, he serves as the Executive Director (Academic) of the Institute for Sustainable Finance (ISF).

Natalia de Assis Brasil Weber is a Doctor in Energy Engineering dedicated to advancing inclusive, resilient, and sustainable energy systems. She currently works as Energy Consultant for the project "Mitigating Climate Change Threats to the Energy Sector in the OSCE Region," supporting the integration of climate data into energy models to inform resilience strategies across national infrastructures. In parallel, she serves as Project Manager at GWNET, leading global mentoring programs that have empowered over 350 women in the energy sector—promoting gender equality and building capacities for clean energy transitions. Her expertise includes energy systems modeling, climate data analysis, and sustainable energy research and capacity building, with experience across Europe and Latin America. She is passionate about driving systemic change through collaborative, multistakeholder approaches to just and equitable energy transitions.

Caroline Zimm is a senior research scholar at the International Institute for Applied Systems Analysis (IIASA). Her research is concerned with the diffusion of technologies and policies for sustainable development, inequalities across countries, and transformative development pathways for humanity within a stable Earth system. Before her career in academia, she worked in development cooperation and on consulting projects in the field of rural electrification, energy access, clean cooking, and climate change. She held positions with the United Nations Industrial Development Organization (UNIDO), a private consulting firm, and the German Development Agency (GIZ) where she was based in Cambodia and India. Zimm received her Ph.D. in Energy Economics from the University of Technology in Vienna, Austria. She holds master's degrees in Environmental and Resource Management (major in Renewable Energies) and in International Business Administration, and a bachelor's degree in Political Science.

List of Figures

List of Tables

CHAPTER 1

Wiring the Future: Financial Strategies, Challenges, and Opportunities in the Sustainable Energy Transition: An Overview

Akram Sadati, Moein Karami, Thomas Walker, and Aphrodite Salas

A. Sadati (✉) · M. Karami · T. Walker
Department of Finance, Concordia University, Montreal, QC, Canada
e-mail: ak_sadat@live.concordia.ca

M. Karami
e-mail: moein.karami@concordia.ca

T. Walker
e-mail: thomas.walker@concordia.ca

A. Salas
Department of Journalism, Concordia University, Montreal, QC, Canada
e-mail: aphrodite.salas@concordia.ca

T. Walker et al. (eds.), *Wiring the Future*, Palgrave Studies in Emerging Risk Management and Sustainable Finance,
https://doi.org/10.1007/978-3-032-14717-2_1

Introduction

The global transition toward sustainable energy systems has emerged as one of the defining challenges of the twenty-first century (Moustakas et al., 2020). Rising greenhouse gas emissions, rapid urbanization, and the intensifying impacts of climate change have heightened the urgency of transforming how societies produce and consume energy (Granjou et al., 2024; Jorgenson et al., 2014; Wang et al., 2021). Cities are expanding rapidly, driving demand for electricity, heating, transportation, and infrastructure; most of which still remain heavily dependent on fossil fuels (Creutzig et al., 2015). Meanwhile, climate change is no longer a distant threat but a pressing and growing challenge. Extreme weather events, rising sea levels, and shifting precipitation patterns are already imposing significant economic and social costs on agriculture, health systems, infrastructure, and migration (Carleton & Hsiang, 2016; Hsiang et al., 2017; Mimura, 2013). If these trends continue unchecked, the long-term economic and societal costs of inaction are projected to exceed the investments required for a decisive transition to a low-carbon energy system.

International frameworks such as the Paris Agreement and the United Nations' Sustainable Development Goals (SDGs) underscore the magnitude of this challenge, calling for rapid and unprecedented changes in the global energy landscape (Paris Agreement, 2015; Rogelj et al., 2016; United Nations, 2015). At the heart of these efforts lies the electrification of society, widely recognized as a cornerstone of sustainable development and a key driver of economic growth, particularly in developing regions (IEA, 2021; Lipscomb et al., 2013). Electrification represents both an environmental necessity and an economic opportunity. It can reduce dependence on fossil fuels in key sectors such as transport, industry, and buildings while simultaneously enhancing energy security and supporting long-term economic growth (Lin & Xu, 2024; Tsao et al., 2018).

Yet achieving large-scale electrification is neither automatic nor straightforward. It requires unprecedented levels of investment, innovative financing mechanisms, and robust regulatory frameworks (Berizzi et al., 2019; Reiche et al., 2006; Sachs et al., 2019). This book examines these financial and regulatory dimensions. It analyzes how governments, financial institutions, corporations, and communities can mobilize the capital, policy instruments, and governance mechanisms necessary to drive the electrification agenda forward.

The central objective of this book is to examine the financial and regulatory dimensions that shape the electrification of society, with a particular focus on mobilizing capital for large-scale green infrastructure projects. It critically examines the interplay between traditional financial metrics such as net present value (NPV), and the pressing need for substantial investments in green infrastructure—a prerequisite for achieving the ambitious targets set by the 2030 Agenda for Sustainable Development. This book addresses the central question of how to mobilize the financial sector and raise the necessary capital to finance the substantial upfront costs associated with long-term investments in environmental sustainability and economic resilience. It explores how governments can encourage investments into a green energy transition and examines both the financial and social returns of sustainable energy projects, such as energy-autonomous buildings, community initiatives, and automated electric vehicle refueling stations.

The uniqueness of this book lies in its collection of scholarly contributions from diverse regions of the world, integrating a wide range of perspectives on the challenges and opportunities inherent in the electrification process. This global collaboration enriches the book's content, ensuring a comprehensive exploration of the subject. The contributors—experts in their respective fields—offer insights that span the spectrum from the impact of competitive pressures and policy mechanisms (such as tax incentives and emission trading schemes) to advanced quantitative analyses of economic variables affected by climate change, including real estate valuations, banking risk assessments, and insurance risk management.

A key feature of the book is its pragmatic approach. It bridges the gap between theoretical frameworks and real-world applications, providing actionable guidance for a diverse set of stakeholders, including corporations, regulatory bodies, and policymakers. By employing case studies and financial modeling techniques, the book sheds light on how investments in green infrastructure can be not only viable but also financially advantageous relative to traditional investments.

Furthermore, the book addresses the challenge of aligning stakeholder interests and designing robust policy incentives for a long-term commitment to green initiatives. It emphasizes the need for an integrated approach that combines financial and environmental models to facilitate effective policy and investment decision-making. This approach is essential for developing governance models that account for the diverse interests of various stakeholders and for mitigating the financial risks associated with climate-related disruptions.

Overview of Content

The book is organized into three thematic parts, each addressing key dimensions of sustainable electrification and providing a comprehensive overview of the financial, governance, and technological aspects of the energy transition. Part I establishes the conceptual and analytical foundations for sustainable electrification, highlighting innovative business models and financing approaches. Part II examines multi-level governance, stakeholder engagement, and policy frameworks that are critical to successful electrification. Part III presents real-world applications, emerging technologies, and lessons from diverse regions, offering practical insights for implementing sustainable energy solutions around the world.

Part I: Frameworks and Financial Strategies

This part establishes the conceptual and financial foundations of sustainable electrification, providing policymakers, corporations, and investors with tools to evaluate risks, design effective business models, and mobilize capital toward the energy transition.

The first chapter in this part—Chapter 2, *Renewable Power Purchase Agreements—An Instrument for Corporate Decarbonization*, by **Castro, Manzolli, Figueiredo, and Silva**—explores the critical role of Power Purchase Agreements (PPAs) in financing renewable energy deployment and enabling corporate decarbonization. The authors examine a variety of PPA structures, risk allocation mechanisms, and pricing arrangements, demonstrating how these contracts provide predictable revenue streams for producers and reduce price volatility for corporations. By integrating real-world examples, the chapter offers practical guidance for decision-makers seeking to implement PPAs that align with financial and sustainability objectives.

Next, Chapter 3—*Business Model Innovation and Business Ecosystem Innovation for the Sustainable Energy Transition*, by **Dehkordi, Ahokangas, Sorvisto, and Mohammadian**—investigates the dynamic interplay between business model innovation (BMI) and business ecosystem innovation (BEI) in supporting sustainable energy transitions.

Drawing on case studies from Finland and Sweden, the authors illustrate how BMI can initiate BEI through novel technologies, cross-sector collaborations, and complementary innovations, while BEI, in turn, reinforces BMI by reshaping market conditions and fostering ecosystem-wide sustainability standards. The chapter emphasizes the strategic advantage of aligning firm-level innovations with broader ecosystem objectives as a means to enable systemic transformation and enhance resource efficiency.

Chapter 4—*Accelerating Distributed Solar and Battery Storage Adoption in India through Tariff Reforms and Community Grids*, by **Krishnan, Kuriachan, Mohan, and Madhu**—focuses on tariff rationalization and community-based strategies to accelerate distributed solar and battery storage adoption. By comparing individual and community-scale approaches and integrating a dynamic feed-in tariff mechanism, the authors demonstrate how policy instruments can enhance financial viability, improve grid resilience, and incentivize prosumer participation. Drawing on empirical data from Indian utilities and consumers, the chapter highlights practical pathways to scaling renewable adoption while balancing economic, social, and operational outcomes.

Chapter 5—*Economic Impacts of Sustainable Electrification*, by **Chapel**—investigates both macro- and microeconomic consequences of electrification. The chapter illustrates how sustainable energy transitions influence energy security, labor markets, and household access to electricity, while reshaping the financial, real estate, and insurance sectors. By examining regional differences and sectoral impacts, the chapter provides a comprehensive assessment of the strategic opportunities and challenges associated with sustainable electrification, highlighting its role in driving broader socioeconomic development.

Part II: Governance, Policy, and Stakeholders

This part emphasizes the importance of governance frameworks, stakeholder inclusion, and equitable participation in shaping sustainable electrification. It highlights how institutions, communities, and marginalized groups influence outcomes and legitimacy in energy transitions.

The section begins with Chapter 6—*The Role of Stakeholders in Sustainable Electrification,* by **Chapel**—which examines how policymakers, private actors, communities, and NGOs shape electrification outcomes. The chapter highlights challenges and emerging engagement strategies across multiple levels of decision-making, emphasizing the importance of coordination to promote inclusive and effective energy transitions.

Chapter 7—*Empowering the Grid: Stakeholder Engagement in Advancing Electrification*, by **Sayegh**—analyzes stakeholder contributions to sustainable energy adoption, considering the roles of governments, private actors, communities, and advocacy groups. The chapter emphasizes integrated governance and corporate leadership, while illustrating how community involvement and advocacy can overcome adoption barriers, improve equity, and drive innovation.

Chapter 8—*Stakeholder Engagement and Governance in the Transition to Renewable Energy Systems*, by **Nyarko and Urmee**—explores decentralized rural electrification in Ghana and Sierra Leone. The authors demonstrate how collaborative decision-making among formal and informal stakeholders ensures project success, and they provide actionable insights for coordinating diverse actors to enhance system reliability, inclusivity, and sustainability.

Chapter 9—*The Multi-Territorial Governance (MTG) Model: Indigenous Nations and Settler-State Collaboration in Energy Transitions*, by **Rioux-Gobeil and Thomassin**—introduces a governance framework designed to respect Indigenous self-determination while enabling decentralized energy solutions in Canada and Australia. The chapter outlines principles for non-hierarchical collaboration, territorial acknowledgment, and strong reciprocal relationships, highlighting lessons for scaling governance models to other regions.

Chapter 10—*Indigenous Ownership in Canada's Energy Transition: A Pathway to Sustainable Infrastructure Development*, by **Poyser**—examines the rise of Indigenous equity participation in energy infrastructure projects. The chapter emphasizes how ownership not only promotes inclusivity and equity but also strengthens long-term sustainability and fosters local economic development. It illustrates mechanisms to integrate Indigenous communities into mainstream energy projects.

Finally, Chapter 11—*Gendered Aspects of Electrification*, by **Zimm and Weber**—addresses the critical gender dimensions of energy access. The authors highlight how limited electricity access disproportionately affects women, with significant impacts on health, education, and economic opportunities. By presenting successful case studies and mentoring programs, the chapter provides strategies to empower women and foster inclusive energy transitions that contribute to broader societal benefits.

Part III: Technological Innovations and Regional Case Studies

The final part of the book applies conceptual and governance insights to real-world technological solutions and regional case studies, illustrating how innovation, policy, and stakeholder engagement interact in practice.

Chapter 12—*A Decentralized eRoaming Business Service Ecosystem to Promote Citizen-Centric Energy Communities*, by **Bokolo**—examines the development of decentralized eRoaming systems for electric vehicles, addressing barriers to interoperability, market fragmentation, and accessibility. The chapter outlines a business ecosystem model that enhances cross-regional coordination, offers practical guidance on stakeholder integration, and facilitates standardized, scalable solutions for citizen-centric energy communities.

Chapter 13—*Imagining Energy Futures: Disrupt, Downsize, Scaleup, or Conserve?* by **Kask, Linton, Carlborg, Hasche, and Klézl**—employs scenario-based analysis to explore Sweden's potential pathways to carbon neutrality. The chapter demonstrates how technological innovation, socio-behavioral change, and adaptive policy interact to shape energy futures, offering insights for utilities, investors, and policymakers to anticipate risks and seize opportunities in energy transitions.

Chapter 14—*Global Perspectives on Electrification: Diverse Approaches and Solutions*, by **Perea-Moreno, Hernández-Escobedo, and Muñoz-Rodríguez**—provides a bibliometric overview of global electrification research. By analyzing trends, key actors, and regional disparities, the authors reveal emerging technologies and highlight socioeconomic impacts, offering a comprehensive lens to inform evidence-based decision-making and policy development.

Chapter 15—*Powering Up Iceland: How Iceland Transformed Itself into a Renewable Energy Powerhouse*, by **Mitchell, Hallgrimsson, and Sigurjonsson**—examines Iceland's near-complete renewable electricity system. The chapter illustrates how resource abundance, long-term planning, collaborative governance, and strategic investments enabled a resilient energy transition, offering transferable lessons for countries seeking rapid decarbonization.

Chapter 16—*Water to Watts: Meghalaya's Electrification Strategy of Turning Rainfall into Power*, by **Sarma and Jaiswal**—investigates India's state-led hydro electrification approach, highlighting strategies for sustainable water resource management, coordinated multi-stakeholder efforts, and socioeconomic development. The chapter demonstrates how policy, finance, and community participation converge to achieve inclusive and environmentally sustainable electrification.

Finally, Chapter 17—*Consumer Behavior in the Energy Transition: The EKB Decision-Making Approach*, by **Kubiczek**—applies the Engel-Kollat-Blackwell model to household energy choices. By analyzing motivations, barriers, and decision-making stages, the chapter provides insights for designing effective interventions, behavioral strategies, and policies that align consumer actions with sustainable energy objectives.

Contribution and Audience

By integrating conceptual frameworks, empirical evidence, and policy analysis, this book addresses a critical gap in the literature on sustainable finance, energy transitions, and electrification. It offers a holistic perspective that bridges theory and practice, providing actionable insights for a wide range of stakeholders involved in the global shift toward sustainable energy systems.

The book will be of interest to a diverse audience. Finance professionals can draw on its analyses to integrate sustainability into investment strategies and evaluate risks and opportunities in renewable energy markets. Policymakers and regulators will find guidance for designing effective frameworks that promote inclusive and resilient electrification. Academics and researchers can leverage its interdisciplinary approach to advance the study of energy transitions, sustainable finance, and sociotechnical innovation. Additionally, students exploring economics, sustainability, and energy policy will gain a comprehensive roadmap for understanding the interactions among governance, technology, finance, and social dynamics in energy transitions.

Ultimately, this volume aspires to serve as both an academic reference and a practical guide. By offering a structured and comprehensive examination of the financial, regulatory, and technological challenges of electrification, it contributes to advancing sustainable development and climate resilience worldwide. Its depth and breadth make it a valuable resource for guiding scholarship, informing decision-making, and supporting the design of inclusive, equitable, and future-ready energy solutions.

References

Berizzi, A., Delfanti, M., Falabretti, D., Mandelli, S., & Merlo, M. (2019). Electrification processes in developing countries: Grid expansion, microgrids, and regulatory framework. *Proceedings of the IEEE, 107*(9), 1981–1994. https://doi.org/10.1109/jproc.2019.2934866

Carleton, T. A., & Hsiang, S. M. (2016). Social and economic impacts of climate. *Science, 353*(6304), aad9837. https://doi.org/10.1126/science.aad9837

Creutzig, F., Baiocchi, G., Bierkandt, R., Pichler, P. P., & Seto, K. C. (2015). Global typology of urban energy use and potentials for an urbanization mitigation wedge. *Proceedings of the National Academy of Sciences, 112*(20), 6283–6288. https://doi.org/10.1073/pnas.1315545112

Granjou, C., Banos, V., Le Berre, S., & Sergent, A. (2024). Greening, climatizing, and decarbonizing: An inquiry into the transformation of productive sectors and activities. *Review of Agricultural, Food and Environmental Studies, 105*(1), 1–16. https://doi.org/10.1007/s41130-024-00210-3

Hsiang, S., Kopp, R., Jina, A., Rising, J., Delgado, M., Mohan, S., Rasmussen, D. J., Muir-Wood, R., Wilson, P., Oppenheimer, M., Larsen, K., & Houser, T. (2017). Estimating economic damage from climate change in the United States. *Science, 356*(6345), 1362–1369. https://doi.org/10.1126/science.aal4369

International Energy Agency (IEA). (2021). *Net zero by 2050: A roadmap for the global energy sector*. OECD. https://doi.org/10.1787/c8328405-en

Jorgenson, A. K., Auerbach, D., & Clark, B. (2014). The (de-) carbonization of urbanization, 1960–2010. *Climatic Change, 127*(3), 56–575. https://doi.org/10.1007/s10584-014-1267-0

Lin, Y., & Xu, C. (2024). Electrification and economic growth: Evidence from China's rural primary electrification program. *China Economic Review, 83*, 102084. https://doi.org/10.1016/j.chieco.2023.102084

Lipscomb, M., Mobarak, A. M., & Barham, T. (2013). Development effects of electrification: Evidence from the topographic placement of hydropower plants in Brazil. *American Economic Journal: Applied Economics, 5*(2), 200–231. https://doi.org/10.1257/app.5.2.200

Mimura, N. (2013). Sea-level rise caused by climate change and its implications for society. *Proceedings of the Japan Academy, Series B, 89*(7), 281–301. https://doi.org/10.2183/pjab.89.281

Moustakas, K., Loizidou, M., Rehan, M., & Nizami, A. S. (2020). A review of recent developments in renewable and sustainable energy systems: Key challenges and future perspective. *Renewable and Sustainable Energy Reviews, 119*, 109418. https://doi.org/10.1016/j.rser.2019.109418

Reiche, K., Tenenbaum, B., & de Mästle, C. T. (2006). Electrification and regulation: Principles and a model law. In *Energy and mining sector board*. The World Bank. https://doi.org/10.1596/17959

Rogelj, J., den Elzen, M., Höhne, N., Fransen, T., Fekete, H., Winkler, H., Schaeffer, R., Sha, F., Riahi, K., & Meinshausen, M. (2016). Paris Agreement climate proposals need a boost to keep warming well below 2 °C. *Nature, 534*(7609), 63–639. https://doi.org/10.1038/nature18307

Sachs, J. D., Woo, W. T., Yoshino, N., & Taghizadeh-Hesary, F. (2019). *Handbook of green finance* (pp. 3–12). Springer Singapore. https://doi.org/10.1007/978-981-10-8710-3

Tsao, J. Y., Schubert, E. F., Fouquet, R., & Lave, M. (2018). The electrification of energy: Long-term trends and opportunities. *MRS Energy & Sustainability*, *5*, E7. https://doi.org/10.1016/j.chieco.2023.102084

United Nations Framework Convention on Climate Change. (2015). *The Paris Agreement*. https://unfccc.int/process-and-meetings/the-paris-agreement

United Nations. (2015). *Transforming our world: The 2030 agenda for sustainable development*. https://sdgs.un.org/2030agenda

Wang, W. Z., Liu, L. C., Liao, H., & Wei, Y. M. (2021). Impacts of urbanization on carbon emissions: An empirical analysis from OECD countries. *Energy Policy*, *151*, 112171. https://doi.org/10.1016/j.enpol.2021.112171

PART I

Frameworks and Financial Strategies

CHAPTER 2

Renewable Power Purchase Agreements: An Instrument for Corporate Decarbonization

Rodolfo Damásio de Castro, Jônatas Augusto Manzolli, Nuno Carvalho Figueiredo, and Patrícia Pereira da Silva

Introduction

The urgent need to combat climate change has accelerated the shift toward cleaner energy sources. Thus, in response to international climate agreements, renewable energy adoption—especially from solar and wind-powered sources—has been driven by financial incentives such as feed-in tariffs, feed-in premiums, and net metering (Enescu et al., 2023). These mechanisms help to ensure steady project revenue, reducing financial risk

R. D. de Castro (✉) · J. A. Manzolli · N. C. Figueiredo · P. P. da Silva
Energy for Sustainability Initiative, University of Coimbra, Coimbra, Portugal
e-mail: rodolfo.castro@dem.uc.pt

INESC Coimbra—Institute for Systems Engineering and Computers at Coimbra, University of Coimbra, Coimbra, Portugal

J. A. Manzolli
e-mail: jonatas.manzolli@deec.uc.pt

T. Walker et al. (eds.), *Wiring the Future*, Palgrave Studies in Emerging Risk Management and Sustainable Finance,
https://doi.org/10.1007/978-3-032-14717-2_2

and attracting investment. However, they often result in higher electricity costs for consumers who share the financial costs. Power Purchase Agreements (PPAs) have gained significant traction as a solution. Businesses, governments, and utilities use PPAs to diversify energy portfolios, stabilize costs, and cut greenhouse gas (GHG) emissions. Unlike traditional support schemes, PPAs enable the development of new renewable projects in a market-based framework without societal sharing costs. PPAs also improve the chances of securing funding from banks and investors by offering predictable revenue streams (Rövekamp et al., 2021). In the long run, they provide an opportunity for stable energy pricing while advancing renewable infrastructure, particularly in solar and windprojects (Table 2.1).

A PPA is a bilateral energy procurement contract between an energy producer and an end consumer (Gabrielli et al., 2022). They are typically long-term contracts, often lasting over ten years, though shorter contracts of as little as three years also exist (Mittler et al., 2025). When the buyer is an energy supplier, it is referred to as a Utility PPA (UPPA). Energy suppliers have widely used UPPAs and are familiar with this kind of contract. Alternatively, end-use corporations can also benefit from PPAs, known as Corporate PPAs (CPPAs), where a company buys energy directly to power its operations rather than for resale. The rise of CPPAs is driven by stricter Environmental, Social, and Governance

N. C. Figueiredo
e-mail: nuno.c.figueiredo@uc.pt

P. P. da Silva
e-mail: patsilva@fe.uc.pt

N. C. Figueiredo
Department of Mechanical Engineering, University of Aveiro, Aveiro, Portugal

P. P. da Silva
CeBER—Centre for Business and Economic Research, University of Coimbra, Coimbra, Portugal

Faculty of Economics, University of Coimbra, Coimbra, Portugal

J. A. Manzolli
IMaTS lab, Department of Civil Engineering, McGill University, Montreal, Canada

Table 2.1 Abbreviations and symbols

Abbreviation	*Full name*
BRP	Balancing Responsible Party
CfD	Contract for Difference
COD	Commercial Operation Date
CPPA	Corporate Power Purchase Agreement
ESG	Environmental, Social, and Governance
ETS	Emissions Trading System
FM	Force Majeure
GHG	Greenhouse Gas
PPA	Power Purchase Agreement
REC	Renewable Energy Certificates
UPPA	Utility Power Purchase Agreement
VRE	Variable Renewable Energy
Symbol	**Description**
t	Index of all periods t in a settlement horizon, $[1, \ldots, T]$
$\mathcal{T}$	Set of all periods, $\mathcal{T} = \{1, \ldots, T\}$
Prod	Production of the power plant during a settlement period (e.g., month)
K	(Strike price) Fixed price of the electricity in the PPA
FP	(Floating Price) Average price captured by the power plant in the wholesale market
DFP	(Discounted Floating Price) Product between the Floating Price and a factor related to the discount (D) on the wholesale price
F	(Floor price) Lower threshold in the wholesale market price
D	(Discount rate) Discount on the wholesale market price
C	(Cap price) Higher threshold in the wholesale market price
WSP	(Wholesale price) Price of electricity in the wholesale market
CR	(Capture rate) Ratio between Floating Price (FP) and the average price of electricity in the spot market

(ESG) requirements and growing consumer demand for sustainable practices. By procuring renewable energy directly from producers, CPPAs help companies reduce their GHG emissions and support the development of new projects without raising consumer electricity costs. However, CPPAs come with risks, and both parties must carefully consider the terms to avoid adverse outcomes.

In this chapter, we will explore the global growth of PPAs, examining their prevalence across significant electricity markets. Furthermore, we present a perspective on policy and regulatory frameworks that shape PPAs and offer sectoral insights into their adoption. We also discuss critical aspects of PPA structures, including delivery schemes, pricing mechanisms, and the role of Renewable Energy Certificates (RECs). The main contribution of this chapter is a comprehensive analysis of the financial and contractual components that characterize PPAs, particularly as regards pricing strategies and the effects of regulation. We aim to provide a deeper understanding of the factors driving the growth of PPAs, and to equip readers with practical insights into their design and implementation in different market contexts.

The remainder of this chapter is organized as follows. Section "A Global Overview of PPA Markets" provides an overview of the global evolution of PPA adoption and regulatory framework across various regions. Section "Structures and Drivers of a PPA" explains the core concepts of a PPA, including delivery schemes, pricing mechanisms, and RECs. In Section "Costs and Other Risks Incurred When Signing a PPA", we discuss the contractual designs of PPAs before Section "Conclusions", where we outline our main conclusions and avenues for further work.

A Global Overview of PPA Markets

The global PPA market has expanded rapidly in the past decade, with substantial increases in the number and capacity of agreements signed. Large corporations have led this surge, motivated by the growing importance of ESG criteria. In the first half of 2023, the total volume of CPPAs exceeded 45 GW, a significant market increase compared to earlier years. This trend reflects the rising commitment of corporations and institutions to decarbonization and sustainability (Briggs et al., 2023). Figure 2.1 depicts the volume of CPPAs per region from 2015 to 2023.

There has been significant growth in adopting CPPAs, with North America taking the lead, followed by Europe and the Asia-Pacific regions. The cumulative capacity is steadily increasing, surpassing 180 GW by 2023, which reflects the growing global interest in renewable energy procurement. The following section provides an overview of different regional perspectives and the adoption of PPAs in various countries.

Regional Overview

North America

Within North America, the USA has emerged as a global leader in adopting CPPAs, with the corporate sector increasingly making long-term

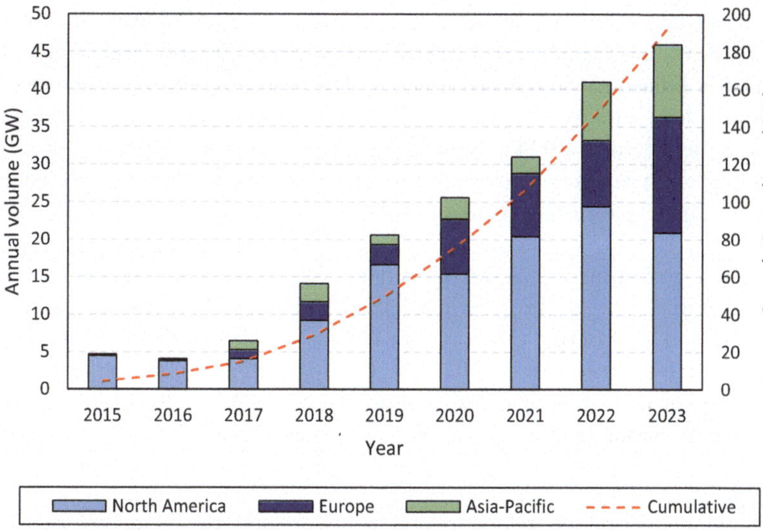

Fig. 2.1 CPPA volume per region (*Data from* Statista [2023])

renewable energy commitments. State and local policy initiatives have fostered further growth of these agreements in the USA. At the same time, corporate sustainability goals and competitive pricing of renewable energy resources, particularly solar and wind power, have played a significant role in the market (Kobus et al., 2021). In 2022, CPPAs in the USA were roughly half of the global market, underlining how the country is the biggest procurer of renewables with such contracts. Driven by their deregulated electricity markets and rich renewable resources, states like Texas and California have become hotspots for CPPA activity. For instance, with significant wind farms and its growing solar sector, Texas has become one of the key drivers in the USA CPPA market. At the same time, companies such as Google, Amazon, and Meta are major PPA contractors, aiming to power their operations with 100% renewable energy. Furthermore, Texas' competitive wholesale electricity market and favorable regulation allows corporations to negotiate directly with renewable energy developers, making it one of the most active regions for CPPAs.

In Canada, the CPPA sector still lags the more established market in the USA. Nevertheless, the adoption of CPPAs in Canada experienced steady growth between 2019 and 2023, particularly in deregulated provinces like Alberta (Brasnett et al., 2023). According to the Business Renewables Center, this growth has created 6,200 jobs and attracted CAD 6.4 billion in capital investments. The resulting contracts generate enough clean energy to power 1.7 million homes annually (Business Renewables Center, 2024). While hydropower has long been the dominant player in Canada's renewable energy landscape, recent years have seen notable expansion in wind and solar developments, particularly in the western provinces. To encourage further renewable energy adoption, the Canadian government has introduced tax incentives that support green energy initiatives. At the provincial level, measures like carbon pricing have created a favorable environment for the growth of PPAs. However, it is essential to note that in 2024 the Canadian PPA market experienced a slowdown, with new projects totaling just 52 MW—a 20-fold decrease from the previous year.

Europe
The European PPA market has become a key driver of renewable energy growth, with corporations and utilities increasingly pursuing long-term agreements to secure clean energy supplies—despite challenges like fluctuating energy prices, economic uncertainty, and regulatory complexities (Kapral et al., 2024). Spain, Sweden, Germany, and the UK lead the market, with significant wind and solar farms as primary energy sources (Harada & Coussi, 2020). The European Union's Green Deal and climate targets, supported by policies like the EU Emissions Trading System (ETS), have encouraged renewable energy procurement. Additionally, mid-sized companies and public institutions are entering the market, recognizing the financial and environmental benefits of PPAs. This movement has led to more flexible PPA structures, such as financial PPAs and multi-buyer agreements, allowing smaller entities to pool resources and participate in renewable energy procurement (Stanitsas & Kirytopoulos, 2023).

In Spain, the deregulated market and favorable climatic conditions for solar energy have positioned it as a hub for solar PPAs (Niculescu, 2024). Additionally, synthetic PPAs for large renewable projects have become increasingly popular, offering Contract for Difference (CfD) arrangements. These contracts help balance financial flows between agreed PPA prices and fluctuating wholesale market prices, promoting renewable energy development within a market-driven framework (World Bank, 2024). Similarly, Sweden's vast wind resources and competitive market framework have made it a top location for wind energy PPAs, with multinational corporations signing significant deals to power their operations. In Germany, PPAs have gained considerable traction as a critical instrument for procuring renewable energy, particularly in the wake of the country's ambitious energy transition goals under the *Energiewende* policy (Rechsteiner, 2021). As subsidies like the feed-in tariffs are gradually phased out, PPAs are becoming increasingly important for financing renewable projects, especially in the wind and solar sectors. The growth of PPAs in Germany is also driven by the rising demand for corporate renewable energy sourcing, with large industrial players and multinational corporations leading the way in signing long-term agreements to meet their energy needs sustainably.

In the UK, PPAs have seen substantial growth, fueled by the rising corporate focus on sustainability and decarbonization. The country's liberalized energy market and transparent regulatory framework, including the RM6289 PPA framework (CCS, 2021), have enabled smaller companies and large energy consumers to engage in CPPA transactions and secure long-term renewable electricity contracts, typically lasting 15–25 years. With the government's ambitious net-zero emissions target for 2050 and the gradual phase-out of subsidies like the CfD scheme, PPAs are increasingly becoming the preferred financing model for renewable energy projects, especially in offshore wind and solar. Major corporations such as BT, Sainsbury's, and Google have signed long-term PPAs.

Asia-Pacific

In Asia, the CPPA market is still developing but is growing steadily, particularly in countries like India, Japan, China, and Vietnam (Rathnasiri & Gunathilaka, 2024). India's PPA framework, developed by the Central Electrical Regulatory Commission, is designed for independent power producers in long-term projects. It offers flexibility in terms of location and fuel sources for power plants. Indian large industrial companies and data centers are increasingly signing PPAs to meet their energy demands while aligning with sustainability goals (Chattopadhyay et al., 2023). Japan's PPA market is also expanding, driven by deregulation in the electricity sector and the government's ambition to reduce reliance on imported fossil fuels. However, challenges remain in terms of regulatory complexity and grid integration for renewables. In China, PPAs are gaining traction as the country accelerates its transition to renewable energy and continues electricity market reforms (Millison et al., 2022). Historically dominated by state-owned companies, the sector is evolving, with policy shifts encouraging direct power transactions between producers and consumers. The government's introduction of green power certificates and a market-based approach to green energy trading, supported by regional power exchange centers, has helped foster PPA growth (Dong et al., 2021). While challenges such as regulatory complexities remain, PPAs demonstrate a growing corporate commitment to renewable energy procurement in China. Vietnam uses PPAs in public tenders for large coal-fired power plants, reflecting the country's state-controlled power sector. The agreements aim to facilitate public–private collaboration while anticipating future market deregulation.

In the Pacific, Australia's PPA market has grown substantially in recent years, driven by corporate demand for renewable energy and abundant solar and wind resources (Briggs et al., 2023). Australia utilizes financial PPAs within its deregulated power market. The agreements enable economic hedging by aligning floating spot market prices with fixed-price contracts, commonly used in financing wind farms. Many companies, including multinational corporations and local businesses, have signed CPPAs to meet their renewable energy targets and reduce exposure to volatile electricity prices. The regulatory environment in Australia is generally favorable, although state-level differences in policy frameworks create some complexity in PPA negotiations.

Africa

Some PPA initiatives are taking place in Africa, with support from the World Bank (Badissy et al., 2020). Africa's PPA market focuses on emergency power and mobile plants as a solution for delivering electricity to rural and underserved areas (Kruger & Eberhard, 2023). These short-term agreements enable purchasing power from mobile plants, often on skids, offering a temporary or emergency supply. Typically drafted by international law firms, these PPAs include an implementation agreement to ensure efficient project execution. Examples of such contracts are found in countries like Namibia, Kenya, and Tanzania. In Namibia, PPAs focus on small-scale projects, offering a short-form agreement and a fuel supply contract. Kenya's PPAs are designed to support hydro, geothermal, and gas-fired power plants, simplifying the process for smaller renewable energy projects with additional model agreements provided by the Energy

Regulatory Commission. In Tanzania, standardized PPAs are available for small power producers, ensuring compatibility with the main grid and isolated microgrids. These agreements also offer consistent tariff methodologies and detailed calculations, facilitating the development of small power projects.

Economic Sector Breakdown and Examples

The IT and heavy industry sectors have primarily dominated the CPPA market, with major tech companies like Amazon, Meta, and Google at the forefront of this movement. Figure 2.2 depicts the main corporate buyers in terms of contracted volume.

Amazon, for instance, claims to have achieved its 100% renewable energy goal by investing in over 240 utility-scale solar and wind projects worldwide, along with 270 on-site solar projects (Amazon, 2024). These contracts have helped create new renewable energy sources, particularly in regions heavily reliant on fossil fuels, such as India, Poland, and South Africa. Amazon's renewable energy portfolio now powers data centers and

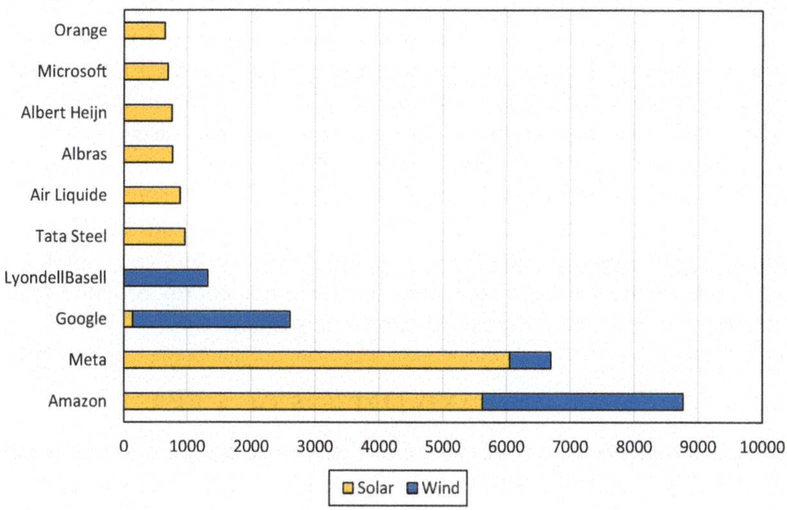

Fig. 2.2 Major corporate buyers of CPPAs in 2024, based on data from the official sustainability reports of Google, Meta, and Amazon (Data for other companies is sourced from BloombergNEF [2024])

facilities globally, significantly decarbonizing the energy grid. Additionally, the company has been a driving force in encouraging other corporations to prioritize emission reductions through impactful renewable energy investments. META claims to be the USA's most prominent corporate buyer of renewable energy by securing 6,700 MW through PPA contracts (Meta, 2025). Noteworthy projects include a 374 MW solar PPA with Silicon Ranch to support its data center operations and a 15-year agreement with the Irish company Highfield Energy to provide 276 MW of solar power. These initiatives are part of Meta's strategy to match 100% of its electricity consumption with renewable energy, aligning with its operational needs and global grid decarbonization efforts. Google, aiming for 24/7 carbon-free energy operations, has also pursued an aggressive PPA strategy. The targets include investments in solar and wind projects, such as a solar farm in Tennessee and wind farms in Sweden. By 2030, Google intends to ensure that all its data centers and campuses operate entirely on renewable energy in real time (Google, 2019).

While the IT and heavy industry sectors have traditionally dominated the CPPA market, the adoption of renewable energy is now expanding into a broader range of industries. Sectors like telecommunications, transportation, consumer goods, and retail are increasingly turning to PPAs as a critical strategy to meet their sustainability targets and reduce their carbon footprints. Figure 2.3 illustrates the breakdown of CPPA volumes across different economic sectors, excluding IT and heavy industry.

By 2023, the telecommunications sector had the largest cumulative renewable energy capacity, exceeding 2,500 MW. Other sectors, including transportation, consumer goods, automotive, and retail, also saw significant increases in PPA capacity, surpassing 500 MW. Additionally, industries such as pharmaceuticals, financial services, and food and beverages are gaining momentum in renewable energy procurement, reflecting the broader shift toward corporate sustainability.

Structures and Drivers of a PPA

Multiple structures of a PPA exist, and the format agreed upon between buyer and seller plays a significant role in the contract's financial performance and the allocation of risks. To better understand the different types of PPAs, this section presents the main concepts that define each type, the pricing structures involved, and how the associated risks are transferred

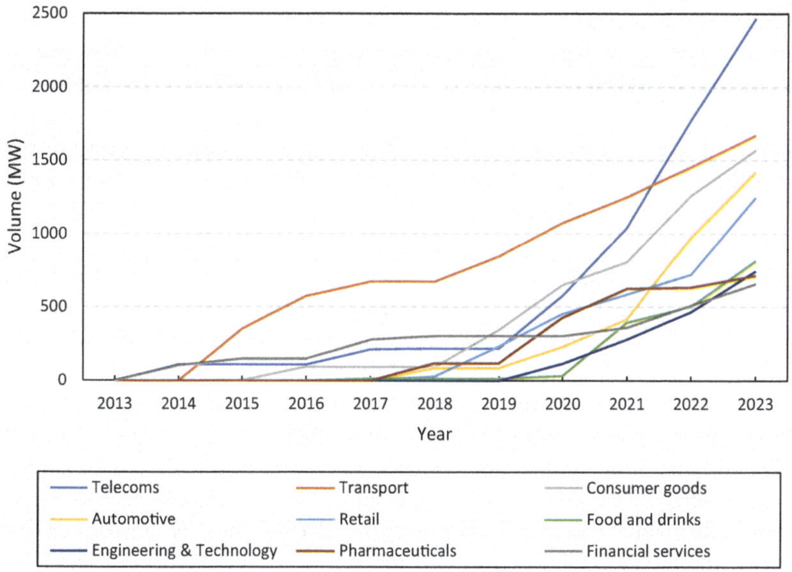

Fig. 2.3 Cumulative CPPA volume by economic sector (excluding IT and heavy industry) in Europe (*Data from* RE-Source Platform [2024])

between the parties. Our focus is on key concepts that enable the financial modeling of PPA contracts and market variables that may affect the pricing structure.

Concepts of PPAs

The main concepts describing a PPA are: (i) the physicality of delivery, (ii) the delivery schemes, and (iii) the pricing structures. We have based these categories on Mittler et al. (2025), but similar ones can be found in other works (Dominy & Zubair, 2021; Mendicino et al., 2019; Simões et al., 2024; Tang & Zhang, 2019).

Physicality of Delivery

The physicality of delivery refers to whether the electricity is delivered to the offtaker.[1] Within this context, the parties can have two types of agreement: physical[2] and financial. In the first category, the energy is physically delivered to the consumer via the electrical grid, a dedicated power line, or a local grid. For this reason, there is a geographic limitation for the supply of energy, dependent on the available grid infrastructure. Physical PPAs can be split into on-site[3] and off-site[4]: in the first, energy is delivered through a dedicated power line or a local grid; in the second, transmission and/or distribution grids are required to transport electricity. In a financial PPA[5] unlike a physical PPA, electricity is not delivered to the offtaker. Instead, the producer sells all the electricity in wholesale markets—day-ahead and intraday—while the consumer buys in the same market or from an energy supplier. The difference between the price agreed upon in the PPA and the spot market price is then settled between the parties. Due to this settlement mechanism, financial PPAs can also be called CfDs (Gabrielli et al., 2022; Mendicino et al., 2019; Tang & Zhang, 2019; Zapparoli, 2023). Figure 2.4 presents a graphical scheme of electricity and cash flows in physical and financial agreements, respectively.

In a financial agreement, cash can flow in both directions, i.e., the buyer can get paid by the seller, depending on the evolution of wholesale market prices. Also, PPAs can include the transfer of RECs to the consumer, providing additional revenues and guaranteeing renewable consumption in the consumer's ESG reports.

[1] The buying party in the contract is commonly referred as offtaker.

[2] Huneke et al. (2018) use CPPA as a synonym of physical PPA. However, this is not a common nomenclature. CPPA is more frequently used as described in Section "Introduction".

[3] On-site PPAs are also called Behind the Meter PPA (Mendicino et al., 2019). Other publications use Private-wire PPA, sometimes not including in the physical PPA category (Natasha, 2019), or differentiating from the On-site in respect to the exact location of the power facility (Labordena, 2020).

[4] Commonly called Sleeved PPA (Mendicino et al., 2019; Natasha, 2019).

[5] Other common names are Virtual PPA or Synthetic PPA (Huneke et al., 2018; Mendicino et al., 2019; Mittler et al., 2025).

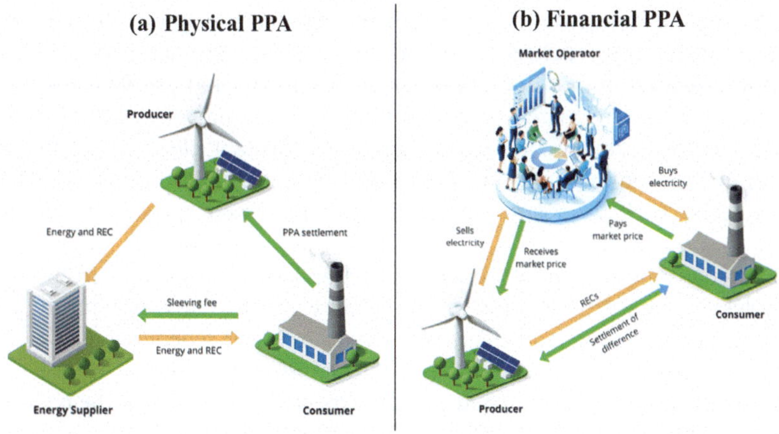

Fig. 2.4 Electricity, REC, and cash flows in physical and financial agreements

Delivery Schemes

Before introducing the concept of delivery schemes, we present some of the risks (volume, shape, and price risks) associated with PPAs. Recognizing these risks is important for comprehending how different delivery schemes allocate them. We discuss additional risks and their mitigation strategies in Section "Costs and Other Risks Incurred When Signing a PPA".

- **Volume risk** has to do with the possibility that the power plant may underperform relative to the contracted amount of electricity over a long-term perspective.
- **Shape risk** (also known as Predefined Profile risk) accounts for generation performance on an hourly basis. While a power plant may meet long-term projections, it can underperform during certain hours.
- **Price risk** involves the uncertainty surrounding the fluctuations of wholesale electricity prices, which can vary both over a long-term and hourly basis, impacting the PPA's financial performance (Brindley et al., 2020). The consumer predominantly bears price risk, as the producer typically has guaranteed revenues under a PPA. If wholesale prices drop below the contracted PPA price, consumers could pay more than the market price for electricity.

In general terms, the delivery schemes refer to the amount of energy the producer must deliver each time during the contract period. This chapter focuses on four delivery schemes: baseload, fixed-profile (or fixed-shape), pay-as-forecasted (or pay-as-nominated[6]), and pay-as-produced. In a baseload scheme, the parties agree to constant electricity delivery during the contract period. For Variable Renewable Energy (VRE) producers, such as solar and wind, this agreement poses significant risks due to uncertain resource availability. If production falls below the contracted amount, the producer must purchase the shortfall from the wholesale market to comply with the agreement. Thus, the producer assumes both volume and shape risks under baseload schemes. In a fixed-profile scheme, instead of committing to constant production, the seller adjusts the hourly delivery to match the expected availability of the energy source. This adjustment significantly transfers shape risk to the consumer. For instance, in a fixed-profile scheme with a solar producer, the seller would not be obliged to deliver energy during the night. Furthermore, since the delivery is based on the source's expected availability, deviations are generally smaller than baseload schemes. As a result, baseload schemes come with higher prices than fixed-profile agreements due to the greater risks involved. On the other hand, in a pay-as-forecasted scheme, the seller periodically determines the delivered volumes based on forecasted generation. Normally, the delivery volume is settled on a day-ahead schedule, reducing the time distance between forecast and actual electricity delivery, which mitigates risks on the seller's side. Therefore, VRE producers are prone to accept lower prices for this type of delivery.

Lastly, the seller transfers most of the risk to the buyer in a pay-as-produced scheme. The delivery is not pre-determined, but the buyer commits to purchasing all the energy produced, regardless of whether their consumption can accommodate it. For this reason, pay-as-produced schemes generally come with lower prices. Figure 2.5 offers a visual representation of each delivery scheme.

Pricing Structures

The pricing structure in a PPA refers to how the electricity price fluctuates during the agreement period. The structures we assess in this section are fixed price, floor price, and collar. Among these, the fixed price is the

[6] As described in (Synertics, 2024) and (SIA Partners, 2023).

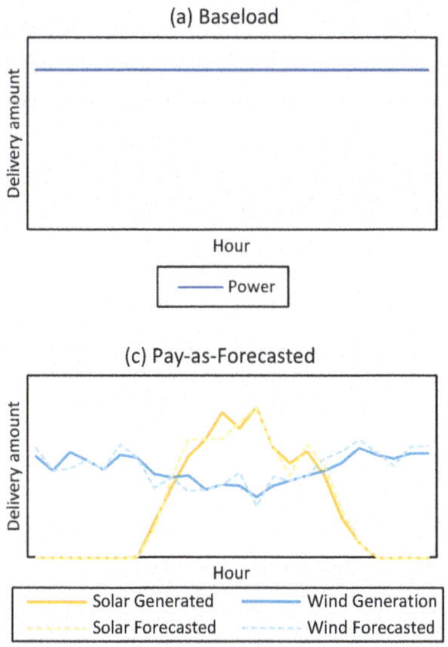

Fig. 2.5 Generation commitment according to the delivery scheme: Baseload; Fixed-Profile; Pay-as-Forecasted; and Pay-as-Produced

most common. Other options beyond these three can be found in the literature (Dominy & Zubair, 2021; Mittler et al., 2025; Simões et al., 2024).

In a fixed-price structure, producers sell electricity at a fixed price for all hours, commonly called the strike price. Variants of a fixed-price structure may include an indexation factor (such as inflation) or other escalation methods. These indexations allow for periodic (e.g., annually) changes in the price, although the price remains the same across all hours within a given period. In the floor structure, the wholesale market price works as the basis of the electricity price at a given hour. The parties apply an agreed discount upon this base price. However, a minimum price (floor) protects producers from extremely low market prices. If the wholesale price drops below this floor, the buyer must pay the difference to ensure the producer receives the minimum agreed amount. In the collar structure, the price also tracks the wholesale market but includes both a floor and a ceiling (cap). While the floor protects producers from low prices, the cap safeguards consumers from excessively high prices. Unlike the floor structure, the collar structure does not always involve a discount on wholesale prices. We provide Eqs. (2.1)–(2.6) to clarify the calculations of the settlements between the parties from a financial delivery contract perspective.

$$\text{Fixed price}: \begin{cases} \text{Seller pays buyer } (FP - K) \times \text{Prod, if } FP \geq K \\ \text{Buyer pays seller } (K - \text{FP}) \times \text{Prod, if } FP < K \end{cases} \tag{2.1}$$

$$\text{Floor price}: \begin{cases} \text{Seller pays buyer } (FP - DFP) \times \text{Prod, if } F \leq \text{DFP} \\ \text{Seller pays buyer } (FP - F) \times \text{Prod, if} DFP < F < FP \\ \text{Buyer pays seller } (F - FP) \times \text{Prod, if } FP \leq F \end{cases} \tag{2.2}$$

$$\text{Collar:} \begin{cases} \text{Seller pays buyer } (FP - C) \times \text{Prod, if} C < FP \\ \text{No settlement if } F \leq FP \leq C \\ \text{Buyer pays seller } (F - FP) \times \text{Prod, if} FP < F \end{cases} \tag{2.3}$$

$$\text{Collar with discount}:$$
$$\begin{cases} \text{Seller pays buyer } (FP - C) \times \text{Prod, if} C < \text{DFP} \\ \text{Seller pays buyer } (FP - DFP) \times \text{Prod, if } F < DFP \leq C \\ \text{Seller pays buyer } (FP - F) \times \text{Prod, if} DFP \leq F < FP \\ \text{Buyer pays seller } (F - FP) \times \text{Prod, if} FP \leq F \end{cases} \tag{2.4}$$

$$FP = \frac{\sum_{t \in T} (\text{Prod}_t \times WSP_t)}{\sum_{t \in T} \text{Prod}_t} \tag{2.5}$$

$$DFP = FP \times (1 - D) \tag{2.6}$$

Equation (2.1) describes the fixed-price structure, where payments between the buyer and the seller are determined by comparing the floating price (*FP*) to the agreed-upon strike price (K). If *FP* is higher than K, the seller should pay the difference to the buyer. If *FP* is lower than K, the buyer should pay the difference to the seller. Equation (2.2) details the Floor price structure, where the buyer benefits from a discounted market price, but if the market price falls below F, the buyer compensates the seller for the shortfall. Equation (2.3) explains the Collar structure, where both parties are protected from extreme market fluctuations by establishing a price range with a floor (F) and a cap (C). Equation (2.4) introduces a variation of the Collar structure that includes a discount (D). In this version, D first adjusts the *FP* before applying

C and *F* limits. Finally, Eqs. (2.5) and (2.6) define the calculation of the *FP* and the discounted floating price (*DFP*), which are essential for determining the settlements under these contracts.

Summary

All PPAs must consider the three critical concepts outlined in this section. While these do not cover every clause of the agreement, they play a significant role during the negotiations. In summary: (i) the physicality of delivery defines whether electricity is physically delivered to the consumer or not, (ii) the delivery schemes define the hourly electricity production commitment outlined in the agreement, and (iii) the pricing structure determines the price of the electricity for each hour during the contract period. Figure 2.6 illustrates the categories and the subcategories explored in this section.

Drivers of PPA Prices

Now that we have covered the fundamental concepts behind a PPA, we will explore the factors that influence the decision of a strike price, as well as the values of the floor and the cap. As in the previous sections, our goal is not to provide an exhaustive analysis or a definitive formula for setting these values, since they are ultimately the result of negotiations between

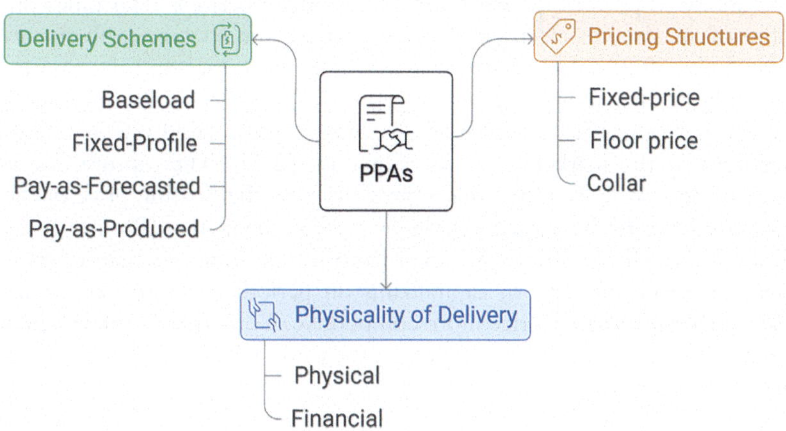

Fig. 2.6 Summary of PPA concepts

the buyer and seller. Instead, we highlight key variables that can drive these parameters higher or lower. Specific incentives, such as tax exemptions, grants, subsidies, and feed-in tariffs, which may also affect pricing, are out of the scope of this chapter. Our focus is on market variables within the competitive landscape that companies face when procuring electricity.

Futures

Energy futures are a type of derivative product in which the parties involved agree to buy and sell energy at a pre-determined price, with delivery set for a future date. This concept is similar to a PPA; however, the main difference between the two is that energy futures are traded on an exchange that predefines the standard features of the product, such as contract size and terms (Hull, 2012). Thus, energy futures markets serve as a tool to predict future prices in the wholesale market (Kim et al., 2024). In this context, the price of energy futures reflects market participants' expectations about market behavior and electricity price fluctuations. If market participants anticipate future prices to rise or fall, electricity generators will consider these projections when setting the prices they are willing to accept in PPAs. Similarly, consumers use future prices as a proxy to define how much the cost of electricity under a PPA.

Floating Price and Capture Rate[7]

The ability to produce power during high-priced periods determines the value of an asset in electricity markets, where the spot price fluctuates frequently (often on an hourly basis). The floating price and the capture rate indicate this ability. The floating price of a power facility represents the revenue generated per unit of energy produced, based on the sale of electricity at the market price (see Equation [2.5]). The capture rate is a key performance indicator that represents how the floating price of the asset compares to the average wholesale price in the market (Blume-Werry et al., 2021; Hirth, 2016). A higher capture rate indicates that a power facility is more effective in capitalizing on periods of higher electricity prices, making the PPA price more competitive. Consequently, developers usually aim to maximize the capture rate when deciding on a project's layout and location.

[7] Floating price can also be referred as capture price or market value. Capture rate can also be referred as value factor.

Equation (2.7)[8] outlines how to calculate the capture rate (CR) of a power facility as a function of the floating price (FP) and the wholesale price (WSP).

$$CR = \frac{FP}{\frac{1}{T}\sum_{t\in T} WSP_t} \tag{2.7}$$

The product of the capture rate and the price of energy futures for the PPA period is a valuable tool for forecasting the floating price of a power plant during the same timeframe. Consequently, buyers and sellers can use it as an initial reference point for setting the PPA price. However, solar and wind technologies rely heavily on weather conditions, which limits the ability of power plant operators and developers to directly influence the floating price and capture rate.

Figure 2.7 illustrates how the production profile of an asset impacts the floating price and the capture rate. The graph shows the average hourly electricity prices in Germany's day-ahead market for 2023, alongside the average hourly solar and onshore wind generation.

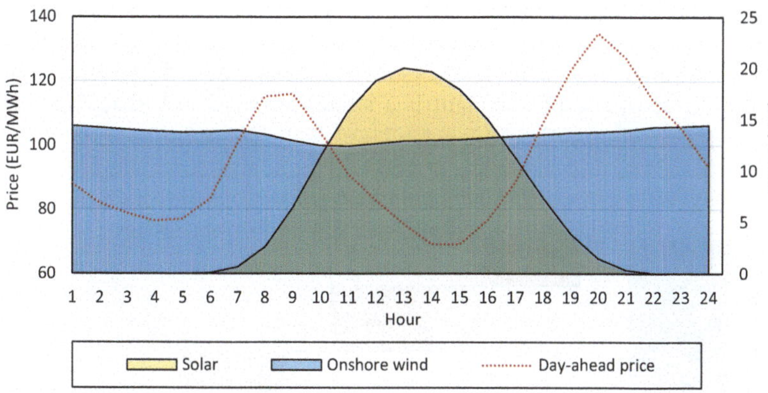

Fig. 2.7 Generation commitment according to the delivery scheme: Baseload; Fixed-Profile; Pay-as-Forecasted; and Pay-as-Produced (*Data from* ENTSO-E [2024])

[8] The average wholesale price can also be referred as base price (Hirth, 2016) or time-weighted average price (Eamon et al., 2016).

As observed in Fig. 2.7, most solar generation occurs during hours when prices are relatively low, with minimal output during the evening price peak. This generation profile leads to a floating price lower than the wholesale price. In contrast, wind generation presents a more consistent profile over the day, allowing assets to capture higher prices during the peak hours in the early morning and late afternoon. This generation profile results in a higher floating price. For instance, in Germany in 2023, solar and onshore wind technologies had average capture rates of about 0.756 and 0.800, respectively (Netztransparenz, 2024). Some strategies can be used to improve capture rates. These include using energy storage systems to provide flexibility or, in the case of solar technology, using trackers to increase production during mornings and evenings. Furthermore, taking advantage of the complementarity of solar and wind can reduce generation uncertainty, mitigating risks for both parties.

Price Composition

The price composition of a PPA is a consequence of how risks are allocated in the agreement. Figure 2.8 depicts how price is composed based on the risks taken by the offtaker, considering a pay-as-produced delivery with a fixed-price agreement.

The price comprises the forecasted floating price and the risks incurred (price, volume, and shape). As previously described, forecasting the floating price of the power plant is a straightforward task that depends

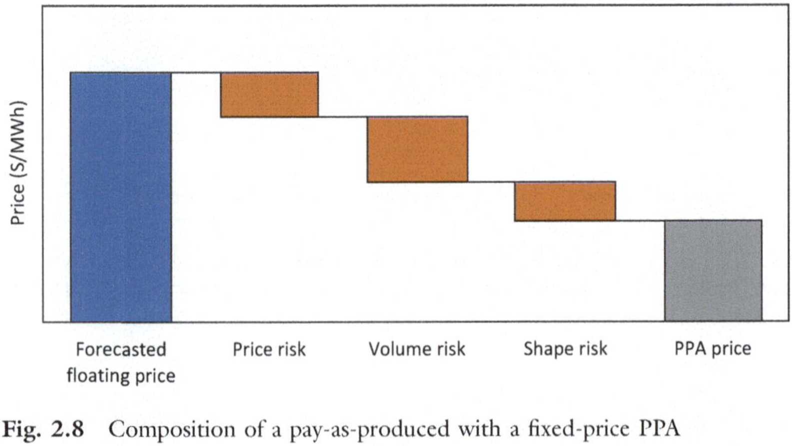

Fig. 2.8 Composition of a pay-as-produced with a fixed-price PPA

only on public data. On the other hand, properly quantifying the risk components requires a thorough analysis of the asset's production profile and the evolution of market conditions. In summary, the more a buyer is willing to take risks, the more a seller is willing to reduce the price.

Other Drivers

In the previous subsections, we presented the market drivers we find most relevant to estimate a reference value for an initial PPA negotiation. However, many other factors may affect the price of a PPA. For instance, including RECs in the agreement would raise the cost of the PPA since offtakers could use the certificate to comply with sustainability standards (e.g., ETS in the European Union). In the same way, governmental and societal pressures on ESG, although hardly quantifiable, would also affect how much a consumer is willing to pay for renewable energy. Lastly, a consumer's credit score also plays a role in the prices. Higher credit scores are more appealing for projects' financing approval with bank institutions, pushing renewable producers' price threshold down.

Costs and Other Risks Incurred When Signing a PPA

We divide this section into two subsections, each outlining critical risks, costs, and mitigation strategies that should be considered during PPA negotiations. In addition to the risk management strategies discussed in the previous section, PPAs usually involve additional risks that can affect the financial attractiveness of the contract. For instance, a CPPA may include performance penalties for the producer if specified maximum and minimum thresholds for energy generation are not achieved (Bruck et al., 2018). Another potential risk involves delays in obtaining necessary permits to build and/or operate the facility, which can prevent the producer from meeting the agreed delivery schedule.

Risks

Previous studies (Badissy et al., 2020; Danquah et al., 2024) have shown that the risks associated with a PPA vary depending on the project phase. Generally, earlier stages of the project development carry higher risks, although other risks persist during the project's lifecycle. In the following, we describe some of them in line with the different phases of a project:

- **Pre-Operational Risks:** Before the project reaches the operational stage, the primary risks involve achieving the planned Commercial Operation Date (COD) during pre-construction or construction phases. Delays in starting construction can lead to a postponed COD, resulting in financial losses. Sellers usually bear these risks since they are generally better positioned to manage them. Potential setbacks such as delays in obtaining construction permits, grid connection approvals, or equipment deliveries can disrupt project timelines. Additionally, the seller must complete proper tests and commissioning before COD to ensure reliable performance. If these requirements are not met, the offtaker may request compensation for financial losses or, in some cases, seek the termination of the contract.
- **Operational Risks:** During the operational phase, risks shift to the seller's ability to deliver the agreed-upon energy output. Parties often include a performance clause to ensure the plant meets generation targets that satisfy the buyer's needs. Failure to achieve these targets can result in compensation to the buyer for any shortfall. In cases of grid congestion, grid operators may request the producer to curtail generation to maintain optimal grid conditions. In this situation, PPAs should clarify whether compensation for lost generation will be provided and if this curtailed amount counts towards minimum performance requirements. Furthermore, negative spot prices present a notable risk. Some offtakers claim the right to request curtailment during these periods, mainly if the PPA includes a fixed or floor price—though producers may resist accepting such terms.
- **Risks Independent of Project Stage:** Certain risks persist independent of the project's development phase. These include changes in legislation, taxation, political risks, asset ownership, and other external factors. For instance, a new law may be introduced, an existing may be changed, or the courts may change the interpretation during the contract duration, potentially impacting the agreement. It is therefore essential to establish a reference point for applicable legislation of the contract, enabling the parties to assess any changes against a fixed baseline. In such cases, the parties

might agree to rediscuss the agreement to restore the initial financial performance of the PPA.

Mitigation Strategies and Costs

Many risks mentioned in the previous subsection can be mitigated with proper insurance or guarantees. When parties use insurance to protect against potential defaults, the costs of these measures should be factored into the financial performance analysis. Parent Company Guarantees can also help to reduce various risks, although not all companies can provide this type of security.

As discussed in Section "Concepts of PPAs", renewable energy producers entering financial PPAs participate in the spot markets, where energy volumes are based on forecasts due to the variability of solar and wind resources. The actual generation may differ from forecasted commitments, leading to potential imbalance costs. Therefore, spot market participation typically requires a third party to operate and manage the power plant's interactions with market operators, grid operators, and fiscal agencies (Mendicino et al., 2019). This third party, the Balancing Responsible Party (BRP), is accountable for making forecasts and placing bids in the daily spot markets. While BRPs are not directly involved in the PPA, their role is essential to ensure smooth operation for renewable energy generators participating in financial PPAs. Since BRPs are responsible for forecasts, they may also assume all the costs of imbalances. However, BRP services come with fees that must be included in the financial performance of a PPA. For off-site physical PPA, where energy is not traded in the spot market, a BRP is not required. Nonetheless, since electricity is transported over a third-party infrastructure, a sleeving party typically charges a fee for grid usage. The sleeving party may also impose specific conditions for connecting to and transporting the energy through their infrastructure, which could lead to additional costs.

PPAs also contain termination clauses outlining the compensation if either party chooses to terminate the agreement before its term. These clauses usually reflect the estimated loss by the non-terminating party to secure a new contract. The parties can agree on a methodology to calculate this amount or appoint an independent third party to estimate the termination values. The termination amount can represent a significant cost, impacting the cash flow of both the seller and the buyer.

Finally, as previously mentioned, some PPAs include performance penalties for the producer if the power plant underperforms or overperforms (Bruck et al., 2018). These costs must be accounted for to ensure proper contract pricing and cash flow projections. A well-defined Force Majeure (FM) clause is also required to protect the parties from unforeseen adverse events. This clause allows each party to be released from obligations if circumstances beyond its control prevent performance. Typical FM events include wars, uprisings, terrorism—whether local or elsewhere but affecting the PPA—and natural disasters such as earthquakes, hurricanes, tsunamis, or severe weather conditions. Given the unpredictable nature of FM events, the terms of the clause should precisely define what constitutes FM and clarify situations in which a claim by either party is not permitted.

Conclusions

PPAs have become an essential mechanism for corporate decarbonization, providing a structured approach for companies to obtain renewable energy, to minimize GHG emissions, and secure stable energy costs. In this chapter, we explored the critical components of PPAs, focusing on their various structures, including physical and financial agreements, distinct delivery schemes, and pricing mechanisms. These elements facilitate risk-sharing between energy producers and corporate buyers. Our chapter also detailed how settlements occur between the parties of a PPA and examined the various factors that influence its pricing.

Our analysis of global markets revealed that PPAs have gained popularity in regions such as North America and Europe, driven by favorable regulatory environments and ESG goals. This chapter also examined the risks associated with PPAs, including market uncertainties and pre-operational delays. It emphasized the importance of contract design, insurance, and balancing services in mitigating these risks. Despite their advantages, PPAs come with inherent risks, such as performance uncertainty, exposure to fluctuations in spot market prices, and potential delays in project execution. These challenges underscore the importance of careful contract structuring, which may involve delivery schemes, pricing mechanisms, and insurance clauses to address unforeseen disruptions.

In this landscape, it is crucial to communicate the definitions and nuances of PPAs, guiding companies in recognizing potential risks and applying effective mitigation strategies when engaging in these agreements. Our intention with this chapter was to equip readers with important concepts and practical insights for designing and implementing PPAs in different market contexts. As renewable technologies progress, PPAs will be central to decarbonizing economic activities and promoting investment in clean energy sources. Emerging PPA models can support the integration of technologies like large-scale energy storage and networks of distributed loads (such as electric vehicles), strengthening the stability and resilience of the electricity grid.

Acknowledgements https://doi.org/10.54499/UIDB/05037/2020, UIDB/00308/2020 with the https://doi.org/10.54499/UIDB/00308/2020, and the PhD grant 2024.18454.PRT; by the RES4CITY project, funded by the Horizon Europe Program (Project: 101075582); and by the Energy for Sustainability Initiative of the University of Coimbra

References

Amazon. (2024, July 15). *How Amazon achieved its 100% renewable energy goal*. Amazon Science. https://www.amazon.science/news-and-features/how-amazon-achieved-its-100-percent-renewable-energy-goal

Badissy, M., Evans, A., Ewelukwa, N., Govender, J., Ketchum, R. T., Loraoui, M., Liebenberg, C., Nagarajan, S., Ndahumba, G. T., Pavry, J., & Vajeth, O. (2020). *Understanding Power Purchase Agreements* (2nd ed.). https://www.booksprints.net/book/understanding-ppa-update/

Blume-Werry, E., Huber, C., Resch, G., Haas, R., & Everts, M. (2021). Value factors, capture prices and cannibalism: Nightmares for renewable energy decision-makers. *The Journal of World Energy Law & Business, 14*(4), 231–247. https://doi.org/10.1093/jwelb/jwab027

Brasnett, G., Hickson, C., & Shor, R. J. (2023). *Mapping techno-economic feasibility of geothermal resources in Alberta, Canada*. In 48th Workshop on Geothermal Reservoir Engineering, Stanford University.

Briggs, C., Nasser, A., McKeon, J., & Prendergast. (2023). *Corporate PPAs in Australia: State of the market 2023*. https://opus.lib.uts.edu.au/bitstream/10453/180722/2/SOM-2023.pdf

Brindley, G., Niklaus, A., Holm, K., Hunt, H., & Ciancibello, V. (2020). *Risk mitigation for corporate renewable PPAs*. Re-Source Platform. https://resource-platform.eu/reports-2/

Bruck, M., Sandborn, P., & Goudarzi, N. (2018). A Levelized Cost of Energy (LCOE) model for wind farms that include Power Purchase Agreements (PPAs). *Renewable Energy, 122*, 131–139. https://doi.org/10.1016/j.renene.2017.12.100

Business Renewables Center. (2024). *Corporate renewable energy deals in Canada (Deal tracker Q3 2024).* https://businessrenewables.ca/sites/default/files/inline_files/BRC%20Deal%20Tracker%20-%202024%20Q3.pdf

CCS. (2021, October 5). *Provision of Power Purchase Agreement.* https://www.crowncommercial.gov.uk/agreements/RM6289

Chattopadhyay, D., Chatterjee, S. K., & Soonee, S. K. (2023). Spotlight on the spot market: A review of the Indian wholesale electricity market. *The Electricity Journal, 36*(1), 107239. https://doi.org/10.1016/j.tej.2023.107239

Danquah, A., Cerqueira, P. A., & Silva, P. P. da. (2024). Risk management in renewable energy power purchase agreements—A bibliographic review. In *Proceedings of the 6th International Conference on Energy & Environment: Bringing together Engineering and Economics* (pp. 69–75). University of Minho.

Dominy, P., & Zubair, S. (2021). *Pricing structures for corporate renewable PPAs.* World Business Council for Sustainable Development (WBCSC).

Dong, J., Liu, D., Zhang, Y., Wang, Y., & Dou, X. (2021). A novel LCOE pricing model for renewable energy with Power Purchase Agreement: A case study in China. *Processes, 9*(10), Article 10. https://doi.org/10.3390/pr9101780

Eamon, B., Kim, H.-G., Kang, Y.-H., & Yun, C.-Y. (2016). Optimizing the electricity price revenue of wind power generation captures in the South Korean electricity market. *Journal of the Korean Solar Energy Society, 36*(1), 63–73. https://doi.org/10.7836/kses.2016.36.1.063

Enescu, D., Ciocia, A., Galappaththi, U. I. K., Wickramasinghe, H., Alagna, F., Amato, A., Díaz-González, F., Spertino, F., & Cocina, V. (2023). Energy tariff policies for renewable energy development: Comparison between selected European countries and Sri Lanka. *Energies, 16*(4), Article 4. https://doi.org/10.3390/en16041727

Gabrielli, P., Aboutalebi, R., & Sansavini, G. (2022). Mitigating financial risk of corporate power purchase agreements via portfolio optimization. *Energy Economics, 109*, 105980. https://doi.org/10.1016/J.ENECO.2022.105980

Google. (2019, September). *Clean Energy: 24x7—Carbon-free Internet—Google Sustainability.* Sustainability. https://sustainability.google/operating-sustainably/stories/24x7/

Harada, L.-N., & Coussi, M. (2020). Power Purchase Agreements: An emerging tool at the centre of the European energy transition: A focus on France. *European Energy and Environmental Law Review,*

29(5). https://www.kluwerlawonline.com/api/Product/CitationPDFURL?file=Journals/EELR/EELR2020043.pdf

Hirth, L. (2016). The benefits of flexibility: The value of wind energy with hydropower. *Applied Energy, 181*, 210–223. https://doi.org/10.1016/j.apenergy.2016.07.039

Hull, J. (2012). *Options, futures, and other derivatives* (8th ed). Prentice Hall.

Huneke, F., Göß, S., Österreicher, J., & Dahroug, O. (2018). *Power Purchase Agreements: Financial model for renewable energies*. https://www.energybrainpool.com/en/power-purchase-agreements-ppas

Kapral, K., Soetaert, K., & Castro, R. (2024). An off-site Power Purchase Agreement (PPA) as a tool to protect against electricity price spikes: Developing a framework for risk assessment and mitigation. *Energies*, *17*(9), Article 9. https://doi.org/10.3390/en17092161

Kim, A., Ryu, D., & Webb, R. I. (2024). Evolving roles of energy futures markets: A survey. *Borsa Istanbul Review, 24*, 1–14. https://doi.org/10.1016/j.bir.2024.05.004

Kobus, J., Nasrallah, A. I., & Guidera, J. (2021). *The role of corporate renewable Power Purchase Agreements in supporting US wind and solar deployment*. Center of Global Energy Policy.

Kruger, W., & Eberhard, A. (2023). The impact of competition, trust and capital on renewable energy auction outcomes in sub-Saharan Africa: Analysing auctions in South Africa, Zambia and Namibia. *Energy Policy*, *178*, 113572. https://doi.org/10.1016/j.enpol.2023.113572

Labordena, M. (2020). *On-site renewable electricity and storage for corporates: Business models & policy framework*. Re-Source Platform. https://resource-platform.eu/reports/

Mendicino, L., Menniti, D., Pinnarelli, A., & Sorrentino, N. (2019). Corporate power purchase agreement: Formulation of the related levelized cost of energy and its application to a real life case study. *Applied Energy*, *253*, 113577. https://doi.org/10.1016/J.APENERGY.2019.113577

Meta. (2025). Energy—Accelerating the transition to renewable energy. *Meta Sustainability*. https://sustainability.atmeta.com/energy/

Millison, D., George, L., Acharya, J., & Wijayatunga, P. (2022). Enabling policy and regulatory environment for solar power development: Lessons in Asia-Pacific region. *Solar Compass*, *2*, 100023. https://doi.org/10.1016/j.solcom.2022.100023

Mittler, C., Bucksteeg, M., & Staudt, P. (2025). Review and morphological analysis of renewable power purchasing agreement types. *Renewable and Sustainable Energy Reviews*, *211*, 115293. https://doi.org/10.1016/j.rser.2024.115293

Natasha, L.-J. (2019, November 12). *Corporate Power Purchase Agreements (PPAs): What are they?* DLA Piper. https://www.dlapiper.com/en/insights/publications/2019/11/what-are-corporate-power-purchase-agreements-ppa

Netztransparenz. (2024). *Market value overview*. https://www.netztransparenz.de/en/Renewable-energies-and-levies/EEG/Transparency-requirements/Market-premium/Market-value-overview

Niculescu, E. (2024). A comparative analysis between the Norwegian and Spanish renewables market, including the level of development and implementation of Power Purchase Agreements. *International Research Journal of Advanced Engineering and Science*, *9*(2).

Rathnasiri, M. S. H., & Gunathilaka, D. H. P. M. (2024). Net zero policies for sustainable future: Issues and challenges in the Asia Pacific region. In R. Singh & D. Crowther (Eds.), *Transition towards a sustainable future: Net zero policies and environmental sustainability* (pp. 253–272). Springer Nature. https://doi.org/10.1007/978-981-97-5756-5_12

Rechsteiner, R. (2021). German energy transition (Energiewende) and what politicians can learn for environmental and climate policy. *Clean Technologies and Environmental Policy, 23*(2), 305–342. https://doi.org/10.1007/s10098-020-01939-3

Rövekamp, P., Schöpf, M., Wagon, F., Weibelzahl, M., & Fridgen, G. (2021). Renewable electricity business models in a post feed-in tariff era. *Energy*, *216*, 119228. https://doi.org/10.1016/j.energy.2020.119228

SIA Partners. (2023). *A practical guide to PPAs*. https://www.sia-partners.com/en/insights/publications/a-practical-guide-ppas

Simões, F., Henriques, C., Figueiredo, N. C., & da Silva, P. P. (2024). Optimizing CPPAs in the renewable energy market: Contract performance assessment. In *Proceedings of the 6th International Conference on Energy & Environment: Bringing together Engineering and Economics* (pp. 19–25). University of Minho.

Stanitsas, M., & Kirytopoulos, K. (2023). Sustainable energy strategies for Power Purchase Agreements (PPAs). *Sustainability*, *15*(8), Article 8. https://doi.org/10.3390/su15086638

Synertics. (2024, July 3). *Pay-as-nominated PPAs*. https://synertics.io/blog/125/pay-as-nominated-ppas

Tang, C., & Zhang, F. (2019). Classification, principle and pricing manner of renewable power purchase agreement. In *IOP Conference Series: Earth and Environmental Science*, *295*(5), 052054. https://doi.org/10.1088/1755-1315/295/5/052054

World Bank. (2024). *Spain Standardized Synthetic PPA*. Spain Standardized Synthetic PPA. https://ppp.worldbank.org/public-private-partnership/library/spain-standardized-synthetic-ppa

Zapparoli, L. (2023). *Multi-Objective Optimal Design of Hybrid PPA Portfolios under Un-certainty Tesi di Laurea Magistrale in Electrical Engineering-Ingegneria Elettrica* [Politecnico di Milano]. https://hdl.handle.net/10589/209902

CHAPTER 3

Business Model Innovation and Business Ecosystem Innovation for the Sustainable Energy Transition

Rashid Dehkordi, Petri Ahokangas, Mika Sorvisto, and Mostafa Mohammadian

Introduction

Climate change, the depletion of fossil fuels, and the growing demand for green energy have made the transition to sustainable energy sources an urgent global priority. International initiatives such as the Paris Agreement (UNFCCC, 2024), along with various policies targeting net-zero emissions by 2050 (European Commission, 2024), underscore the critical need to shift to renewable energy solutions and enhance energy

R. Dehkordi (✉) · P. Ahokangas
Oulu Business School, Martti Ahtisaari Institute, University of Oulu, Oulu, Finland
e-mail: Rashid.Dehkordi@oulu.fi

P. Ahokangas
e-mail: petri.ahokangas@oulu.fi

T. Walker et al. (eds.), *Wiring the Future*, Palgrave Studies in Emerging Risk Management and Sustainable Finance,
https://doi.org/10.1007/978-3-032-14717-2_3

efficiency across all sectors. However, this transition faces significant challenges, including the high initial costs of renewable energy technologies, the need for robust infrastructure such as energy storage systems, and complex regulatory barriers, all of which require intricate coordination and collaboration across industries and stakeholders (Crichton et al., 2024; Tolonen, 2024). For instance, the upfront investment required to deploy utility-scale solar or wind farms remains prohibitively high for many regions, especially where financing mechanisms or government incentives are limited.

Companies are increasingly recognizing that transitioning to sustainable energy requires more than just technological advances; it demands a holistic approach that integrates the three pillars of social, economic, and environmental sustainability. This transition requires innovative approaches to business operations, value creation, and value capture, as well as an engagement with their broader business ecosystem (Evans et al., 2017a, 2017b). Companies thus need to rethink and innovate their business models (BMs) to align with sustainable practices. Business Model Innovation (BMI) can be defined as introducing "novel business models to the product market space in which the firm competes" (Snihur & Zott, 2020, p. 556). Recent examples highlight the importance of BMI in advancing the sustainable energy transition. A notable example is Tesla, the electric vehicle and energy company, which is innovating its BM by integrating electric vehicles (EVs), energy storage solutions, and its own Supercharger network into a unified business model (Chen & Perez, 2018). This unified model enabled Tesla to control multiple elements of the energy and mobility, from energy storage to vehicle use and charging infrastructure. Tesla's BMI has extended beyond the company itself and embraced innovation in the broader ecosystem by prompting

M. Sorvisto
Ahola Group, Kokkola, Finland
e-mail: mika.sorvisto@aholagroup.com

M. Mohammadian
University of Colorado, Boulder, CO, USA
e-mail: mostafa.mohammadian@colorado.edu

complementary innovations (e.g., renewable energy and EV infrastructure) and collaboration across industries such as energy, transportation, and technology.

Companies striving for a successful sustainable energy transition need to adopt an ecosystemic perspective, understanding that they are part of a larger interconnected network of actors, including energy producers, technology firms, policymakers, grid operators, and consumers (Dehkordi et al., 2024a). The concept of a business ecosystem highlights how these networks of interconnected actors collaborate and compete to co-create and co-capture value (Adner, 2017). In the context of a business ecosystem, BMI must go beyond the boundaries of individual firms (Snihur & Bocken, 2022a, 2022b), recognizing the critical interdependencies that demand collaboration, co-creation, and, at times, coopetition (Ahokangas et al., 2012; Björkdahl & Holmén, 2013). These interdependencies often arise from the complementary resources, capabilities, and activities that diverse actors must align to co-create and capture value, especially when addressing complex sustainability challenges no single entity can solve alone (Jacobides et al., 2024; Thomas et al., 2022). In practice, recognizing these interdependencies involves adopting a collective mindset and engaging in cross-boundary collaboration that facilitate mutual adaptation and co-evolution.

Despite the potential of BMI to drive the sustainable energy transition, existing research on BMI predominantly focuses on technological innovations (Baden-Fuller & Haefliger, 2013; Loock, 2020; Wu et al., 2022) and economic impacts at the firm level (Teece, 2017), with a strong emphasis on profitability and value creation for customers (Foss & Saebi, 2017). This firm-centric focus often overlooks the broader implications of BMI on business ecosystems. While some studies acknowledge the link between BMI and business ecosystem innovation (BEI), they often fail to examine the effects of BMI on the broader business ecosystem (Stubbs & Cocklin, 2008). The neglect of BMI's wider implications limits the understanding of its role in facilitating collaborative innovation across interdependent firms and stakeholders. In addition, an innovative ecosystem has an impact on the BMI of firms that are part of the ecosystem, a relationship that remains unexplored in current research. To address these gaps, this study aims to explore how BMI and BEI relate to each other in the context of sustainable energy transition. Specifically, this chapter aims to answer the following research question: How does BMI within firms impact and embrace innovation within the broader business ecosystem and vice versa?

This study applies a business model framework, following Amit and Zott (2001), who define business models as "the content, structure, and governance of transactions designed so as to create value through the exploitation of business opportunities" (Amit & Zott, 2001, p. 511). The framework is built on three business model elements—content, structure, and governance—that go beyond interdependencies among the activities of firms. This framework allows us to explore the mutual impacts of BMI and BEI in the sustainable energy transition, providing insights into how BMI within individual firms can drive broader BEI, and reciprocally, how BEI influences BMI. This study adopts a qualitative case study of the business ecosystem of sustainable energy transition. Data were collected from primary and secondary sources. The primary data was chosen from 26 semi-structured interviews in Finland and Sweden. To enhance the credibility and contextual richness of the findings, the study also draws on an extensive set of secondary sources, including peer-reviewed academic literature, business case analyses, corporate communications, and media publications.

This chapter contributes to the theoretical understanding of BMI and BEI in the context of sustainable energy transitions by offering a structured analysis of content, structure, and governance-related impacts. By categorizing BMI and BEI impacts, this research advances existing frameworks, which traditionally focus on firm-level innovation, expanding these insights to the ecosystem level. It also advances the understanding of how BMI and BEI impact each other, addressing calls for perspectives on BMI's ecosystemic effects (Snihur & Zott, 2020). In addition, this study highlights practical implications for firms in the energy sector by noting specific impacts of BMI and BEI that can guide strategy and operational adjustments for enhanced collaboration. The findings offer a framework for policymakers and industry leaders to design supportive regulations and infrastructure that foster alignment, interoperability, and adaptability across diverse participants.

The rest of this chapter is structured as follows. Section "Background" provides relevant context and background, introducing key concepts related to BMI and BEI, followed by a focused discussion of BMI and BEI in the context of the sustainable energy transition. Section "Method" outlines the research methodology. Section "Results" presents the results, first examining BMI's influence on BEI and then analyzing BEI's influence on BMI. Finally, Sect. "Conclusion" offers concluding insights, summarizes implications, and suggests directions for future research.

Background

Business Model Innovation and Business Ecosystem Innovation

In recent years, business models (BMs) have garnered significant interest among practitioners and researchers, representing effective tools to understand how companies conduct business in the market and how they create value (Casadesus-Masanell & Ricart, 2010; Teece, 2010). Traditionally, scholars conceptualized BMs as firm-level concepts that aim to articulate how a firm creates, delivers, and captures value (Osterwalder & Pigneur, 2010; Teece, 2010). Although the concept of BM initially focused on the firm level, many scholars increasingly recognize the systemic nature of BMs. Zott and Amit (2010) conceptualize BMs as "boundary-spanning activity systems" that are centered on the focal firm but span boundaries with customers and partners in the pursuit of value creation and capture (Zott et al., 2011). They highlight BMs' span across organizational boundaries, linking firms with external stakeholders, including customers, suppliers, and partners, in a coordinated network of value co-creation. This perspective positions BMs as a unit of analysis that lies between the individual firm and the broader ecosystem, reflecting their role in connecting firms with the external environment for value creation.

In this study, we employ the business model conceptualization of Zott and Amit (2010) because of its rich theoretical foundation and its capacity to consider a firm's entire activity system. They defined business models as depicting "the content, structure, and governance of transactions designed to create value through the exploitation of business opportunities" (Zott & Amit, 2010, p. 219). Zott and Amit (2010) propose three design elements for BM: content, structure, and governance. Content refers to the selection of activities (e.g., what is being designed, including the factors that shape the business model and characterize its outcomes). Structure describes how the activities are linked (e.g., the sequencing between them) and captures their importance for the business model. Governance refers to who performs the activities.

As a result of their boundary-spanning nature, business models have evolved to serve as tools for driving sustainability within and beyond organizational boundaries, reflecting the growing emphasis on integrating sustainable practices into core business strategies (Evans et al., 2017a, 2017b; Lüdeke-Freund et al., 2018). Companies are increasingly adopting BMs that extend beyond targeting profitability, enabling them to describe, analyze, manage, and communicate their sustainable value propositions to customers and stakeholders (Doganova & Eyquem-Renault, 2009; Lüdeke-Freund et al., 2018). These BMs guide firms in creating and delivering value through sustainable methods, while ensuring that they capture economic value in ways that preserve or regenerate natural, social, and economic capital beyond their organizational boundaries (Evans et al., 2017a, 2017b; Schaltegger et al., 2016). Recent research further highlights the role of BMs in creating not only economic value, but also social and environmental value as BMs are encouraged to transcend profit-driven objectives by addressing societal challenges through circular practices, social inclusion, and environmental stewardship (Snihur & Zott, 2020).

The concept of business ecosystem was introduced by James F. Moore in 1993. It stems from the insight that innovative companies do not operate in isolation; instead, they depend on a range of external resources—they need capital, partners, suppliers, and customers to create cooperative networks. The business ecosystem draws parallels with biological ecosystems, wherein diverse organisms interact and depend on one another for survival and development. Similarly, in business ecosystems, firms collaborate and compete simultaneously, relying on each other's capabilities of innovating and advancing shared goals (Adner, 2017). Previous studies identified various categories of ecosystems, such as business ecosystems (Iansiti & Levien, 2004), innovation ecosystems (Autio & Thomas, 2020), and entrepreneurial ecosystems (Wurth et al., 2022). However, we focus on business ecosystems, which can be defined as a network of interconnected actors, including firms, suppliers, customers, competitors, and regulatory entities that both collaborate and compete to co-create value. The success of each actor within this ecosystem is inherently tied to the collective functioning and health of the entire system, highlighting the deep existing interdependencies within these ecosystems (Moore, 2006; Iansiti & Levien, 2004). For example, in the smartphone industry, device manufacturers, app developers, service providers, and consumers all form a complex ecosystem that co-evolves. Each actor's success is influenced by the ability of others to innovate and adapt, demonstrating the importance of collaboration alongside competition within such ecosystems. In addition, complementors play a crucial role in these ecosystems by offering complementary products or services that enhance the value of primary offerings, thereby enriching the overall value proposition (Jacobides et al., 2024). They contribute to positive network effects, where the addition of more complementors or users increases the value of the ecosystem for all participants (e.g., app developers in the smartphone industry).

As industries evolve, business model innovation (BMI) has emerged as a crucial tool for firms to remain competitive and relevant. Foss and Saebi (2017) define BMI as "designed, novel, nontrivial changes to the key elements of a firm's BM and/or the architecture linking these elements" (Foss & Saebi, 2017, p. 201) and maintaining the focal firm perspective. BMI can occur in several ways. For example, Amit and Zott (2012) highlight that BMI is the process of reconfiguring a firm's BM by innovating its core elements: content; structure; and governance. In their view, firms can innovate their BM by adding activities (novel content), linking activities in novel ways (novel structure), or bringing in partners to perform specific activities (novel governance) (Amit & Zott, 2012). Taking a system-level view, Snihur and Zott (2020, p. 556) describe BMI as introducing "novel BMs (in terms of its content, structure, or governance) to the product market space in which the firm competes."

BMI has been identified as a significant driver of business ecosystem innovation (BEI), reshaping not only individual firms but also the broader networks of interconnected actors within their ecosystems (Snihur & Zott, 2020; Yi et al., 2022).BEI refers to the broader changes that occur when innovations introduced by a firm, such as new BMs, trigger innovation among other ecosystem participants and not only the focal firm. As Snihur and Zott (2020) highlight, BMI can instigate or stimulate BEI, as new BMs create opportunities for complementary innovations across the ecosystem. These innovations often reshape relationships, roles, and dependencies in the ecosystem. For example, when Airbnb launched its accommodation-sharing platform, it not only disrupted the traditional hospitality industry, but also facilitated the emergence of new relationships between hosts and guests, fostering innovations such as short-term rental management services and real estate ventures in the cities where Airbnb operates (Snihur & Bocken, 2022a, 2022b). Similarly, Tesla innovated its business model through Vertical integration (owning the supply chain and selling directly to consumers), bundling products (integrating

EVs, energy storage, and solar), and Product-service integration (continuous software updates and energy services) through these innovations. These business model innovations spurred the development of complementary infrastructure such as charging stations, while stimulating further innovation among energy providers and suppliers. As a result, BMI and BEI are deeply interdependent: The successful implementation of a BMI can create ripple effects, generating new opportunities and challenges for other ecosystem actors. This interdependence underscores the importance of firms not only focusing on their internal strategies but also on understanding their role within the larger ecosystem.

While companies like Airbnb and Tesla are often celebrated for their innovations, they also generate negative externalities that challenge sustainability goals. Airbnb has been linked to reduced housing availability and unsustainable tourism practices, while Tesla's contributions are overshadowed by concerns over labor practices and its CEO's opposition to Diversity, Equity, and Inclusion (DEI) initiatives, despite evidence linking DEI to positive climate action. These examples underscore the need to critically assess BMI not only for its economic and technological impact but also for its broader social and environmental consequences.

BMI and BEI in Sustainable Energy Transition

The global transition to sustainable energy created significant challenges and opportunities for firms in the energy sector, leading to a growing body of research on business model innovation and its role in this transformation. BMI emerged as a strategic tool for companies aiming to create, deliver, and capture value while incorporating renewable energy technologies and aligning with sustainability goals (Evans et al., 2017a, 2017b). In the context of energy transition, Richter (2013a) is among the first to examine the role of BMI in the distributed solar/photovoltaic generation market, highlighting how utilities have largely failed to capitalize on this opportunity. Similarly, his later work analyzes the BMs of utilities for large-scale renewable energy generation, revealing a gap in BMs that cater to small-scale renewable technologies (Richter, 2013b).

Other researchers expanded this focus on firm-level BMI to include a systems-based approach. Bolton and Hannon (2016) examine how local authorities in the UK developed innovative business models for decentralized energy systems such as combined heat and power (CHP) and district heating (DH). They argue that a systems approach is necessary to understand the dynamic interactions between BMs, energy infrastructure, and institutions. Further, Wadin et al. (2017) explore joint business model innovation between large multinational utilities and smaller solar energy firms, illustrating how alliances between different-sized actors can foster knowledge dissemination and accelerate the adoption of sustainable technologies. While large firms may benefit more from such collaborations, Wadin's study emphasizes the importance of co-creation in scaling up sustainability innovations across industries.

The literature highlights the crucial role of collaboration, digitalization, and social innovation in driving BMI and BEI within the energy sector. Wadin et al. (2017) underscore the importance of collaboration and co-creation for BEI in the energy sector. Through a case study of a joint business model innovation effort between a multinational utility and a solar energy company, they demonstrate how such alliances can accelerate learning and the diffusion of sustainable technologies. While large firms often dominate the learning outcomes in technological collaborations, the authors argue that business models oriented toward sustainability promote a more cooperative mode of innovation, thereby fostering ecosystem-wide transformations. Loock (2020) adds another critical dimension by examining the role of digitalization as a catalyst for BMI. He develops a typology of digital business model innovations that respond to key bottlenecks in the European energy transition. These include barriers related to both the integration of sustainable technologies into incumbent systems and the creation of independent, decentralized energy configurations. Loock categorizes BMI into incremental and transformative innovations, showing how digital technologies such as smart grids, real-time data analytics, and energy platforms facilitate value creation beyond firm boundaries, thus acting as a core enabler of BEI.

In addition, the rise of peer-to-peer (P2P) energy trading communities further illustrates the transformative impact of BMI on energy ecosystems. Plewnia and Guenther (2021) explore how P2P energy trading, enabled by digital platforms, allow producers and consumers to transact energy directly, challenging the conventional centralized energy market structure. Their study reveals that these energy communities not only empower consumers and decentralize energy governance but also require active engagement from grid operators, regulatory authorities, and technology providers, thereby expanding and reshaping the broader energy ecosystem. Costa et al. (2022), in turn, shift the focus to the convergence of EVs and renewable energy, illustrating how cross-sectoral integration drives BMI. Through empirical analysis, they show how innovations such as vehicle-to-grid (V2G), second-life battery use, and EV-based home energy systems blur the boundaries between the mobility and electricity sectors. This convergence creates new ecosystem configurations, enabling firms to co-develop solutions that address shared sustainability goals and opening up new opportunities within an expanded energy-mobility ecosystem.

Further, Mihailova (2023) emphasizes the role of social innovation in shaping sustainable business models in Europe's decentralizing energy system. Her analysis of citizen collectives and municipal initiatives reveals how these actors contribute to the emergence of Business Models for Sustainability (BMfS), which prioritize community engagement, social value, and inclusivity over purely financial returns. These BMs often challenge traditional utility-based business models by fostering bottom-up participation, democratizing energy ownership, and addressing local needs and values—thereby complementing technological and organizational innovation and enriching the ecosystem-level dynamics of energy transitions.

Method

This research is based on a qualitative case study of the business ecosystem of sustainable energy transition. This business ecosystem encompasses a highly complex network of interdependent actors and relationships, including energy producers, distributors, technology providers, electric vehicle manufacturers, logistics companies, charging infrastructure providers, and policymakers. Each of these actors plays an active role in advancing the sustainable energy transition by contributing specialized resources, knowledge, and technologies essential for developing, implementing, and scaling innovative energy solutions. For example, electric vehicle manufacturers integrate energy storage technologies and charging infrastructure into mobility solutions, while technology providers develop smart grid systems and digital platforms that enable real-time energy management and peer-to-peer trading. Policymakers, on the other hand, create regulatory frameworks and financial incentives that align stakeholder interests and accelerate the deployment of renewable energy across sectors.

The qualitative case study is a research method that involves the in-depth investigation of a specific case, such as an organization, network, or ecosystem, within its real-life context using multiple sources of data. This method is especially well-suited for process-oriented research aimed at understanding how and why complex phenomena evolve over time, rather than simply testing predefined hypotheses (Yin, 2009). It is particularly effective when studying underexplored topics where theoretical foundations are still emerging, as is the case with the intersection of BMI and BEI (Eisenhardt & Graebner, 2007). The qualitative case study enables the integration of diverse data sources and perspectives, which is essential to capture the dynamic and systemic nature of business ecosystems (Yin, 2009). Moreover, it supports stakeholder-relevant inquiry by generating insights that are both contextually grounded and practically meaningful (Langley et al., 2013).

This ecosystem is a pertinent case study in which to explore BMI and BEI in the context of the energy transition, as it allows the capture of the multifaceted interactions and interdependencies among the actors involved in driving the transition to sustainable energy. The rationale behind our selection of the qualitative case study method stems from its suitability for exploring complex phenomena within their real-world contexts (Yin, 2009). Qualitative methods allow a detailed exploration (Hancock et al., 2021), providing rich insights into the underlying processes, motivations, and behaviors that shape the energy transition. Qualitative research is also suitable for addressing complex research questions that necessitate the understanding of context, meaning, and social processes (Yin, 2009).

Data were collected from primary and secondary sources. The primary data source was chosen from 26 semi-structured interviews conducted with professionals, policymakers, and regulators involved in energy transition in Finland and Sweden. These interviews involved management-level experts specializing in electrification within their organizations, including CEOs, directors, and strategic advisors. Participants represented a broad spectrum of companies, ensuring a diverse and comprehensive understanding of the energy transition and the impact of innovative business models in driving this transition. To complement and validate the data collected through interviews, the study also incorporated a range of secondary data sources, including peer-reviewed research papers, business case studies, scholarly conference proceedings, company releases, and news and press releases.

The sample of interview participants was mainly purposeful sampling to select information-rich respondents, followed by snowball sampling. We selected Initial participants through a search of relevant companies in Finland and Sweden, and further contacts were generated through referrals from interviewees. Respondents were selected based on their expertise and management-level roles in strategy and business model development. To capture diverse viewpoints, participants were drawn from a range of organizations across the ecosystem.

This study employed thematic analysis to analyze the data collected from interviews, focusing on BMI and BEI in the context of the sustainable energy transition and. This method allowed the systematic identification of recurring themes and patterns within the qualitative data, offering deep insights into how BMI and BEI shaped the interactions between key actors in the energy ecosystem (Saldana, 2016; Terry et al., 2017). The data analysis followed the six-step process, including familiarization with the data; generating initial codes; searching for themes; reviewing themes; defining and naming themes; and finally, producing the report (Braun & Clarke, 2006). As part of the coding phase, initial codes were derived from the theoretical framework of business model design elements, content, structure, and governance, to capture how firms reconfigured their business models in response to the demands of the sustainable energy transition. Content-related codes included discussions around the introduction of new value propositions, such as bundling electric mobility with renewable energy supply, integrating smart charging technologies, or offering energy-as-a-service solutions. Structure-related codes focused on the novel sequencing and linking of activities, including coordination between energy providers and logistics companies and the integration of digital platforms for customer interaction. Governance-related codes addressed changes in the allocation of responsibilities, such as shifting key activities to external partners and forming partnerships. These theoretically informed codes provided a foundation for understanding the mutual influence of BMI and BEI, helping clarify its role in facilitating the sustainable transition and transforming the wider energy ecosystem.

To ensure the trustworthiness of the research, several steps were taken to enhance the credibility, transferability, and reliability of the findings. This research reinforced credibility through respondent-checking, where selected interviewees were given the opportunity to review and comment on the preliminary analysis, ensuring that the interpretations accurately reflected their perspectives and experiences. This study also supported transferability by providing detailed contextual information about the research setting, the structure of the ecosystem, and the background of participants, allowing future researchers to assess the applicability of the findings to other contexts. In addition, the research process was documented in detail, including the participant selection criteria, data collection procedures, and analytic decisions, enabling a clear audit trail. All interviews were transcribed in full and systematically validated to ensure consistency and transparency throughout the analysis. These measures collectively contributed to the overall rigor and robustness of the qualitative inquiry.

To ensure the trustworthiness of the research, we implemented several steps that enhanced the credibility, transferability, and reliability of our findings. We reinforced credibility by conducting respondent-checking, where we invited selected interviewees to review and comment on the preliminary analysis. This step ensured that our interpretations accurately reflected their experiences and insights. To support transferability, we provided detailed contextual information about the research setting and participant backgrounds, enabling other researchers to assess the relevance of our findings in different contexts. We ensured methodological transparency by thoroughly documenting our research process, including participant selection and data collection. We also transcribed and validated all interviews to ensure consistency and transparency throughout the analysis. Together, these actions strengthened the overall rigor and integrity of the qualitative research.

Results

This study explores how BMI and BEI relate to each other in the context of sustainable energy transition (Fig. 3.1). This section presents the results, starting with the impact of BMI on BEI, organized into content, structure, and governance. Each dimension reveals distinct ways through which BMI influences BEI. Following this, we explore the reverse impact of BEI on BMI, analyzing how BEI reciprocally impacts the BMI of individual firms. Table 3.1 in the Appendix provides selected quotes from interviews.

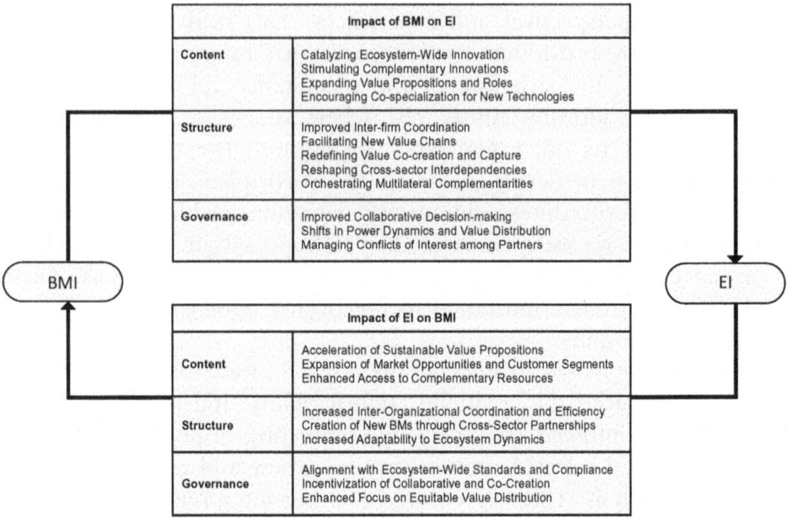

Fig. 3.1 The reciprocal relationship between business model innovation (BMI) and business ecosystem innovation (BEI)

The Impact of Business Model Innovation on Business Ecosystem Innovation

Impact of Content-Related BMI on BEI: Content-related BMI acts as a catalyst for ecosystem-wide innovation by introducing new activities that compel other ecosystem actors to evolve and develop complementary solutions. This finding, which aligns with Amit and Zott (2012) and Snihur and Zott (2020), highlights that when firms bring innovative products or services to market, they often create ripple effects across the ecosystem, prompting partners, suppliers, and even competitors to adapt (Sundberg & Holmström, 2024). For example, home battery storage systems spurred energy providers, solar panel manufacturers, and smart home solution companies to innovate and integrate their offerings with this new technology. This triggered a wave of innovation that spilled over across the energy ecosystem (Gambardella et al., 2021), illustrating how a single firm's content innovation can lead to widespread technological development (Key examples are presented in Table 3.2 in the Appendix). The success of such innovations relies on interdependencies within the ecosystem, highlighting the need for collaborative responses to innovation and the ability of ecosystems to accept and scale new ideas (Dehkordi et al., 2024a).

Moreover, content-related BMI stimulates complementary innovations as firms introduce new products or services that create opportunities for ecosystem participants to develop aligned solutions (Adner, 2017; Teece, 2017). For example, the rise of solar rooftop installations encouraged companies to create complementary technologies such as solar inverters and residential energy management systems, which became essential to the success of the solar market. This process fosters a symbiotic relationship between innovators and their ecosystem, as each firm's innovation builds upon and expands the ecosystem's infrastructure. This dynamic enhances the ecosystem's overall value proposition, improving the customer experience and encouraging further innovation across the network of participants.

Content innovation can also alter value propositions within an ecosystem, redefining the roles of participants and creating opportunities around new technologies. By expanding the functionality of offerings, content innovation shifts the dynamics between ecosystem actors, making some activities obsolete while introducing new opportunities for participation (Massa & Tucci, 2013). For example, introducing solar leasing models transformed homeowners from passive consumers to active participants in renewable energy generation, fundamentally changing their role in the ecosystem. This shift illustrates how content innovation can empower new participants and alter the traditional value exchange, driving ecosystem evolution and introducing new dynamics between established and emerging actors.

Furthermore, Content-related BMI often enables co-specialization (Jacobides et al., 2018), where ecosystem participants develop expertise around new technology, fostering collaborative partnerships, and creating differentiated joint offerings. A prime example is the development of proprietary battery technology by automakers, which led to partnerships with battery producers, allowing both industries to co-specialize and advance battery innovation. This co-specialization enhances the capabilities of the entire ecosystem, as multiple actors collaborate to improve their technological offerings. The partnerships formed through content innovation drive progress in a specific field while fostering adaptability in the ecosystem as participants become interdependent on one another's success.

Impact of Structure-Related BMI on BEI: Structure-related BMI is essential to improve inter-firm coordination and facilitate the formation of new value chains across ecosystems. When companies restructure their activities, they can streamline interactions among ecosystem participants, leading to more effective collaboration and efficient value delivery. For example, integrating blockchain-based smart contracts in supply chain logistics has allowed suppliers, shippers, and manufacturers to automate and verify transactions, thereby reducing friction and enhancing trust across the network. This example demonstrates how restructuring activities can increase the smoothness of collaboration, fostering stronger interdependencies between firms.

Structure-related BMI can give rise to new value chains by connecting industries that were previously unlinked, enabling the creation of more comprehensive offerings (Dehkordi et al., 2024b). Energy-as-a-service (EaaS) BMs illustrate this by integrating solar panel manufacturers, battery storage providers, and service consultants into a unified solution for industrial clients. This structural innovation spans multiple sectors, creating an entirely new value chain. By connecting disparate industries, these innovations open new market opportunities while fostering deeper ecosystem interdependence. The EaaS example underscores the potential of structural innovations to drive ecosystem-wide collaboration and deliver holistic solutions that benefit all the participants involved.

Structure-related BMI also redefines how value is co-created and captured in the ecosystem. By altering the structure of activities, firms enable ecosystem participants to engage more actively in creating shared value (Snihur & Bocken, 2022a, 2022b), which can increase the total value generated by the ecosystem. For example, forming energy cooperatives allows community members to jointly invest in renewable energy projects and share in the benefits of green electricity production. This structural innovation empowers stakeholders to become active contributors to value creation rather than passive consumers, fostering a more equitable and participatory ecosystem. Through such democratization, structural innovations enhance the ecosystem's social and economic sustainability, creating more inclusive growth opportunities.

Structure-related BMI can reshape interdependencies across sectors, requiring collaboration and co-evolution within the ecosystem. This reshaping fosters a collaborative environment that supports ecosystem growth, extending Iansiti and Levien's (2004) view of the impact of ecosystem interactions on business relationships. The development of smart city solutions exemplifies this by establishing new connections between tech companies, utilities, municipalities, and transportation companies. These sectors now depend on each other to provide integrated urban services, which highlights how structural innovations can transform the relationships between industries. This restructuring facilitates the exchange of value and increases the complexity and resilience of the ecosystem, driving further innovation and enabling growth. By linking once separated sectors, structural innovations foster a collaborative environment that benefits all ecosystem participants.

Another impact of Structure-related BMI is the orchestration of multilateral complementarities among ecosystem actors and the reduction of bottlenecks that hinder ecosystem growth. By restructuring the links between activities, structural innovation can create complementarities, where the success of one firm's activities depends on the complementary actions of others (Thomas et al., 2024). An example of this is the collaboration between utility companies, EV manufacturers, and charging network providers, which has resulted in an efficient EV charging network. Here, the coordinated activities of these firms enhance energy distribution and load balancing across the grid. This case shows how structural innovations can foster interdependent relationships that improve both the technological capabilities and operational efficiency of the entire ecosystem.

Impact of Governance-Related BMI on BEI: Governance-related BMI influences BEI by enhancing collaborative decision-making, shifting power dynamics and value distribution, and managing conflicts of interest among partners. In large-scale renewable energy projects, firms adopt governance models that foster coordinated efforts among diverse stakeholders. This collaborative decision-making, as seen in offshore wind farm joint ventures, enables resource sharing and collective responsibility, streamlining project execution while leveraging the unique capabilities of each participant. This collaborative governance, while accelerating the sustainable energy transition, also sets a precedent for coordinated efforts across the ecosystem, inspiring other firms to adopt similar governance models that enhance ecosystem efficiency and resilience.

Furthermore, governance innovations in BMI shift power dynamics and redistribute value in the ecosystem (Bohnsack et al., 2024), fostering a balanced and competitive environment that supports broader BEI. For example, in the EV industry, the rise of independent charging network providers has transformed value distribution by creating equitable opportunities for smaller ecosystem participants. This governance shift compels automakers to collaborate with charging providers, thereby expanding infrastructure accessibility and promoting EV adoption. Managing conflicts of interest among ecosystem partners is another impact of governance innovation. In complex environments such as VPPs, operators must mediate contracts between renewable energy producers, consumers, and energy storage facilities to balance competing interests and ensure smooth operations. Effective governance in such environments ensures that all participants' goals are aligned, fostering long-term collaboration. This is particularly important in ecosystems with diverse actors and objectives, ensuring that governance structures support equitable value distribution and sustained engagement.

The Impact of Business Ecosystem Innovation on Business Model Innovation

Impact of Content-Related BEI on BMI: Content-related BEI influences BMI by accelerating sustainable value propositions, expanding market opportunities and customer segments, and enhancing access to complementary resources. As ecosystems increasingly adopt sustainable technologies such as renewable energy solutions, firms are pressured to align their BMs with these new standards to maintain their relevance and competitiveness. This shift accelerates the development of sustainable value propositions within BMIs (Ahokangas et al., 2024; Baldassarre et al., 2017), positioning sustainability as a core element rather than an add-on (Lüdeke-Freund et al., 2020). For example, as energy ecosystems adopt solar and wind power, firms operating within these networks are incentivized to incorporate these renewable solutions, reconfiguring their BMIs to highlight environmental stewardship. This alignment reinforces the firm's commitment to sustainability and creates a competitive advantage, as businesses that embrace ecosystem-driven standards often become preferred partners for stakeholders who prioritize sustainable practices.

Content-related BEI also opens new market opportunities and customer segments, pushing firms to expand their BMI in ways that capture emerging demands. For example, in the automotive ecosystem, the rise of electric mobility has compelled traditional automakers to adapt their business models to engage with the growing EV market, creating new service-based BMs related to EV maintenance and charging infrastructure. Additionally, content-related BEI provides firms with enhanced access to complementary resources (Bohnsack et al., 2024) such as specialized suppliers or technology platforms that may have been too costly or complex to develop independently. This access enables firms to integrate advanced capabilities into their BMIs, enriching their offerings and enhancing customer value. Consequently, content-related innovations within ecosystems drive firms to continuously refine their business models, expanding both the scope and depth of their value propositions to align with the evolving ecosystem.

Impact of Structure-Related BEI on BMI: Structure-related BEI influences BMI by increasing inter-organizational coordination and efficiency, enabling the creation of new business models through cross-sector partnerships and collaborations, extending Pedersen's (2021) view of the importance of cross-sector collaborations for the transition to new sustainability-driven business models. Structural innovations such as integrating digital platforms promote enhanced coordination among firms in the ecosystem, optimizing supply chains, and improving operational efficiencies. This increased coordination drives BMI by creating opportunities for firms to refine their internal processes and develop more efficient BMs, particularly in logistics and retail, where operational streamlining is critical.

Additionally, structure-related BEI enables the formation of cross-sector partnerships (Lüdeke-Freund et al., 2024), which create opportunities for firms to explore entirely new business models that span multiple industries. For example, collaborations between technology companies and energy providers for smart city solutions prompt firms to innovate their BMs around new services that capitalize on shared digital infrastructure and data. This structural expansion encourages firms to design BMs that leverage these cross-industry resources, capturing value from services that go beyond their traditional offerings. Moreover, structural innovations make ecosystems more adaptable, allowing firms to adjust their BMIs to accommodate shifts in market demand or technological innovations (Baden-Fuller & Haefliger, 2013). In ecosystems with modular production or flexible supply chains, firms can quickly incorporate emerging technologies or adapt their BMs to meet evolving market needs. This adaptability enhances resilience, enabling firms to respond swiftly to ecosystem dynamics while maintaining alignment with structural innovations that support long-term growth and competitiveness.

Impact of Governance-Related BEI on BMI: Governance-related BEI influences BMI by ensuring alignment with ecosystem-wide standards and compliance, incentivizing collaboration and co-creation, and enhancing the focus on equitable value distribution. Governance innovations within ecosystems often establish stringent standards such as sustainability benchmarks that firms must integrate into their BMIs to maintain compliance and secure their position within the ecosystem. For example, ecosystems that emphasize environmental responsibility push firms to embed eco-friendly practices into their business models, thereby promoting alignment with shared sustainability goals. This governance-driven compliance encourages firms to innovate responsibly, reshaping their business models to prioritize environmental or social criteria in addition to profitability.

Additionally, governance-related BEI that incentivizes collaboration and co-creation prompts firms to integrate shared value creation into their BMs, adapting their BMs to leverage collective resources and capabilities. Ecosystem-level incentives for open innovation such as shared intellectual property (IP) frameworks drive firms to reconfigure their business models to incorporate collaborative research and development (R&D), creating synergies that benefit the entire ecosystem. Furthermore, governance frameworks increasingly emphasize equitable value distribution, leading firms to design BMIs prioritizing inclusivity and fairness.

Conclusion

This chapter aims to examine the mutual relationship between BMI and BEI ecosystems in the sustainable energy transition through content, structure, and governance. Innovating the content of BMs introduces new technologies and value propositions that generate a ripple effect across the ecosystem, motivating co-specialization and complementary innovations among various ecosystem participants. Structural and governance-related BMI is shown to be integral in sustaining innovations, providing the organizational and regulatory frameworks necessary for collaboration and adaptability across diverse actors. More specifically, structural innovations such as cross-sector partnerships enable the creation of new value chains, fostering interdependencies among diverse industries within the ecosystem. Governance innovations establish collaborative frameworks, promoting shared decision-making processes that enhance ecosystem cohesion. In the reverse, BEI drives BMI by equipping firms with critical resources and reshaping market conditions, necessitating adaptations within business models. This ecosystem-driven influence compels firms to adopt sustainable value propositions, implement ecosystem-compatible structures, and develop governance frameworks that align with the broader ecosystem objectives.

In theoretical terms, this study contributes to the ongoing discussion about BMI by integrating ecosystem-level impacts into traditionally firm-focused perspectives on BMI. The findings challenge conventional views that confine BMI's effects within the boundaries of individual firms, suggesting instead that BMI catalyzes ecosystem-wide innovation. Additionally, these insights bridge gaps in the literature on sustainability, BMI, and ecosystem innovation, revealing that ecosystem and firm-level dynamics interact synergistically to foster sustainable transformations more effectively. This perspective encourages a shift in thinking about BMI, positioning it not only as a firm-specific concept but as a transformative force within business ecosystems.

This research provides actionable insights for firms in the energy sector aiming to strengthen their competitive position in the ecosystem. It highlights that BMI and BEI can serve as catalysts for the sustainable energy transition by enabling firms and ecosystems to develop and implement novel solutions that promote sustainability. By rethinking traditional business models, companies can integrate renewable energy sources, enhance energy efficiency, and reduce environmental impacts, while creating economic value. Firms can also gain a strategic advantage by aligning their BMI with broader ecosystem goals, allowing them to build cross-sector partnerships and collaborations, thereby facilitating co-development of complementary solutions and enhancing the overall value proposition. Firms are encouraged to adopt an ecosystemic perspective, fostering partnerships with diverse stakeholders to facilitate knowledge exchange and co-creation, thereby strengthening their response to market demands and regulatory pressures. In addition, by remaining adaptable to ecosystem-driven opportunities, companies can continually refine their business models to reflect technological advances and sustainability standards, enhancing their competitive positioning. For policymakers, this study highlights the importance of designing supportive regulatory frameworks that establish ecosystem-wide standards and promote interoperability among stakeholders. By fostering collaboration across diverse ecosystem participants, policymakers can accelerate sustainable innovation and encourage an integrated approach to achieve sustainable energy transition goals.

While this study provides valuable insights about BMI and BEI in the sustainable energy transition, it has some limitations. The study's scope is limited to specific geographical and industrial contexts, which may restrict the generalizability of its findings to other regions and sectors with varying regulatory, economic, and environmental factors. Future research could broaden these findings by examining BMI-BEI relationships across diverse industries to capture context-specific variations. Additionally, this study relies on qualitative assessments, leaving room for future studies to incorporate quantitative analyses to provide a more robust understanding. Longitudinal studies could also offer valuable insights into the long-term effects of these innovations on ecosystem resilience, shedding light on how BMI and BEI contribute to sustained ecosystem adaptability over time.

Appendix

See Tables 3.1 and 3.2.

Table 3.1 Selected quotes from interviews

Themes	*Selected quotes*
Impact of Content-related BMI on BEI	Complementary Innovations • "More and more new platforms are being created by different companies. What's interesting is how companies will use other platforms. They're not just working on their own anymore, they're trying to build something that fits together." Expanding Value Propositions and Roles • "We need different kinds of business model ideas from different players. It's not just one company anymore, everyone involved needs to think about how they can do things in a new way." Encouraging Co-specialization for New Technologies: • "Does one company need to do everything, or could someone else take the lead? For example, one company, maybe the battery maker, could own the batteries and then offer them as a service. So, you'd have one player focusing on that part, while others do something else."
Impact of structure-related BMI on BEI	Improved Inter-firm Coordination • "You often need around five companies to work together to deliver the complete solution. These services have to be aligned, and companies need to develop interfaces so that everything connects smoothly." Facilitating New Value Chains • "To make electrification a reality, you need a full value chain. That means multiple companies working together, each handling a different part." Redefining Value Co-creation and Capture • "The best reason for building shared interfaces and combining APIs is when there's a business opportunity that no single company can reach on its own." Reshaping Cross-sector Interdependencies • "You've got energy companies, data providers, logistics firms, and even municipalities all playing a role." Orchestrating Multilateral Complementarities • "Few companies act as a kind of middleman, trying to stay in between the others. They want to play an interface role, basically trying to connect the different parts of the ecosystem."
Impact of governance-related BMI on BEI	Improved Inter-firm Coordination • "When you start using renewables and electric vehicles, you need real-time monitoring and control. That's why we're working with ICT companies to handle the data transfers related to energy use." Shifts in Power Dynamics and Value Distribution • "If we set the rules for how things work, and especially if we're the ones providing the key interface, then others have to follow." • "The ones who manage the data and the platforms are starting to have more power and control in deciding how the whole system is running." Managing Conflicts of Interest among Partners • "It's easier when the companies have the same business goals. Then they're more willing to integrate their cloud systems and APIs. If their goals don't match, integration becomes much harder."

(continued)

Table 3.1 (continued)

Themes	*Selected quotes*
Impact of Content-related BEI on BMI	Acceleration of Sustainable Value Propositions • "Many farms in Finland are leading the way in adopting local energy solutions. For example, they've been among the first to install solar panels." • "There's also a growing need to replace diesel generators with more sustainable backup power sources." Expansion of Market Opportunities and Customer Segments • "Customers now expect certain services as standard. For instance, people going to gas stations or grocery stores expect to find fast chargers for electric vehicles." Enhanced Access to Complementary Resources • "Right now, taking a futures position on electricity is extremely costly. But with providing services based on battery storage, we can offer some hedging capacity by shifting loads within the same day."
Impact of structure-related BEI on BMI	Increased Inter-Organizational Coordination and Efficiency • "Our company is dealing with the systemic change and that is why we build up new ways of working with both our customers and our partners." Creation of New BMs through Cross-Sector Partnerships • "We need to develop new platforms and technology solutions that allow the rapid increase in the use of green electricity in decarbonization of various sectors." Increased Adaptability to Ecosystem Dynamics • "We have to manage this complexity and the new dynamics that comes with fast increase of renewable electricity on the supply side."
Impact of governance-related BEI on BMI	Alignment with Ecosystem-Wide Standards and Compliance • "Emission standards in Europe are getting stricter, so moving from diesel to electric trucks is necessary. You can make the switch without changing your business model, but it will cost more. That's why you have to rethink how you run your business with electric trucks." Enhanced Focus on Equitable Value Distribution • "The first question is always whether the change makes economic sense. Then it's about who's willing to invest, and how the value gets shared among all the ecosystem players."

Table 3.2 Examples of the mutual impacts between business model innovation and business ecosystem innovation in the sustainable energy transition

BM element	*Examples*
Impact of Content-related BMI on BEI	Catalyzing Ecosystem-Wide Innovation: A utility's launch of bundled solar-plus-storage services prompted adjacent industries—such as smart home tech and building automation—to develop interoperable products Stimulating Complementary Innovations: An introduction of energy management services led suppliers and manufacturers to innovate sensor technologies and data platforms compatible with the new services Expanding Value Propositions and Roles: A solar energy provider partnered with a mobility-as-a-service (MaaS) company to offer bundled services combining EV leasing, solar panel installation, and dynamic charging tariffs Encouraging Co-specialization for New Technologies: An automotive manufacturer investing in battery technologies formed long-term partnerships with chemical companies and lithium mining operations, advancing joint development of advanced materials
Impact of structure-related BMI on BEI	Improved Inter-firm Coordination: A logistics company introduced a BM that synchronizes EV fleet management with smart charging infrastructure. This BM required close coordination between the logistics firm, charging station operators, and electricity providers, leading to shared scheduling systems and improved alignment in energy demand forecasting Facilitating New Value Chains: The deployment of EaaS BMs replace ownership with service-based energy access and enabling collaboration across traditionally separate sectors such as Utilities, IT companies, telecom providers, and financial institutions Redefining Value Co-creation and Capture: A utility company invited local users to invest in and co-manage microgrids, turning passive consumers into active value co-creators Reshaping Cross-sector Interdependencies: Smart city projects integrating transportation, energy, and ICT infrastructure led stakeholders from formerly separate sectors to redesign their activities around shared mobility and energy solutions Orchestrating Multilateral Complementarities: An EV infrastructure provider facilitated collaboration between auto manufacturers and grid operators to build a shared charging network, where each actor's contribution enhanced the others' offerings

(continued)

Table 3.2 (continued)

BM element	*Examples*
Impact of governance-related BMI on BEI	Improved Collaborative Decision-making: Offshore wind consortia established joint governance to align engineering standards, risk-sharing mechanisms, and project phasing among multiple stakeholders Shifts in Power Dynamics and Value Distribution: The rise of platform-based energy trading redistributed influence from large utilities to decentralized aggregators, altering bargaining dynamics and revenue flows Managing Conflicts of Interest among Partners: A VPP operator introduced smart contracts to transparently govern roles between energy producers, storage owners, and retailers, minimizing disputes over dispatch priorities and remuneration
Impact of Content-related BEI on BMI	Acceleration of Sustainable Value Propositions: The widespread adoption of green energy targets across Europe led firms to embed lifecycle carbon tracking and traceability into their core offerings Expansion of Market Opportunities and Customer Segments: The emergence of prosumer communities incentivized firms to develop tailored solutions for home-based generation, storage, and peer-to-peer energy trading Enhanced Access to Complementary Resources: Growing ecosystem partnerships enabled energy startups to access satellite-based weather forecasting data, enriching their solar planning models and business services
Impact of structure-related BEI on BMI	Increased Inter-Organizational Coordination and Efficiency: The rollout of interoperable energy data standards enabled firms to streamline service delivery and cut integration costs, reshaping their pricing and service architecture Creation of New BMs through Cross-Sector Partnerships: Collaborations between automotive OEMs and utilities around V2G technologies gave rise to BMs for bidirectional energy services Increased Adaptability to Ecosystem Dynamics: The modularization of energy hardware (e.g., EV batteries, inverters) allowed firms to adapt quickly in response to ecosystem-level innovation, adjusting their BMs and offerings
Impact of governance-related BEI on BMI	Alignment with Ecosystem-Wide Standards and Compliance: ESG-driven procurement rules in the ecosystem required firms to reconfigure their BM around verified sustainability practices Incentivization of Collaborative and Co-Creation: Funding programs that rewarded inter-firm co-development led firms to embed shared IP ownership and joint models directly into their BM design Enhanced Focus on Equitable Value Distribution: Municipal energy communities pressured participating firms to design BMs that prioritized local ownership, shared profits, and inclusive governance mechanisms

References

Adner, R. (2017). Ecosystem as structure: An actionable construct for strategy. *Journal of Management, 43*(1), 39–58.

Ahokangas, P., Atkova, I., Yrjölä, S., & Matinmikko-Blue, M. (2024). Business model theory and the becoming of new mobile communications technologies. *Business Model Innovation: Game Changers and Contemporary Issues* (pp. 263–293). Springer International Publishing.

Ahokangas, P., Matinmikko, M., Myllykoski, J., Okkonen, H. (2012). Future scenarios, ecosystems and business models for cognitive radio systems. *Future Times, 5*(5).

Amit, R., & Zott, C. (2012). Creating value through business model innovation. *MIT Sloan Management Review.*

Baden-Fuller, C., & Haefliger, S. (2013). Business models and technological innovation. *Long Range Planning, 46*(6), 419–426.

Baldassarre, B., Calabretta, G., Bocken, N. M. P., & Jaskiewicz, T. (2017). Bridging sustainable business model innovation and user-driven innovation: A process for sustainable value proposition design. *Journal of Cleaner Production, 147*, 175–186.

Björkdahl, J., & Holmén, M. (2013). Business model innovation–the challenges ahead. *International Journal of Product Development, 18*(3/4), 213–225

Bohnsack, R., Rennings, M., Block, C., & Bröring, S. (2024). Profiting from innovation when digital business ecosystems emerge: A control point perspective. *Research Policy, 53*(3), 104961.

Bolton, R., & Hannon, M. (2016). Governing sustainability transitions through business model innovation: Towards a systems understanding. *Research Policy, 45*(9), 1731–1742.

Braun, V., & Clarke, V. (2006). Using thematic analysis in psychology. *Qualitative Research in Psychology, 3*(2), 77–101.

Casadesus-Masanell, R., & Ricart, J. E. (2010). From strategy to business models and onto tactics. *Long Range Planning, 43*(2–3), 195–215.

Chen, Y., & Perez, Y. (2018). Business model design: Lessons learned from Tesla Motors. *Towards a Sustainable Economy: Paradoxes and Trends in Energy and Transportation*, 53–69.

Costa, E., Wells, P., Wang, L., & Costa, G. (2022). The electric vehicle and renewable energy: Changes in boundary conditions that enhance business model innovations. *Journal of Cleaner Production, 333*, Article 130034.

Crichton, R., Shrivastava, P., Walker, T., Farhidi, F., Renwick, D., & Ellegate, N. (2024). Going green in the Norwegian fossil fuel sector? The case of sustainability culture at Equinor. *German Journal of Human Resource Management, 38*(2), 140–158.

Dehkordi, R., Ahokangas, P., Evers, N., & Sorvisto, M. (2024a). Assessing business model performance using scalability and replicability as performance indicators: A Case of an electric commercial vehicle ecosystem. *IEEE Transactions on Engineering Management, 71*, 5798–5814.

Dehkordi, R., Ahokangas, P., Evers, N., & Sorvisto, M. (2024b). Business model design for electric commercial vehicles (ECVs): An ecosystemic perspective. *Energy Policy, 186*, Article 113971.

Doganova, L., & Eyquem-Renault, M. (2009). What do business models do?: Innovation devices in technology entrepreneurship *Research Policy, 38*(10), 1559–1570.

Eisenhardt, K. M., & Graebner, M. E. (2007). Theory building from cases: Opportunities and challenges. *Academy of Management Journal, 50*(1), 25–32.

European Commission. (2024). 2050 long-term strategy. Climate Action — European Commission. https://climate.ec.europa.eu/eu-action/climate-strategies-targets/2050-long-term-strategy_en

Evans, N. J., Phua, J., Lim, J., & Jun, H. (2017a). Disclosing Instagram influencer advertising: The effects of disclosure language on advertising recognition, attitudes, and behavioral intent. *Journal of Interactive Advertising, 17*(2), 138–149.

Evans, S., Vladimirova, D., Holgado, M., Van Fossen, K., Yang, M., Silva, E. A., & Barlow, C. Y. (2017b). Business model innovation for sustainability: Towards a unified perspective for creation of sustainable business models. *Business Strategy and the Environment, 26*(5), 597–608.

Foss, N. J., & Saebi, T. (2017). Fifteen years of research on business model innovation: How far have we come, and where should we go? *Journal of Management, 43*(1), 200–227.

Gambardella, A., Heaton, S., Novelli, E., & Teece, D. J. (2021). Profiting from enabling technologies? *Strategy Science, 6*(1), 75–90.

Hancock, D. R., Algozzine, B., & Lim, J. H. (2021). Doing case study research: A practical guide for beginning researchers.

Iansiti, M., & Levien, R. (2004). Strategy as ecology. *Harvard Business Review, 82*(3), 68–81.

Jacobides, M. G., Cennamo, C., & Gawer, A. (2018). Towards a theory of ecosystems. *Strategic Management Journal, 39*(8), 2255–2276.

Jacobides, M. G., Cennamo, C., & Gawer, A. (2024). Externalities and complementarities in platforms and ecosystems: From structural solutions to endogenous failures. *Research Policy, 53*(1), Article 104906.

Langley, A. N. N., Smallman, C., Tsoukas, H., & Van de Ven, A. H. (2013). Process studies of change in organization and management: Unveiling temporality, activity, and flow. *Academy of Management Journal, 56*(1), 1–13.

Loock, M. (2020). Unlocking the value of digitalization for the European energy transition: A typology of innovative business models. *Energy Research & Social Science, 69*, Article 101740.

Lüdeke-Freund, F., Carroux, S., Joyce, A., Massa, L., & Breuer, H. (2018). The sustainable business model pattern taxonomy—45 patterns to support sustainability-oriented business model innovation. *Sustainable Production and Consumption, 15*, 145–162.

Lüdeke-Freund, F., Rauter, R., Pedersen, E. R. G., & Nielsen, C. (2020). Sustainable value creation through business models: The what, the who and the how. *Journal of Business Models, 8*(3), 62–90.

Lüdeke-Freund, F., Wells, P., & Aagaard, A. (2024). The catalytic role of sustainability transitions for business models. *Business Model Innovation: Game Changers and Contemporary Issues* (pp. 127–155). Springer International Publishing.

Massa, L., & Tucci, C. L. (2013). Business model innovation. *The Oxford Handbook of Innovation Management, 20*(18), 420–441.

Mihailova, D. (2023). Redefining business models for the energy transition: Social innovation and sustainable value creation in the European energy system. *Energy Research & Social Science, 100*, Article 103114.

Moore, J. F. (2006). Business ecosystems and the view from the firm. *Antitrust Bulletin, 51*(1), 31–75.

Ocasio, W., & Radoynovska, N. (2016). Strategy and commitments to institutional logics: Organizational heterogeneity in business models and governance. *Strategic Organization, 14*(4), 287–309.

Osterwalder, A., & Pigneur, Y. (2010). *Business model generation: A handbook for visionaries, game changers, and challengers* (Vol. 1). John Wiley & Sons.

Pedersen, E. R. G., Lüdeke-Freund, F., Henriques, I., & Seitanidi, M. M. (2021). Toward collaborative cross-sector business models for sustainability. *Business & Society, 60*(5), 1039–1058.

Plewnia, F., & Guenther, E. (2021). The transition value of business models for a sustainable energy system: The case of virtual peer-to-peer energy communities. *Organization & Environment, 34*(3), 479–503.

Richter, M. (2013a). German utilities and distributed PV: How to overcome barriers to business model innovation. *Renewable Energy, 55*, 456–466.

Richter, M. (2013b). Business model innovation for sustainable energy: German utilities and renewable energy. *Energy Policy, 62*, 1226–1237.

Saldana, J. (2016). *The Coding Manual for Qualitative Researchers* (2nd ed.). Sage Publications.

Schaltegger, S., Hansen, E. G., & Lüdeke-Freund, F. (2016). Business models for sustainability: Origins, present research, and future avenues. *Organization & Environment, 29*(1), 3–10.

Snihur, Y., & Bocken, N. (2022a). A call for action: The impact of business model innovation on business ecosystems, society and planet. *Long Range Planning, 55*(6), Article 102182.

Snihur, Y., & Bocken, N. M. P. (2022b). Sustainable business model innovation: Exploring evidences in sustainability-oriented and circular start-ups. *Business Strategy and the Environment, 31*(2), 926–942.

Snihur, Y., & Zott, C. (2020). The genesis and metamorphosis of novelty imprints: How business model innovation emerges in young ventures. *Academy of Management Journal, 63*(2), 554–583.

Stubbs, W., & Cocklin, C. (2008). Conceptualizing a "sustainability business model". *Organization & Environment, 21*(2), 103–127.

Sundberg, L., & Holmström, J. (2024). Innovating by prompting: How to facilitate innovation in the age of generative AI. *Business Horizons*.

Teece, D. J. (2010). Business models, business strategy and innovation. *Long Range Planning, 43*(2–3), 172–194.

Teece, D. J. (2017). Profiting from innovation in the digital economy: Standards, complementary assets, and business models in the wireless world. *Research Policy, 47*(8), 1367–1387.

Terry, G., Hayfield, N., Clarke, V., & Braun, V. (2017). Thematic analysis. *The SAGE Handbook of Qualitative Research in Psychology, 2*(17–37), 25.

Thomas, L. D., Autio, E., & Gann, D. M. (2022). Processes of ecosystem emergence. *Technovation, 115*, Article 102441.

Thomas, L. D., Ritala, P., Karhu, K., & Heiskala, M. (2024). Vertical and horizontal complementarities in platform ecosystems. *Innovation*, 1–25.

Tolonen, E. (2024). Keeping up with the decarbonization: Conceptualizing and investigating incumbents' responses to transition pressures in the post-Paris world. *Journal of Cleaner Production, 476*, Article 143520.

UNFCCC. (2024). *The Paris Agreement*. https://unfccc.int/process-and-meetings/the-paris-agreement

Wadin, J. L., Ahlgren, K., & Bengtsson, L. (2017). Joint business model innovation for sustainable transformation of industries–A large multinational utility in alliance with a small solar energy company. *Journal of Cleaner Production, 160*, 139–150.

Wu, O. Q., Yücel, Ş, & Zhou, Y. (2022). Smart charging of electric vehicles: An innovative business model for utility firms. *Manufacturing & Service Operations Management, 24*(5), 2481–2499.

Wurth, B., Stam, E., & Spigel, B. (2022). Toward an entrepreneurial ecosystem research program. *Entrepreneurship Theory and Practice, 46*(3), 729–778.

Yi, Y., Chen, Y., & Li, D. (2022). Stakeholder ties, organizational learning, and business model innovation: A business ecosystem perspective. *Technovation, 114*, Article 102445.

Yin, R. K. (2009). *Case study research: design and methods* (Vol. 5). Sage.

Zott, C., & Amit, R. (2010). Business model design: An activity system perspective. *Long Range Planning, 43*(2–3), 216–226.

Zott, C., Amit, R., & Massa, L. (2011). The business model: Recent developments and future research. *Journal of Management, 37*(4), 1019–1042.

CHAPTER 4

Accelerating Distributed Solar and Battery Storage Adoption in India Through Tariff Reforms and Community Grids: A Utility-Prosumer Perspective

Anith Krishnan, Bino Kuriachan, Vivek Mohan, and Nimal Madhu M.

Introduction

India has made significant commitments to develop renewable energy and reduce carbon emissions under the Paris Agreement and other climate pacts. At COP26 in 2021, India revised its renewable energy targets, aiming to install 500 gigawatts (GW) of renewable energy capacity and to generate 50% of its total electricity from renewable sources by 2030. Since 2015, the country has directed 80% of future investments toward

A. Krishnan · B. Kuriachan · V. Mohan (✉) · N. Madhu M.
Department of Electrical Engineering, National Institute of Technology Calicut, Kozhikode, India
e-mail: vivekmohan@nitc.ac.in

A. Krishnan
e-mail: anith_p230024ee@nitc.ac.in

T. Walker et al. (eds.), *Wiring the Future*, Palgrave Studies in Emerging Risk Management and Sustainable Finance,
https://doi.org/10.1007/978-3-032-14717-2_4

solar energy and implemented various policies and initiatives to promote renewable energy growth to achieve these ambitious goals. The latest initiative, the PM Surya Ghar: Muft Bijli Yojana scheme, subsidizes the installation of on-grid solar PV systems in 10 million households (*National Portal for Rooftop Solar—Ministry of New and Renewable Energy*, n.d.).

This renewable energy transition has come with both technical and social challenges, which the Indian Ministry of Power (MoP) and the Indian Government have actively addressed at all levels. For instance, during solar hours, the grid faces a major technical challenge: an increase in grid frequency caused by uncontrolled generation from solar PV (SPV) systems installed across the country (Report on High Frequency Operation in Indian Power System on 4, 11, and 25 August 2024, 2024). One effective strategy to manage high grid frequency involves storing excess energy in Battery Energy Storage Systems (BESS), which function as a load while charging and as a generator while discharging (Maghami et al., 2024). Additionally, a dynamic tariff structure that offers cheaper electricity during peak solar hours —and charges more during non-solar hours—encourages consumers to use electricity when it is most available. This helps keep energy demand and supply balanced and supports the stability of the grid's frequency.

As for social challenges, regulatory uncertainties, financial constraints, and land acquisition issues have plagued India's energy transition process. Stakeholders (such as the Ministry of Power, utility companies, local communities, etc.) can mitigate these challenges to an extent through community grids, where participating entities share resources. These grids offer several benefits, including localized energy management (matching supply and demand), pooled physical and financial resources, and simplified regulatory approvals, since the control area is much smaller than that of the national grid. Figure 4.1 presents the installed capacity in GW of the Indian power sector as of September 31, 2024.

B. Kuriachan
e-mail: bino_p230054ee@nitc.ac.in

N. Madhu M.
e-mail: nimalmadhum@nitc.ac.in

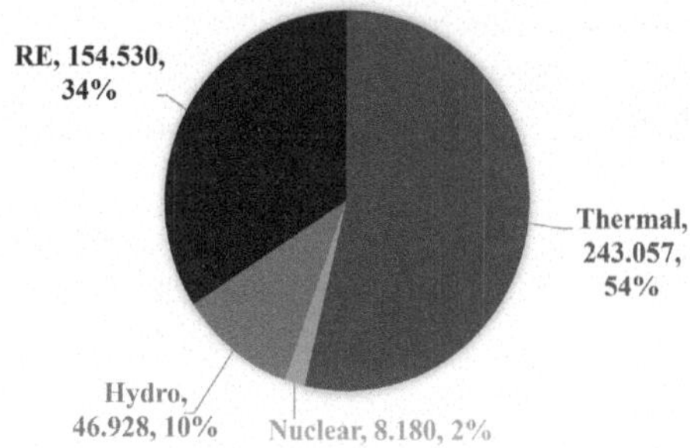

Fig. 4.1 Installed capacity (in GW) of the Indian power sector

The MoP implemented the Renewable Purchase Obligation (RPO) in India as a policy tool under the Electricity Act 2003 to promote renewable energy sources and reduce dependence on fossil fuels for electricity generation. The RPO ensures that entities such as distribution companies and industries must use a certain percentage of the grid's supply which comes from renewable sources (see Table 4.1).

The MoP introduced the Energy Storage Obligation (ESO) in India to support the country's growing renewable energy capacity, especially in light of its aggressive target of achieving 500 GW of non-fossil fuel-based power capacity by 2030 (Ministry of power, Government of India. (2022)). The ESO serves as a mechanism to address the variability of renewable energy sources, such as solar and wind, whose generation fluctuates with changes in weather and time of day. Table 4.2 lists the renewable energy storage obligations mandated by the Government of India.

To complement the ESO and further enhance grid stability, utility companies increasingly adopt dynamic pricing mechanisms such as ToU

Table 4.1 Renewable purchase obligation in India

Year	*Wind RPO*	*Hydro RPO*	*Other RPO*	*Total RPO*
2024–25	2.46%	1.08%	26.37%	29.91%
2025–26	3.36%	1.48%	28.17%	33.01%
2026–27	4.29%	1.80%	29.86%	35.95%
2027–28	5.23%	2.15%	31.43%	38.81%
2028–29	6.16%	2.51%	32.69%	41.36%
2029–30	6.94%	2.82%	33.57%	43.33%

Table 4.2 Energy storage obligation in India

Financial Year	*Energy Storage Obligation*
2023–24	1%
2024–25	1.5%
2025–26	2%
2026–27	2.5%
2027–28	3%
2028–29	3.5%
2029–30	4%

tariffs. In India, ToU tariffs and their adjustments play an increasingly vital role in the country's energy transition These tariffs are designed to encourage more efficient patterns of energy consumption and production. They benefit both consumers and prosumers while tackling challenges posed by the evolving renewable energy sources (RES) used by the power grid.

ToU tariffs typically divide the day into different time periods with varying electricity rates, encouraging consumers to shift their usage to off-peak hours when electricity consumption rates are lower (Oprea & Bara, 2020). This approach can lead to peak shaving and shifting, which is crucial for grid stability and cost-effectiveness. In India, the implementation of ToU tariffs can reduce peak demand by 5–25% by shifting interruptible loads to non-peak hours (Gupta, 2021).

Some researchers have suggested a consumer-centric ToU tariff design which offers distinct on-peak and off-peak unit prices tailored to each customer's consumption habits. This approach aims to address externality issues and rectify flawed pricing mechanisms (G et al., 2022). Although ToU tariffs offer a promising solution to optimize electricity consumption, they are only one approach to addressing the challenges posed by the evolving energy landscape. As India moves toward a greater integration of RES, there is an increasing need for more localized solutions that complement these tariff structures. Another strategy is community microgrids, which can play a crucial role in furthering these objectives, with their decentralized nature and capacity for integrating renewable energy.

Community microgrids offer numerous benefits, including improved reliability, energy savings, consumption reduction, and environmental protection. They provide communities with decentralized grid access to distributed energy resources and enhance their reliability and resiliency

(Kaczmarski, 2022). These systems also contribute to meeting the development targets set out in the Paris agreement by mitigating greenhouse gas emissions through an increased implementation of renewable energy sources and by adapting to climate-related disturbances (Warneryd et al., 2020 As India's energy transition progresses, community microgrids are likely to play an increasingly important role in decentralizing power systems and promoting renewable energy adoption. The UI-ASSIST initiative in Uttar Pradesh, for instance, involved the setting up of two interconnected microgrids in different hamlets of a village based on a socio-economic survey and local resource assessment (Singh et al., 2021). The microgrid design incorporates various renewable energy sources and technologies, including Solar PV systems, Lithium-Ion Battery Energy Storage Systems (BESS), and hybrid biomass systems.

For utility companies, integrating energy storage with SPV systems can help enhance grid stability by mitigating the variability of renewable energy sources. This synergy supports services such as peak shaving, operational flexibility, and voltage regulation, which are crucial for maintaining grid security (Hou et al., 2022; Peprah et al., 2022). Additionally, utilities can adopt the concept of a Virtual Battery (VB), wherein prosumers export surplus PV outputs to the grid and later receive the same energy back, while paying a premium for storage services (Laoharojanaphand & Ongsakul, 2021).

Energy storage systems and feed-in tariffs (FiTs) serve as another essential tool to encourage the adoption of SPV systems (Tanaka et al., 2021). While Lithium-Ion Battery Energy Storage Systems (BESS) addresses the temporal mismatch between energy production and consumption by storing excess solar power for later use, FiTs improve the economic viability of grid-connected PV systems, particularly in terms of providing financial incentives for prosumers. By leveraging the benefits of energy storage and FiT mechanisms, prosumers can optimize their energy independence while capitalizing on favorable policies that support the growth of renewable energy systems. In Romania, for instance, a program offering subsidies for the procurement and installation of SPV systems has made these systems economically feasible, especially those with a smaller energy production capacity (Cristea et al., 2020). FiTs can be modified in several ways to benefit prosumers with installed on-grid SPV systems. Lin et al. (2022) suggest that peak-valley tariffs can increase cost savings for prosumers, although this may come at the expense of grid revenue. Their findings underscore the significance of striking a balance between storage installation and variable renewable energy deployment to sustain the profitability of SPV system-generated electricity in local electricity markets. Additionally, Ilham et al. (2022) compare different compensation schemes in Malaysia, and find that the Net Energy Metering (NEM) 2.0 model shows higher accrued bill savings compared to other models. This suggests that policymakers should consider implementing similar one-on-one offset price mechanisms in FIT designs to maximize prosumer benefits.

Market-linked FiTs offer more dynamic and potentially higher compensation to prosumers, particularly during periods of high electricity demand. This approach aligns prosumer incentives with market conditions, potentially leading to more efficient energy production and consumption patterns (Dynge et al., 2022). However, the shifting from fixed to market-linked tariffs may reduce overall revenue for prosumers compared to the traditional FiT scheme (Dynge et al., 2022; Ilham et al., 2022). Despite this, the increased flexibility allows prosumers to make informed decisions about when to export energy to the grid, which can offset potential revenue declines. Interestingly, the effectiveness of market-linked tariffs depends on several factors such as price caps, consumption/production ratios, and local market conditions. For instance, participation in organized power market segments becomes economically attractive when day-ahead market prices exceed 130 EUR/MWh, regardless of household tariff rates (Zhou et al., 2023). Additionally, the transition from FiT to net billing can significantly promote prosumer development under high-price caps (Zhou et al., 2023). Mohan et al. (2022) further explore a consumer-centric business model for rooftop solar PV in India, formulating a multi-objective optimization framework where FiT and solar PV capacity are decision variables aimed at balancing utility profits and prosumer savings.

Current ToU tariff and FiT models predominantly rely on static pricing structures based on predefined time periods, yet they fail to capture real-time fluctuations in grid and market conditions. This lack of responsiveness limits the effectiveness of these models in optimizing energy consumption and generation. In addition, there is no comprehensive decision-making framework that evaluates the trade-offs between remaining an individual prosumer or joining a community energy setup, leaving a critical gap in assessing the financial and operational advantages of community-based energy solutions. Furthermore, while some researchers have explored FiT structures, there is a notable scarcity of studies focusing on market-responsive FiT models that incentivize prosumers to export energy during periods of high electricity market prices— a critical factor in improving the economic viability of renewable energy systems in India. Researchers have also given limited attention to strategies for enhancing the profitability of Battery Energy Storage Systems (BESS) for both prosumers and utilities. In particular, the potential for prosumers to store energy during peak solar hours and discharge it during high-price periods—thereby boosting financial returns and support grid stability—remains unexplored. These research gaps underscore the need to develop more adaptive and integrated models that align renewable energy generation, storage, and consumption across individual and community scales.

In this chapter, we focus on four key areas and propose actionable strategies to support the integration of solar photovoltaic (SPV) systems and battery energy storage. First, we adjust Time-of-Use (ToU) tariffs to benefit both utilities and consumers/prosumers by aligning pricing with solar generation patterns (Sect. "Time-of-Use Tariff"). Secondly, we evaluate the feasibility of implementing community-scale SPV projects by comparing individual and community-based electricity pricing models (Sect. "Evaluating Individual vs. Community-Grid Approaches"). Thirdly, we assess the economic benefits of Battery Energy Storage Systems (BESS) for prosumers to propose targeted subsidies and incentives to improve their financial viability (Sect. "Incentive and Subsidy for BESS"). Finally, we revise the Feed-in Tariff (FiT) structure based on Time-of-Generation (ToG), ensuring a minimum revenue guarantee while promoting wider SPV adoption among prosumers (Sect. "Modification of Feed-In-Tariff Based on Time-of-Generation").

The following section outlines our general methodology for developing the proposals detailed above. Then, Sect. "Results and Discussion" presents specific-use cases based on data from India. Finally, Sect. "Conclusion" reviews our findings and draws key inferences.

Methodology

This section proposes a framework to assist prosumers and utility companies in making informed decisions about the financial viability of the integration of distributed renewable energy sources, both individually and through targeted policy interventions. We adapt the existing ToU tariff structure to respond to dynamic electricity market conditions, generation capacity, and real-time solar availability, while ensuring cost-effective energy procurement for the utility. We also evaluate and compare the economic viability of individual solar PV installations and participation in a community grid to identify the most cost-effective path to achieving net-zero status.

Time-of-Use Tariff

Balancing demand and generation pose a critical challenge in modern power systems, particularly with the growing integration of rooftop solar PV sources and fluctuating electricity market prices. In India, utility companies meet their power demand by entering into bilateral agreements and energy exchanges, such as the Indian Energy Exchange (IEX), Power Exchange India Ltd. (PXIL), and Hindustan Power Exchange (HPX). These bilateral agreements typically involve central generating stations (CGS) and long-term, medium-term, and short-term open access (LTOA, MTOA, and STOA). Procurement costs vary significantly between off-peak hours—when the demand is much lower than the available generation capacity—and peak hours. A utility's available generation capacity refers to the specific power it can deliver to the load, including both its own generation capacity and the power secured through term-based contracts.

To address this dynamic environment, we propose a new ToU tariff designed to optimize grid efficiency, promote renewable energy adoption, and offer financial incentives to consumers. This tariff leverages parameters such as electricity market rates, available generation capacity (in MW), and solar hours (09:00 to 15:00) to classify time periods into Normal, Off-Peak, and Peak zones—ensuring operational and economic benefits for both utilities and consumers. We focus on the Day Ahead Market (DAM) and Real Time Market (RTM) for energy trading, as they reflect the real-time dynamics in Indian Rupees (INR) per unit of electricity.

Utility companies usually purchase a major portion of their peak demand from the markets (Pramanick et al., 2021). In India, the utility companies have options to place a demand bid in the Term Ahead Market (TAM), Day Ahead Market (DAM), Green Day Ahead Market (GDAM), and Real Time Market (RTM). TAM are submitted in advance for a specific term, such as a week or a month. DAM and GDAM bids are placed a day before the scheduled delivery, while RTM bids are submitted just one hour prior. The GDAM exclusively supports the trading of green energy.

We classify time periods systematically using three key variables—DAM/RTM rates, available generation capacity, and peak solar hours—to propose the new ToU tariff for electricity. Since the DAM and RTM rates in the electricity exchanges follow a similar trend (see Sec. 3), we consider them together. Table 4.3 shows all eight possible combinations using the three variables and their proposed representative zones:

Condition 1 is a Normal Zone because it occurs outside of peak solar hours, the DAM/RTM rate is lower than the Average Cost of Supply (ACoS), and although central generating stations (CGS) and bilateral agreements do not provide enough generation capacity to meet the load, the shortfall can be filled by buying cheaper power from the energy exchange.

In contrast, Condition 2 classifies an Off-Peak Zone as it occurs during peak solar hours, when SPV systems can meet a significant portion of the demand. With DAM/RTM rates still below the ACoS, this condition incentivizes consumers to shift their demand and enables utility companies to integrate more renewable energy into the grid efficiently.

Conditions 3 and 4 also represent Off-Peak Zones. In these cases, the DAM/RTM rate remains below the ACoS, and generation capacity is sufficient to meet demand—either during non-solar hours (Condition 3) or solar hours (Condition 4). This encourages consumers to shift their demand to these times, offering financial benefits while helping utility

Table 4.3 Classification of zones based on electricity market rates, cost of supply, available generation capacity, and peak solar hours

Condition	*DAM/RTM Rate*	*Available Generation Capacity*	*Peak Solar Hours*	*Zone*
1	< ACoS	Insufficient	No	Normal
2	< ACoS	Insufficient	Yes	Off-peak
3	< ACoS	Sufficient	No	Off-peak
4	< ACoS	Sufficient	Yes	Off-peak
5	> ACoS	Insufficient	No	Peak
6	> ACoS	Insufficient	Yes	Peak
7	> ACoS	Sufficient	No	Normal
8	> ACoS	Sufficient	Yes	Off-peak

companies avoid the cost of surrendering excess generation capacity and supporting greater renewable energy integration.

Conditions 5 and 6 represent Peak Zones, where DAM/RTM rates exceed the ACoS, and the generation capacity is insufficient to meet demand—whether during non-solar hours (Condition 5) or solar hours (Condition 6). In these scenarios, the utility must procure additional power at a higher cost to meet the demand.

Conditions 7 and 8 state that when the DAM/RTM rate exceeds the ACoS, the generation capacity is sufficient. The classification then depends on the time of day: during non-solar hours (Condition 7), it is a Normal Zone because demand is met without incurring additional costs. During solar hours (Condition 8), we classify the period as an Off-Peak Zone to encourage demand shifting, allowing consumers to benefit financially and helping utility companies increase renewable energy by raising daytime demand.

Evaluating Individual vs. Community-Grid Approaches

This section outlines our proposed methodology for evaluating the financial implications for both individual and community-based consumers. This methodology addresses the following questions:

1. Should consumers continue to receive bills from utility companies as individuals, or should they become part of a community that functions as a single entity from the utility's perspective?
2. Should consumers install an on-grid SPV system individually, or should a community-led initiative install a larger on-grid SPV system?

Evaluation of Individual and Community-Based Electricity pricing Approaches

We propose assessing individual and community-based electricity pricing through the following three scenarios:

Scenario 1: The Utility Bills Each Consumer Individually.

The electricity bill for an individual consumer is based on the monthly electricity consumption, the type of connection (e.g., domestic, commercial, etc.), and other charges—such as fixed charge, duty, fuel surcharge and meter rent. $E_{i,m}$ is the monthly electricity consumption of the i th consumer for the m th month, and let the utility apply the tariff function $T\left(E_{i,m}, u_{i_{type}}, u_{i_{oc}}\right)$ to calculate the monthly electricity bill, where $u_{i_{type}}$ is the type of consumer, and u_{oc} indicates the other charges.

The bill amount $B_{i,m}$ for the i th consumer for the m th month is calculated as

$$B_{i,m} = T(E_{i,m}, u_{i_{type}}, u_{i_{oc}}) \tag{4.1}$$

Therefore, the annual electricity bill for a consumer is calculated as

$$B_{ia} = \sum\nolimits_{m=1}^{12} B_{i,m} \tag{4.2}$$

Scenario 2: The Utility Bills the Community of Consumers (Called Community Grid) as Those in Scenario 1, but as a Single Entity.

The utility companies calculate the monthly electricity bill of the community as follows:

$$B_{c,m} = T(E_{comm,m}, u_{type}, u_{oc}) \tag{4.3}$$

$E_{comm,m} = \sum\nolimits_{i=1}^{n} E_{i,m}$ represents the total electricity consumption of the community grid for the m th month, while u_{type} and u_{oc} are the consumer type and the other charges respectively, according to applicable tariff. Thus, the utility companies calculate the annual electricity bill of the community grid by summing up the monthly bills:

$$B_{ca} = \sum\nolimits_{m=1}^{12} B_{c,m} \tag{4.4}$$

Scenario 3: Consumers Form a Community Grid and Purchase Electricity from the Electricity Market.

In the current scenario, utility companies distribute energy to their consumers primarily through their own generation resources or via other generating companies under fixed-term purchase agreements. They purchase a very small portion of the distributed energy from the electricity market—typically around 13%, although this share has steadily increased over recent months (Economics Division, Central Electricity Regulatory Commission, 2024). Global experience from modern electricity markets suggests that this share will continue to grow as they evolve. In this context, we propose incorporating market dynamics by enabling the community grid to procure 100% of its electricity consumption from the market.

We calculate the annual electricity bill of the community grid (B_{cm}) which purchases electricity from the electricity market, as follows:

$$B_{cm} = \sum\nolimits_{i=1}^{n} \sum\nolimits_{t} E_i(t) P_m(t) \tag{4.5}$$

where, $P_m(t)$ and $E_i(t)$ are the market price and the electricity consumption of the i th consumer respectively, during the period t.

While we do not explicitly quantify the financial benefit for each individual consumer, we emphasize that if a consumer joins a community grid and the total community bill is lower than the sum of individual bills, the consumer receives a proportional benefit based on their consumption. Conversely, if the community bill exceeds the total of individual bills, the consumer saves more by remaining independent, with savings again aligned to their usage.

Let $B_{total} = \sum_{i=1}^{n} B_{ia}$ be the sum of the individual electricity bills of all consumers. If $B_{total} > B_{ca}$ or if $B_{total} > B_{cm}$, then forming a community grid is financially advantageous.

Once the decision to form a community grid is made, the choice between purchasing electricity from the utility or the market is determined by comparing B_{cm} and B_{ca} i.e., if $B_{cm} < B_{ca}$, purchasing from the electricity market is preferable; otherwise, purchasing from the utility company is a better option.

Assessing Financial Viability of On-Grid Solar PV Systems: Individual vs. Community Approaches

To ascertain whether consumers should install individual on-grid solar photovoltaic (SPV) systems or participate in a larger community-level on-grid SPV system, we size SPV systems for both scenarios to reach annual net-zero energy consumption and assess their financial viability based on the simple payback period (SPP) for each case, which is calculated using Eq. (4.6).

$$\text{SPP} = \frac{\text{Initial investment of SPV system}}{\text{Annual savings in electricity bill}} \tag{4.6}$$

We size the SPV system for annual net-zero energy consumption using peak solar hours, following Hussain et al.'s method (2022). The size of the on-grid SPV system (P_{PV}) is determined as

$$P_{PV} = \frac{E_{annual}}{H_{sun} \times 365} \tag{4.7}$$

where E_{annual} is the annual energy generated by the SPV systems and H_{sun} is the peak solar hours, which represent the average number of hours per day during which sunlight is at or near its peak (approximately 1000W/m^2).

Incentive and Subsidy for BESS

Utility companies can fulfill their Energy Storage Obligation (ESO) either through Grid-Connected Storage (GCS) or a Distributed Storage System (DSS). In a GCS configuration, the storage system is directly connected to the main grid, whereas in a DSS, it is installed closer to the end-user. In a GCS system, utility companies must account for Transmission and Distribution (T&D) losses, transmission costs, and distribution costs. By contrast, DSS allows end users to utilize stored energy without incurring T&D losses or associated costs. Utility companies can pass these avoided costs to end users as an incentive to consume stored energy during its peak demand periods.

Here, we suggest installing a BESS on the prosumer's property, which they can own and operate themselves. Ideally, the BESS would be charged during solar hours and use the stored energy during the utility company's peak hours, typically between 18:00 and 22:00 hours.

Sizing of BESS

We size the BESS to compensate for the prosumer's usage during peak hours, when solar PV generation is unavailable. We focus exclusively on-peak hours, as electricity prices are typically highest during this period. In the context of this study, we assume that the peak hours have consistent daily duration, such as 18:00 to 22:00.

Let $E_{t,m}$ denote the prosumer's energy consumption during time interval t for the m th month. To ensure full utilization of the battery every month, we size the BESS based on the minimum energy consumption recorded during the specified time interval t across all the twelve months. Therefore, the minimum monthly peak-hour consumption of a prosumer, denoted as E_{min}, is defined as

$$E_{min} = min(E_{t,1}, E_{t,2}, E_{t,3}, \dots, E_{t,12}) \tag{4.8}$$

where $E_{t,1}$, $E_{t,2}$, etc., are the prosumer's consumptions during time interval t in months 1, 2, and so on, respectively. Thus, the size of the battery in kWh can be calculated as follows:

$$E_{battery} = \frac{E_{min}}{DoD_{max} \times \eta} \tag{4.9}$$

where DoD_{max} is the maximum permissible depth-of-discharge of the battery and η is the round-trip efficiency of the battery.

Economic Evaluation

The proposal to install a BESS alongside an SPV system on a prosumer's premises is evaluated by using two financial metrics: Net Present Value (NPV) and Discounted Payback Period (DPP). NPV is used to assess the financial viability of the combined BESS and SPV project by comparing the present value of expected cash inflows (electricity bill savings) to the present value of anticipated cash outflows (initial investment, replacement costs, maintenance expenses, etc.). This approach accounts for the time value of money by discounting future cash flows to their present value.

Key factors in calculating NPV include the initial costs of the SPV and BESS systems, the annual savings generated by the prosumer, the discount rate, and the operational lifespans of both the SPV and BESS systems. Equation 4.10 defines the NPV:

$$\text{NPV} = \sum_{t=0}^{n} \frac{C_t}{(1+r)^t} \tag{4.10}$$

Here, C_t= Net savings during year t, in INR, where the prosumer's savings in electricity bill is taken as positive, the initial investment of the SPV and battery, the replacement cost of the battery, etc. are taken as negative, r is the percentage discount rate and n is the life of the BESS and the SPV system in years.

Discounted Payback period

The Discounted Payback Period (DPP) is a financial measure used to determine the time it takes for the cumulative discounted cash flow—representing electricity bill savings—from the SPV and BESS to recover the initial investment. Unlike the simple payback period, DPP incorporates the time value of money by discounting future cash flows to their present value. The steps for calculating the DPP are as follows:

1. Calculate the present value of annual cash flows (discounted cash flow) using the discount rate, see Eq. (4.11).
2. Accumulate the discounted cash flows year by year, as per Eq. (4.12).
3. Determine the point at which the total discounted cash flow equals or exceeds the initial investment, see Eq. (4.13).

$$\text{Discounted Cash Flow (DCF)} = \frac{C_t}{(1+r)^t} \tag{4.11}$$

$$\text{Cumulative discounted cash flow (CDCF)} = \sum\nolimits_{t=1}^{k} \frac{C_t}{(1+r)^t} \tag{4.12}$$

$$\text{DPP} = max(t : CDCF_t \leq C_0) + \frac{C_0 - CDCF_t}{DCF_{t+1}} \tag{4.13}$$

Here, C_0 is the cost of the SPV and BESS, and is taken as negative. A shorter DPP indicates a quicker recovery of investment in the SPV and BESS, whereas a longer DPP may suggest slower returns, potentially making the SPV and BESS less attractive.

Modification of Feed-in-Tariff Based on Time-of-Generation

FiTs play a pivotal role in promoting rooftop solar PV adoption and ensuring fair compensation for decentralized energy producers. Utility companies in India have traditionally implemented FiTs as fixed-rate schemes, offering predictable returns but lacking responsiveness to real-time market dynamics. This static approach often fails to reflect fluctuations in demand, grid requirements, or the dynamic pricing of solar energy, potentially limiting the full potential of roof top solar PV integration. To address these shortcomings, we propose a dynamic FiT model that links the tariff to the green energy market price, ensuring that it aligns with current grid conditions while guaranteeing a minimum revenue for producers.

The proposed FiT (see Eq. (4.14)) also incentivizes renewable energy generation by ensuring a minimum revenue.

$$C_{FiT}(t) = \max\left(\alpha, \alpha + \left[C_{gm}(t) - \alpha\right] \times \beta\right) \tag{4.14}$$

Here $C_{FiT}(t)$ is the feed-in-tariff during time duration (15-minute block in India) t(in INR/kWh), α is the incentive (in INR/kWh) for renewable energy exported to the grid, $C_{gm}(t)$ is the price of electricity (in INR/kWh) in the green energy market under consideration and β is the Green Energy Support Factor (GESF).

The proposed FiT (see Eq. (4.14)) ensures that the prosumer gets an incentive when the price of electricity in the green energy market is less than this incentive. When the price of electricity exceeds the incentive, the prosumer receives a fraction (GESF) of this difference, in addition to the incentive, which encourages the export of green energy to the grid during periods of high market price. The utility company can fix the value of GESF based on administrative and policy considerations, ensuring that the proposed FiT remains a regulated fraction of the market price.

Results and Discussion

This section uses data from consumers/prosumers, utility company, and electrical markets, adhering to the approach outlined in Sect. "Methodology" to derive the results and conclusions.

Time-of-Use Tariff

The Kerala State Electricity Board Limited (KSEBL) implements a comprehensive tariff structure designed to address the diverse needs of its consumer base while maintaining operational sustainability and economic efficiency. The Kerala State Electricity Regulatory Commission (KSERC) approves and regulates this structure annually. KSEBL's tariff framework is segmented based on consumer categories, including residential, commercial, industrial, agricultural, and public utility services. Each of these categories is further subdivided, enabling differential pricing that reflects usage patterns, consumer demand profiles, and load characteristics.

Residential consumers typically benefit from a progressive block tariff system, where unit charges increase with higher levels of consumption. This structure incentivizes energy conservation while protecting low-income households through lower rates for essential electricity use. Utility companies often charge commercial and industrial consumers based on a two-part tariff model that includes a fixed charge (based on contracted demand or connected load) and a variable charge (based on actual energy consumption). Utility companies also implement Time-of-Use (ToU) tariff structures, particularly for high-demand categories, to encourage load shifting and manage peak demand more effectively. Table 4.4 presents the existing ToU tariff in KSEBL (Kerala Tariff, 2024). Here, 150% indicates that the applicable electricity rate during the zone is 150% of the normal tariff (in INR/kWh).

Figure 4.2 displays the actual demand curve of the KSEBL, the DAM/RTM rates for a typical day in January 2024, and the applicable Average Cost of Supply (ACoS). The exchange provides DAM/RTM rates for each 15-minute time block, resulting in 96 time blocks per day. The ACoS for the KSEBL is INR 7.16 ($0.086) for FY 2023–2024. The DAM/RTM rate is greater than the ACoS from block 29 (07:15) to 47 (11:45) during morning time and from block 67 (16:45) to 84 (21:00) during the evening time, reflecting the peak demand periods in the Indian power grid.

During the solar peak hours—when maximum irradiance is incident on the Earth's surface—the generation from solar PV systems across the

Table 4.4 Existing ToU tariff zones of KSEBL

Tariff	*Particulars*		
	Normal Period (06:00 to 18:00) (Zone 1)	*Peak Period (18:00 to 22:00) (Zone 2)*	*Off-peak Period (22:00 to 06:00) (Zone 3)*
Industrial	100%	150%	75%
Domestic & Commercial	100%	120%	90%

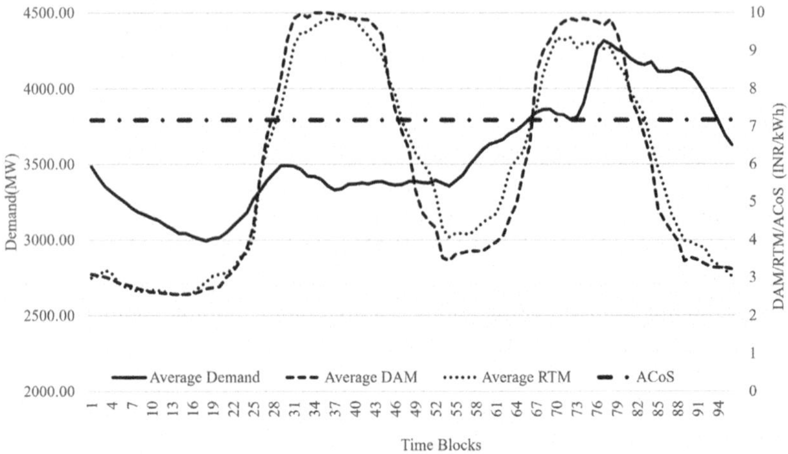

Fig. 4.2 Demand, Average DAM/RTM price and ACoS of January 2024 vs Time blocks

country will be at its peak which drives down the DAM/RTM rates during these periods.

During nighttime (23:00 to 06:00) to early morning, demand is generally low across the Indian power grid, resulting in lower DAM/RTM rates. However, Kerala shows a different demand profile, with two demand peaks; the morning peak from block 24 (06:00) to block 34 (08:30) and the evening peak from block 72 (18:00) to block 92 (23:00).

Although the DAM/RTM rates are higher than ACoS from 7:15 till 11:45, indicating a peak across Indian power grid, Kerala offsets this peak due to the generation from prosumers connected to the utility grid (installed capacity of 1009.30 MW as on 31-03-2024) (KSEBL, 2024). Solar generation ramps up significantly from 9:00 onwards.

Kerala's base load is primarily met by thermal and nuclear power stations. From 9:00 to 15:00, substantial solar energy generation leads to an output exceeding the demand of the utility company. To maintain the load-generation balance, the utility company must either surrender the power from thermal stations or increase demand during this period. However, surrendering thermal power incurs fixed cost (INR/kWh) placing a financial burden on the utility company. This causes an additional financial burden to the utility company and reduces revenue for the thermal generation plants. Furthermore, these stations must also

Table 4.5 The proposed ToU zones

Proposed Zones	*Time of the Day*
Morning peak	06:00 to 09:00
Day off-peak	09:00 to 15:00
Normal	15:00 to 19:00
Night peak	19:00 to 23:00
Night off-peak	23:00 to 06:00

run on a minimum load called the technical minimum, beyond which the power plant cannot reduce their generation. We propose introducing a new tariff zone, the Day Off-Peak Zone, during this period (as shown in Table 4.5) to encourage consumers to increase electricity usage at a reduced rate. This approach financially benefits consumers for shifting their load, reduces the burden on utilities, and aligns with India's Renewable Purchase Obligation (RPO) goals—requiring 33.57% of annual energy consumption to come from other RPO sources (mainly solar) by FY 2029–30 (Ministry of Power, Government of India, 2022).

Evaluating Individual vs. Community-Grid Approaches

We select a set of six representative consumers (as shown in Table 4.6) for the analysis of the community grid. Consumers C1 and C5 represent domestic users with low energy consumption, while C4 represents a domestic consumer with medium energy consumption, and C6 represents a domestic consumer with high energy consumption. Consumers C2 and C3 are representative commercial consumers. We calculate the annual electricity bill based on the latest tariff revision order of the Kerala State Electricity Regulatory Commission (KeralaTariff, 2024).

Here, we analyze six different scenarios, as explained in Sect. "Evaluating Individual vs. Community-Grid Approaches", and the analysis is performed using the on-site load data from consumers C1 to C6.

In Scenario 1, the total annual bill for the six consumers amounts to INR 17,88,897 ($20,734.47) when they are treated individual consumers and billed by the utility company as per the existing electricity Low Tension (LT) tariff, using Eqs. (4.1) and (4.2). For the tariff calculations, $u_{i_{oc}}$ is assumed to be zero.

In Scenario 2, we club all six consumers into a hypothetical micro-community. When the utility company treats this community as a single

Table 4.6 The annual consumption & annual electricity bills of selected consumers

Consumer	*Annual Consumption (kWh)*	*Electricity Bill (INR)*
C1	3,080	18,423($215.51)
C2	92,273	867,366($10,146.45)
C3	24,463	229,952($2,689.90)
C4	18,730	168,330($1,969.12)
C5	3,148	21,310($249.28)
C6	54,785	483,516($5,658.24)
Total		1,788,897($20,926.52)

entity and applies the existing High Tension (HT-IV) tariff (u_{type}) of KSEBL (Kerala Tariff, 2024), resulting in total annual bill of INR 16,55,712 ($19,375). This calculation is based on the following assumptions:

(i) The distribution losses are zero.
(ii) Infrastructure and maintenance cost of the community grid is zero
(iii) The community grid operator takes zero profit
(iv) Other charges (u_{oc}) are zero

In Scenario 3, the community grid opts to purchase electricity from the open market. Our analysis in this section is based on data from the Indian Energy Exchange (IEX). We use Day Ahead Market (DAM) rates during FY 2023–2024 to calculate B_{cm} (see Eq. (4.5)), where t represents a 15-minute time block. Based on these DAM rates, the total electricity cost for the community grid electricity amounts to INR 15,58,143 ($18,227.16).

A comparison of the original load data of the six consumers across all three scenarios reveals that scenario 3 offers the best return, followed by Scenario 2 as shown in table 4.7.

Scenarios 2 and 3 show that forming a community grid is more economical than operating as individual prosumers. In Scenario 4, we size SPV systems to achieve annual net-zero energy consumption for both individual households and the community as a whole. Table 4.8 presents detailed results.

Table 4.7 Annual electricity bills of consumers/community in various scenarios

Scenario 1	*Scenario 2*	*Scenario 3*
Total Electricity Bill of All Consumers (INR)	**Bill as per HT IV(B) (INR)**	**Bill as per DAM (INR)**
1,788,897 ($20,926.52)	1,655,712 ($19,368.52)	1,558,143 ($18,227.16)

Table 4.8 Energy & cost of prosumers/community

Scenario 4					
Entity	*Annual Energy Consumption (kWh)*	*Size of SPV for Net Zero Energy, P_{PV} (kW)*	*Cost of SPV System (INR)*	*Annual Savings (INR)*	*Simple PaybackPeriod (Years)*
Community	196,479	135	5,271,750 ($61,660.53)	1,788,897 ($20,929.27)	2.95
C1	3,080	2	98,186 ($1,148.52)	18,423 ($215.51)	5.33
C2	92,273	63	2,741,319 ($32,051.18)	867,366 ($10,144.92)	3.16
C3	24,463	17	739,721 ($8,653.26)	229,952 ($2,689.98)	3.22
C4	18,730	13	565,669 ($6,617.20)	168,330 ($1,969.12)	3.36
C5	3,148	2	98,186 ($1,148.58)	21,310 ($249.28)	4.61
C6	54,785	38	1,865,534 ($21,823.02)	483,516 ($5,656.17)	3.86

Incentive and Subsidy for BESS

The increasing penetration of renewable energy (RE), primarily solar energy, in the Indian power system has led to the Duck **curve** effect on the demand curve. The duck curve effect is characterized by a reduction in net load during peak solar generation hours, illustrating high penetration of distributed solar power. As solar generation diminishes after sunset, net load rapidly increases in the late afternoon (around 15:00) or early evening (around 16:00), driven by increasing consumer demand.

In Kerala, the Kerala State Electricity Regulatory Commission (KSERC, 2024) currently limits the maximum installed capacity of the on-grid solar PV systems to 90% of the rated capacity of the distribution transformer. This restriction encourages prosumers to install additional capacity for charging the BESS, thereby enhancing power supply reliability and reducing dependence on a utility company's electricity during peak pricing periods, especially at night when SPV output is zero.

The steep gradient in netload during the late afternoon necessitates the development of power systems to deploy fast-ramping conventional generation assets. However, this poses challenges for baseload plants that are not designed for such flexible operation. The resulting variability increases demand for spinning reserves and ancillary services to maintain grid stability and ensure adequate response during rapid transitions. Frequent cycling and ramping of conventional plants lead to increased operational and maintenance costs, affecting its lifespan and economic viability. The need for grid flexibility has driven investments in technologies such as energy storage systems (ESS), demand-side management (DSM) programs, and grid modernization to handle the dynamic load changes.

The annual consumption of the KSEBL is 30,937.883 MU. According to the Ministry of Power, Government of India, the ESO for the current fiscal year is 1.5%, with an annual increment of 0.5% in subsequent years.

In the KSEBL, peak demand mostly occurs during the summer months of April and May between 22:00 and 23:00 hours. In the past few years, during this period, most of the transmission lines and distribution systems were experiencing overload. If the growth in demand continues, it will necessitate a huge investment in transmission and distribution infrastructure to cater to the increased load present for a shorter period, which is in fact not a financially sound decision. Furthermore, the development of additional infrastructure will take more time for implementation. One solution to mitigate this problem is to introduce the distributed storage system. These will help in reducing the investment cost of the utility company in distribution and transmission, as there is stored energy available in the vicinity of the load thereby decreasing their dependence on the transmission and distribution system. Additionally, the use of distributed storage will result in demand reduction during the peak time, thereby improving the stability of the grid.

Currently, KSEBL must purchase additional power to meet the peak demand from the energy exchange, often at a higher rate, due to supply constraints. The prosumers with installed BESS can help in lowering system demand during peak hours. This, in turn, decreases the utility company's dependence on the energy exchange and it can also generate financial savings. This financial benefit of the utility company can be transferred partially to the prosumers by way of offering incentives/subsidies for installing BESS and using stored energy during peak hours.

We use the annual consumption patterns of two prosumers, C1 and C6 to analyze the Battery Energy Storage Systems (BESS) in low voltage systems. C1 is a low-consumption household (<300 units consumption per month) and C6 is a high-consumption household. The utility company currently bills C1 with a cross-subsidy rate whereas C6 is billed at higher rates (without cross-subsidy). The average billing rate (ABR), calculated by dividing total revenue from electricity sales by the total energy sold is INR 4.84 ($0.057) per kWh for C1 and INR 8.80 ($0.10) per kWh for C6.

The Average Power Purchase Cost (APPC) refers to the weighted average price at which a power distribution company (DISCOM) or utility purchases electricity from different sources, including power generators, renewable energy providers, and power exchanges. During peak hours, the APPC borne by the utility is higher than the ABR of C1 due to cross subsidies. It is important to note that the APPC is higher than the ACoS, leading to revenue loss for the utility company. This financial loss can be mitigated by introducing incentives for the self-stored energy that the consumers use in the peak hours. However, in the case of the prosumer C6, which is not cross-subsidized, the utility company has no revenue gain and hence no incentives shall be given to C6 for the self-stored energy.

We analyze the effectiveness of the Battery Energy Storage System (BESS) by considering four different scenarios involving two prosumers, C1 and C6 (see Table 4.9). C1 has an installed SPV capacity of 8.25 kW, while C6 has 25.6 kW. Based on the installed SPV capacity and annual energy consumption, C1 acts as a net exporter of electricity, whereas C6 remains a net importer.

Scenario 1 represents the baseline case, where neither prosumer has an installed SPV system. Under the current KSEBL tariff, C1 pays an annual electricity bill of INR 18,522 ($216.68), while C6 pays INR 483,524 ($5,654.92) in Scenario 1. Scenarios 2, 3, and 4 represent cases where the prosumers have installed SPV systems on their premises. Scenarios 1 and 2 follow electricity billing under the existing tariff structure, while Scenarios 3 and 4 offer more realistic conditions from both the prosumer's and the utility company's perspectives.

In Scenario 2, we calculate the annual bills for C1 and C6 as INR –24,303 (-$284.17) and INR 16,674 ($195.08), respectively, based on

Table 4.9 Financial analysis of BESS for C1 and C6 with and without subsidy and incentive

Prosumer	*With BESS (no subsidy & incentives)*		*With BESS (with 40% subsidy on CAPEX)*		*With BESS (with 40% subsidy on CAPEX & incentive of INR, 4/kWh)*	
	NPV (INR)	*Discounted Payback Period (Yrs)*	*NPV (INR)*	*Discounted Payback Period (Yrs)*	*NPV (INR)*	*Discounted Payback Period (Yrs)*
C1	−297,370 (−$3,478.77	NA	−88,769 (−$1,038.59)	NA	−74,279 (−$868.92)	NA
C6	787,292 ($9,209.74)	7.8	1,511,372 ($17,680.03)	4.71	NA	

net metering and a 100% energy banking facility. The negative value for C1 indicates that the prosumer earns income during the billing period by exporting excess electricity to the utility company, which pays a FiT of INR 3.15/kWh (0.037 $/kWh). The banking facility under net metering allows the prosumer to feed surplus electricity from their renewable energy system—such as an SPV system—back into the grid and store it as energy credit. The consumer can later use this credit when their energy consumption exceeds their system's generation.

In Scenario 3, we calculate the annual bill based on Time-of-Use (ToU) net metering (see Table 4.4), where the prosumer can utilize exported energy only within the same time zone, with no inter-zone carry-forward allowed. We envision this scenario as a likely development in the near future, where the utility company avoids returning energy to the prosumer during the peak zone (non-solar hours) that was originally exported during the normal zone (solar hours). As a result, C6 faces a significant increase in energy expenses compared to Scenarios 1 and 2, especially since they remain a net importer of electricity. In contrast, the reduced Feed-in Tariff (FiT) lowers C1's income from energy exports.

In scenario 4, BESS stores energy during solar hours, which C1 and C6 use during non-solar peak hours. We size the BESS as detailed in Sect. "Incentive and Subsidy for BESS", and this is done only for representation of the benefits that can be derived from its installation. This setup results in financial savings for both C1 and C6. However, we can determine the actual financial benefit only after accounting for the substantial investment required for purchasing and installing the BESS.

To further assess the financial viability of a BESS, we calculate the Net Present Value (NPV) and Discounted Payback Period (DPP). As shown in Table 4.9, C6 achieves a positive NPV and recover the capital expenditure (CAPEX) for the BESS installation within 7.8 years, even without subsidies or incentives. This is primarily due to their higher Average Billing Rate (ABR). In contrast, C1 has a negative NPV, C1 cannot recover the CAPEX even with a 40% subsidy and an incentive of INR 4 per kWh ($0.047 per kWh) consumed during peak hours. For low-consumption domestic consumers, it becomes essential to offer higher subsidies and greater incentives to make the BESS financially viable. The decision to grant higher subsidies highlights the need for appropriate policy interventions, while the feasibility of offering higher incentives depends on the utility's Average Power Purchase Cost (APPC). The data and assumptions taken for the calculations are given below.

(i) The cost of the SPV system is INR 42,000 ($491.19)/ kWp (Boruah and Chandel (2023)) and the cost of BESS (Including PCS and EMS) is 35,000 INR ($405.65)/ kWh,
(ii) We assume that the BESS will be replaced every 10 years
(iii) We assume the minimum state-of-charge maintained in the BESS as 20%
(iv) We consider the number of useful charge–discharge cycles as 5000
(v) We assume the round-trip efficiency of BESS as 90%,
(vi) We apply a discounting factor of 10.0%,
(vii) We calculate the electricity bills based on the latest applicable tariff of KSEBL

From the utility company's perspective, significant savings can be made when consumers use locally stored energy instead of drawing power from the grid during periods— particularly when the utility company purchases electricity from the exchange. These savings include transmission and distribution (T&D) costs as well as reduced losses—which account for 12.84% in FY 2023–24. KSEBL currently incurs a combined cost of INR 1.82/kWh (0.0213 $/kWh) for transmission, distribution, and energy exchange. Therefore, for every unit of electricity the consumer draws from their own battery storage system, the utility company saves INR 1.82 ($0.0213).

Our analysis of C6 is relevant for commercial consumers, as their per-unit electricity cost is higher according to the existing tariff. We anticipate analogous outcomes for consumers with higher per-unit electricity costs.

Modification of FiT Based on Time-of-Generation

We analyze the proposed dynamic FiT scheme using C5, a low-consumption prosumer with a 3.54 kWp on-grid SPV system. We calculate the Levelized Cost of Energy (LCoE) of the SPV system as INR 2.2 ($0.0257) over a 20-year lifetime, factoring in the replacement of the inverter every 10 years.

Our proposed FiT for each time block is defined in Eq. (4.14). We set the incentive α equal to the LCoE of the SPV system to ensure that the prosumer does not have any financial loss. The variable $C_{gm}(t)$ represents the Green Day Ahead Market (GDAM) rate. Thus, Eq. (4.14) becomes,

$$C_{FiT}(t) = \max(2.2, 2.2 + [GDAM(t) - 2.2] \times \beta)$$

Table 4.10 shows the revenue generated by C5 for different values of GESF (β) ranging from 0.1 to 1 along with the corresponding financial gain or loss. The highest financial benefit happens when GESF equals 1. Unlike current FiT scheme, which pays a fixed rate of INR 3.15 ($0.037)/kWh, the proposed FiT scheme allows each prosumer to maximize revenue by exporting more and consuming less during time blocks when GDAM rates exceed the prosumer's LCoE. Additionally, the proposed FiT benefits the utility by reducing its reliance on the electricity market during high-price periods. Under this scheme, prosumers like C5 gain the flexibility to export electricity strategically during peak-price intervals, resulting in improved financial returns.

Table 4.10 Variation of prosumer's profit with GESF (β)

Green Energy Support Factor, β (GESF)	*Prosumer's Revenue as per new FiT (INR)*	*Present Revenue of Prosumer (INR)*	*Loss (INR)*
0.1	87($1.0177)	119	−32(−$0.37)
0.2	90($1.0528)		−29(−$0.34)
0.3	94($1.0998)		−25(−$0.2925)
0.4	97($1.134)		−22(−$0.2572)
0.5	101($1.18)		−18(−$0.21)
0.6	104($1.22)		−15(−$0.1755)
0.7	108($1.26)		−11(−$0.1287)
0.8	111($1.30)		−8(−$0.0936)
0.9	115($1.35)		−4(−$0.0468)
1	118($1.38)		−1(−$0.0117)

Conclusion

Our chapter explores strategies to improve roof top solar PV(RE) integration, increase prosumer participation, and promote community-based energy solutions, with a focus on the Kerala and Indian electricity grid. We propose a modified Time-of-Use (ToU) tariff that introduces a day off-peak zone during peak solar hours, which addresses the significant gap between electricity generation and demand during midday. By lowering electricity costs during this period, the day off-peak zone encourages increased daytime consumption and higher RE integration.

We also highlight the economic advantages of forming community grids. As shown in Sect. "Evaluating Individual vs. Community–Grid Approaches", the total electricity cost drops from INR 1,788,897 ($20,926.52) to INR 1,558,143 ($18,227.16), resulting in an annual savings of INR 230,754 ($2,700). Our case study confirms that installing community-scale solar PV systems is more economical than deploying individual systems to achieve net-zero energy.

In addition, we analyze the financial viability of using Battery Energy Storage Systems (BESS) under different subsidies and incentive structures. Our findings indicate that the adoption of BESS is economically feasible for high-consumption prosumers (using at least 300 kWh per month), even without subsidies. However, low-consumption prosumers require subsidies of 40–60% and incentives up to INR 4 (0.047 $)/kWh to make BESS adoption economically advantageous. Utility companies also benefit when prosumers use stored energy during peak hours, as this reduces transmission, distribution, and energy exchange costs.

Finally, we develop a green electricity market-linked FiT model that offers a flexible pricing mechanism. This model guarantees a minimum revenue per unit exported to the grid while allowing prosumers to earn more by exporting during high-price intervals. Although our analysis indicates that a fixed FiT currently offers better returns in the short term, the dynamic FiT model empowers prosumers to optimize their exports and increase revenue based on market signals over time.

Policy Suggestions:

We recommend several policy measures to support rooftop solar PV initiatives, enhance energy storage adoption, and improve the economic sustainability of prosumers and utility companies. Firstly, utility companies can introduce a ToU tariff adjustment by designating peak solar hours as a day off-peak zone. This encourages consumers to shift their electricity usage to daytime hours, which improves roof top solar PV integration.

Secondly, we recommend the implementation of targeted strategies to promote the financial viability of BESS adoption. For low-consumption prosumers ($\leq$300 kWh/month), offering subsidies between 40 and 60%, along with incentives up to INR 4 ($0.047) /kWh, can significantly boost BESS adoption in Kerala. In contrast, high-consumption prosumers (>300 kWh/month) can justify BESS installations without subsidies, promoting energy self-sufficiency and decentralized energy use. Moreover, when prosumers rely on stored energy during peak demand periods, utility companies benefit through reduced transmission, distribution, and energy exchange costs—ultimately improving overall grid economics.

We also emphasize the importance of supporting community grids through policy frameworks that attract investment in localized energy generation, leading to long-term cost savings and enhanced energy security.

Finally, implementing a green electricity market-linked feed-in tariff (FiT) model provides prosumers with flexible pricing. By allowing prosumers to earn higher revenue by exporting electricity during high-price intervals, while still ensuring a minimum guaranteed return for each unit of energy exported, the dynamic FiT incentivizes strategic exports and aligns prosumer behavior with grid needs.

Overall, our findings suggest that encouraging community-based energy models, revising tariff structures, and incentivizing energy storage adoption can lead to more efficient electricity use, increased rooftop solar

PV integration, and financial benefits for both consumers and utility companies. However, the effectiveness of these strategies depends on consumer consumption patterns and the prevailing tariff structures. This highlights the importance of designing tailored solutions and implementing supportive policy measures to ensure successful outcomes across different user categories.

References

Ahmed, S. D., Al-Ismail, F. S. M., Shafiullah, M., Al-Sulaiman, F. A., & El-Amin, I. M. (2020). Grid integration challenges of wind energy: A review. *IEEE Access, 8*, 10857–10878. https://doi.org/10.1109/access.2020.2964896

Alam, M. S., Al-Ismail, F. S., Salem, A., & Abido, M. A. (2020). High-level penetration of renewable energy sources into grid utility: Challenges and solutions. *IEEE Access, 8*, 190277–190299. https://doi.org/10.1109/access.2020.3031481

Alnawafah, H., Sarrias-Mena, R., Harb, A., Fernández-Ramírez, L. M., & Llorens-Iborra, F. (2023). Evaluating the inertia of the Jordanian power grid. *Computers & Electrical Engineering, 109*, Article 108748. https://doi.org/10.1016/j.compeleceng.2023.108748

Hussein, A. -M. M. K., Al-Asemi, T., Sabr, N., Almawari, K., Abuelrub, A., & Ehsani, M. (2022). *On-grid photovoltaic energy system-A case study*. In 2022 3rd International Conference on Smart Grid and Renewable Energy (SGRE) (pp. 1–6). IEEE. https://doi.org/10.1109/SGRE53517.2022.9774138

Boruah, D., & Chandel, S. S. (2023). A comprehensive analysis of eight rooftop grid-connected solar photovoltaic power plants with battery energy storage for enhanced energy security and grid resiliency. *Solar Energy, 266*, Article 112154. https://doi.org/10.1016/j.solener.2023.112154

Commitments made under CoP 26. https://pib.gov.in/PressReleaseIframePage.aspx?PRID=1882840

Cristea, C., Cristea, M., Birou, I., & Tîrnovan, R. (2020). Economic assessment of grid-connected residential solar photovoltaic systems introduced under Romania's new regulation. *Renewable Energy, 162*, 13–29. https://doi.org/10.1016/j.renene.2020.07.130

Dynge, M. F., Halden, U., Klæboe, G., & Cali, U. (2022). LCOE-Based pricing for DLT-Enabled local energy trading platforms. *Frontiers in Energy Research, 10*. https://doi.org/10.3389/fenrg.2022.901009

Economics Division, Central Electricity Regulatory Commission. (2024). *Monthly report on short-term transactions of electricity in India*. https://www.cercind.gov.in/2024/market_monitoring/MMC%20Report%20August%202024.pdf

Gupta, U. (2021, April 14). *Implementing time-of-use tariff for peak load management*. Pv Magazine India. https://www.pv-magazine-india.com/2021/04/14/implementing-time-of-use-tariff-for-peak-load-management/

Hou, P., Yang, G., Hu, J., Douglass, P. J., & Xue, Y. (2022). A distributed transactive energy mechanism for integrating PV and storage prosumers in market operation. *Engineering, 12*, 171–182. https://doi.org/10.1016/j.eng.2022.03.001

Ilham, N. I., Dahlan, N. Y., Hussin, M. Z., Sintuya, H., & Setthapun, W. (2022). An optimal compensation schemes decision framework for solar PV distributed generation trading: Assessing economic and energy used for prosumers in Malaysia. *Energy Reports, 8*, 536–544. https://doi.org/10.1016/j.egyr.2022.05.220

Kaczmarski, J. I. (2022). Public support for community microgrid services. *Energy Economics, 115*, Article 106344. https://doi.org/10.1016/j.eneco.2022.106344

Kerala State Electricity Regulatory Commission. (2024). Kerala state electricity regulatory commission (renewable energy and net metering) (second amendment) regulations, 2024. In *Kerala Gazette* [Regulation]. https://dev.erckerala.org/api/storage/draft-regulations/HELssGmIIdUTJtcA8bZtmID9W0e0jOdAORuigBjW.pdf

KSEBL. (2023). *Tariff Order*. https://kseb.in/uploads/Downloadtemsuppy/eogfiledownload%20(5)-1698992147936773737.pdf

KSEBL, (2024), *Annual administration report*. https://kseb.in/uploads/Downloadtemsuppy/AAR2023-24-1717151134433353632.pdf

Laoharojanaphand, V., & Ongsakul, W. (2021). Virtual battery storage service using hydropower plant with co-located floating solar and wind generation. *Sustainable Energy Technologies and Assessments, 47*, Article 101531. https://doi.org/10.1016/j.seta.2021.101531

Lin, B., Chen, J., & Wesseh, P. K. (2022). Peak-valley tariffs and solar prosumers: Why renewable energy policies should target local electricity markets. *Energy Policy, 165*, Article 112984. https://doi.org/10.1016/j.enpol.2022.112984

Maghami, M. R., Pasupuleti, J., & Ekanayake, J. (2024). Energy storage and demand response as hybrid mitigation technique for photovoltaic grid connection: Challenges and future trends. *Journal of Energy Storage, 88*, Article 111680. https://doi.org/10.1016/j.est.2024.111680

Ministry of power, Government of India. (2022). *Renewable_Purchase_Obligation & Energy_Storage Obligation_Trajectory_till_2029_30*. https://powermin.gov.in/sites/default/files/Renewable_Purchase_Obligation_and_Energy_Storage_Obligation_Trajectory_till_2029_30.pdf

MNRE. (2024). *All India installed capacity (In Mw) of power stations (As on30.09.2024)*. https://cea.nic.in/wpcontent/uploads/installed/2024/09/IC_Sep_2024_allocation_wise.pdf

Mohan, V., DS, H. R., Patil, G., M, J., Unni, R. C., K, N., Thirumala, K., & Bu, S. (2022). Policy assistance for adoption of residential solar PV in India: A stakeholder-centric approach for welfare optimization. *Energy Reports*, *8*, 197–206. https://doi.org/10.1016/j.egyr.2022.05.109

National Portal for Rooftop Solar—Ministry of New and Renewable Energy. (n.d.). https://pmsuryaghar.gov.in/

Oprea, S., & Bara, A. (2020). Setting the time-of-use tariff rates with NoSQL and machine learning to a sustainable environment. *IEEE Access, 8*, 25521–25530. https://doi.org/10.1109/access.2020.2969728

Peprah, F., Gyamfi, S., Effah-Donyina, E., & Amo-Boateng, M. (2022). Evaluation of reactive power support in solar PV prosumer grid. *e-Prime—Advances in Electrical Engineering Electronics and Energy*, *2*, 100057. https://doi.org/10.1016/j.prime.2022.100057

Pramanick, S., Ahmad, N., & Jain, D. K. (2021). Real Time Electricity Market (RTM) in India. In *IEEE International Conference on Smart Grid and Smart Cities (ICSGSC)* (5th ed., pp. 112–117). IEEE. https://doi.org/10.1109/icsgsc52434.2021.9490431

Report on high frequency operation in Indian power system on 4, 11 and 25 August 2024. (2024). *Grid Controller of India Limited.* https://posoco.in/en/download/report-on-high-frequency-operation-for-4th-11th-and-25th-aug-2024/

Singh, K., Meena, R. S., Kumar, S., Dhyani, S., Sheoran, S., Singh, H. M., Pathak, V. V., Khalid, Z., Singh, A., Chopra, K., Bajar, S., Ansari, F. A., Gupta, S. K., Varjani, S., Kothari, R., Tyagi, V. V., Singh, B., & Byun, C. (2023). India's renewable energy research and policies to phase down coal: Success after Paris agreement and possibilities post-Glasgow Climate Pact. *Biomass and Bioenergy, 177*, Article 106944. https://doi.org/10.1016/j.biombioe.2023.106944

Singh, S. K., Schulz, N., Srivastava, A., Srivastava, S. C., Mishra, S., Sharma, A., & Mukherjee, D. (2021, October 19). *A rural microgrid field pilot in india ensuring reliable electricity supply and social upliftment*. https://doi.org/10.1109/ghtc53159.2021.9612517

Tanaka, K., Wilson, C., & Managi, S. (2021). Impact of feed-in tariffs on electricity consumption. *Environmental Economics and Policy Studies, 24*(1), 49–72. https://doi.org/10.1007/s10018-021-00306-w

Warneryd, M., Håkansson, M., & Karltorp, K. (2020). Unpacking the complexity of community microgrids: A review of institutions' roles for development of microgrids. *Renewable and Sustainable Energy Reviews, 121*, Article 109690. https://doi.org/10.1016/j.rser.2019.109690

Zhou, M., Pysmenna, U., Kubatko, O., Voloshchuk, V., Sotnyk, I., & Trypolska, G. (2023). Support for Household Prosumers in the Early Stages of Power Market Decentralization in Ukraine. *Energies, 16*(17), 6365. https://doi.org/10.3390/en16176365

CHAPTER 5

Economic Impacts of Sustainable Electrification

Capucine Chapel

Introduction

Electrification through sustainable energy sources represents a major transformation for modern economies, redefining economic dynamics at both global and local levels. By harnessing renewable energies such as solar, wind, and hydro power, nations and citizens are not only meeting growing demands for environmental sustainability; they are also embarking on a process of economic restructuring, generating significant opportunities as well as major challenges.

This chapter examines the economic impacts of sustainable electrification, analyzing its overall economy-wide effects, its targeted implications for households and businesses, and the strategic transformations it engenders in key sectors such as real estate, finance, and insurance. The

C. Chapel (✉)
Université Clermont Auvergne, INRAE, VetAgro Sup, UMR Herbivores, Saint-Genès-Champanelle, France
e-mail: capucine.chapel@inrae.fr

T. Walker et al. (eds.), *Wiring the Future*, Palgrave Studies in Emerging Risk Management and Sustainable Finance,
https://doi.org/10.1007/978-3-032-14717-2_5

first section analyzes macroeconomic impacts, addressing energy security, labor market transformations, and the relationship between energy and economic growth. The second section focuses on microeconomic effects, highlighting the transformations experienced by households and businesses, with a particular focus on fuel poverty in developed countries and energy poverty in developing countries. Finally, a third section analyzes sectoral impacts, focusing on the adaptation of strategic sectors such as real estate, finance, and insurance, and their role in this transition. Through a literature review, concrete examples, and statistical analyses, this chapter offers an integrated vision of the economic opportunities and challenges posed by the energy transition.

Macroeconomic Implications: Energy Security, Jobs, and Growth

Energy Security

The energy transition is not just about environmental considerations. One of the fundamental issues it raises is that of energy security. In a world where dependence on fossil fuels leads to economic and geopolitical vulnerabilities, electrification via renewable energy sources appears to be a strategic solution for guaranteeing long-term energy stability. This section explores how the energy transition is redefining global energy balances and helping to strengthen the energy independence of nations, while at the same time posing new challenges.

Energy security is a key strategic concept for ensuring the stable, affordable, and sustainable supply of energy that is essential to the functioning of modern economies. This security is based on several key dimensions: access to reliable energy sources, diversification of supplies, infrastructure resilience, and effective management of geopolitical and environmental risks.

Dependence on fossil fuels is a key issue for energy security. Global energy consumption continues to grow, with fossil fuels still accounting for a significant share. In 2023, fossil fuels accounted for approximately 80% of global energy consumption (IEA, 2024b). The International Energy Agency (IEA) predicts that, without additional policies, this proportion would likely only decrease by 5% by 2030, indicating a gradual but insufficient transition to meet international climate targets.

In this context, the lack of diversification of energy sources compromises energy stability and exposes many countries to major risks. In addition to their high environmental impact, fossil fuels such as oil and gas are finite resources whose reserves will inevitably run out. This poses a major challenge to meeting growing energy needs in a sustainable way, especially when their extraction and transport require costly investments and specific infrastructures. Moreover, dependence on fossil fuels is all the more worrying given that their production is highly concentrated in a small number of regions around the globe. Just a few countries hold the majority of the world's reserves, particularly in the case of oil. Indeed, the three biggest oil producers—the USA, Saudi Arabia, and Russia, in descending order—together accounted for over 40% of world oil production and 37% of exports by 2022 (Fattouh & Economou, 2023). This uneven distribution limits the energy autonomy of importing countries, making them vulnerable to price fluctuations and geopolitical tensions.

The recent energy crisis, exacerbated by Russia's invasion of Ukraine in 2022, exposed the vulnerabilities of countries reliant on energy imports. While energy markets were already under strain due to post-COVID demand recovery and supply constraints, the conflict significantly disrupted global energy flows, particularly in Europe, where Russia plays a key role in the supply of oil and gas. Following the invasion, Western countries—including the European Union, the United States, the United Kingdom, and others—imposed economic sanctions on Russia. These measures led to a significant reduction in Russian gas exports to Europe, which had previously depended on Russia for around 40% of its natural gas (IEA, 2022a). The reduction in Russian gas flows has put a strain on Europe's energy infrastructure. Natural gas is a key energy source for power generation, particularly in Europe, where many power plants run on gas. This reduction in supplies has caused prices to soar on energy markets, directly impacting electricity and heating costs for households and businesses in many countries (IEA, 2022b).

In some cases, increased demand to compensate for Russian gas deficits has led to heightened global competition for liquefied natural gas (LNG) cargoes, further amplifying price increases and tensions between importers. Faced with this situation, governments have had to implement energy rationing policies and mobilize financial resources to subsidize high-energy costs, putting further pressure on public finances.

In addition to increased competition for LNG cargoes, the strategic routes used by these ships are also sources of vulnerability, as disruptions there can further amplify tensions on energy supplies (IEA, 2024b). These so-called "chokepoints" include narrow passages such as the Suez Canal, the Strait of Malacca, and the Panama Canal, where geopolitical, security, and environmental conditions can restrict the transit of ships.

For example, the Panama Canal recently suffered capacity restrictions due to a prolonged drought, limiting the number of ships allowed through and causing supply delays. Similarly, the Suez Canal, which was already blocked for six days in 2021 by an incident involving a container ship, is crucial for LNG routes and remains sensitive to security tensions in the region. The Bab al-Mandab Strait, close to conflict zones, has recently seen a drop in the number of cargo ships and tankers in transit due to the activity of armed groups. If these various passages were temporarily inaccessible, ships would be forced to follow longer routes, such as the Cape of Good Hope, increasing journey times and costs. These constraints on strategic routes not only lengthen delivery times, but they also create greater price volatility and increase the pressure on importing countries to secure their energy supplies.

This energy crisis and the various vulnerabilities it has created have highlighted the need for importing countries to strengthen their security. The energy transition, which relies on both improved energy efficiency and greater diversification of sources via renewable energies, thus appears to be an essential solution for strengthening energy security in many countries. Indeed, thanks to their decentralized nature, renewable energies offer local production that limits dependence on imported fossil fuels. Unlike fossil resources, which are concentrated in certain regions of the globe, renewable energy sources are more equitably distributed, enabling each country to take advantage of its own natural resources, whether wind, sun, or water. This enables countries to develop their domestic energy capacities, strengthening their autonomy and reducing their vulnerability to international market fluctuations and geopolitical tensions.

In Europe, this vision took shape with the REPowerEU plan, launched in 2022 by the European Commission in response to the energy crisis worsened by the war in Ukraine (European Commission, 2022). The plan aims to significantly accelerate the integration of renewable energies to achieve sufficient levels of renewable energy production in the medium and long term, thereby reducing energy dependence on imported gas, oil, and coal. REPowerEU sets ambitious targets, including reaching 480 GW of wind and 420 GW of solar power capacity by 2030—representing a tripling of current levels—as well as deploying up to 15 TWh of rooftop solar annually and installing 30 million heat pumps by 2030. These measures are expected to deliver substantial gas savings and enhance Europe's energy security.

In addition, REPowerEU encourages the deployment of innovative energy solutions, such as energy storage infrastructures and smart grids, which optimize the integration of renewable energies into power systems and increase grid resilience. By aiming for a significant reduction in fossil fuel imports, the REPowerEU plan does more than simply increase green energy production capacity: it reorients European energy policy to strengthen the continent's energy autonomy and encourage a more sustainable, less carbon-intensive economic model. According to European Commission forecasts, these initiatives should reduce natural gas consumption in electricity and heating production by around 40% by 2030 (European Commission, 2022), thus stimulating the green economy and limiting energy costs for European citizens.

Beyond this example, economic literature highlights the macroeconomic benefits of the energy transition and its potential to trigger a virtuous circle. Gökgöz and Güvercin (2018) show that the development of renewable energies reduces the need to import fossil fuels, thereby contributing to greater energy security. This improvement in energy security in turn has positive repercussions on key macroeconomic variables, such as trade balance and current account deficits. For example, Kaushik and Shastri (2024), analyzing data from India, a country heavily dependent on oil imports, between 1985 and 2019, demonstrate that a 1% increase in the price of oil leads to a 0.003% decrease in the trade balance, while a 1% increase in renewable energy consumption improves it by 0.011%. Ozkan and Okay (2024) confirms this trend for 15 OECD countries between 1990 and 2021: countries using a higher share of renewable energy in their electricity production see an improvement in their current account deficits.

However, this shift creates new challenges. Countries that export hydrocarbons must rethink their economic models to address declining fossil fuel revenues. A United Nations Development Programme (UNDP) report (Jensen, 2023) highlights that 40 fossil fuel-dependent economies, where these resources contribute an average of 14.3% to GDP and over 60% to exports, could face a revenue loss of more than 60% from oil by 2040 under a scenario aligned with the Paris Agreement. This loss, estimated at between \$12 and \$14 trillion, would threaten economic stability and essential public spending. These figures demonstrate the urgent need for these countries to diversify their economies and prepare for rapid global decarbonization.

Moreover, the transition to renewable energies, while reducing dependence on fossil fuels, is creating new dependencies on the critical minerals needed for technologies such as batteries, solar panels, and wind turbines. According to the IEA (2021), demand for critical materials such as lithium, cobalt, and rare earths is set to at least quadruple by 2040 to meet the needs of energy transitions. The International Renewable Energy Agency (IRENA) points out that the production and refining of these materials to meet this growing demand is geographically concentrated, with China dominating global supply (IRENA, 2023). For example, China provides 100% of the refined supply of natural graphite and dysprosium, 70% of cobalt, and nearly 60% of lithium and manganese. Consequently, in a recent report, the World Economic Forum warns of the geopolitical vulnerabilities arising from this concentration, which could reproduce the strategic dynamics once seen with oil (World Economic Forum, 2024). To mitigate these risks, it is essential to diversify supply chains, develop the recycling of critical materials, and encourage technological innovations aimed at reducing dependence on these resources.

The energy transition, while promising greater resilience, requires a complex reorganization of supply chains and renewed international cooperation. This transformation is not limited to energy supply; it also has profound implications for the labor market.

Job Creation and Sectoral Transformation

The energy sector is a major employer worldwide. According to the IEA (2024), this sector added almost 2.5 million jobs over the year, bringing the total to over 67 million, an increase of 3.8%, outstripping the 2.2% average recorded for the economy as a whole. This growth is attributed to record levels of investment in various energy sources, following the global health and energy crises. In addition, the energy sector encompasses a wide variety of jobs and industries, from fossil fuel power generation to renewable energies such as wind, solar, and hydro, as well as jobs in the distribution, maintenance, and management of energy infrastructures.

Renewable energies are increasingly driving employment growth in the energy sector. Jobs in renewables span various activities, including the manufacturing of solar panels, wind turbines, and batteries, as well as the construction, installation, and maintenance of these infrastructures. While the oil and gas industries continue to play a significant role in energy sector employment, their growth is slowing due to the shift toward clean energy. In 2023, clean energy sectors created 1.5 million new jobs, significantly outpacing the 940,000 jobs added by fossil fuel industries (IEA, 2024a).

The energy transition, by encouraging the switch from fossil fuels to renewable sources, is having a significant impact on the job market. While the energy transition is creating new jobs, it is also destroying jobs in traditional fossil fuel sectors, such as coal mining, oil refining, and gas production. This trend is particularly marked in regions where the economy is heavily dependent on these activities. It is therefore essential to consider the prospects for the net effect of the energy transition on the job market.

Recent studies point to an overall trend toward net job creation, although this dynamic varies by sector and region (Fragkos & Paroussos, 2018; Zhang et al., 2024). According to projections from the IEA (2023) current policies should lead to the creation of 8 million clean energy jobs by 2030, while the fossil fuel sectors could lose 2.5 million jobs, a net gain of 5.7 million jobs worldwide. This figure could even rise to 17 million if countries align their energy transition policies in line with net-zero emissions targets by 2050. Despite these promising prospects in terms of net job creation, there are still some grey areas, notably regarding regional inequalities and the structural challenges involved in meeting the demands of this transition.

Indeed, the energy transition does not affect all regions uniformly, with its impact varying according to economic dependence on carbon-intensive industries and the development potential of renewable energies. In Europe, McDowall et al. (2023) highlight increased vulnerability in regions where employment is concentrated in sectors such as coal, oil, and energy-intensive industries. These areas, characterized by heavy dependence on these industries and high structural unemployment rates, are likely to be the hardest hit, and require specific policies to diversify their economies and reduce their dependence. In China, Zhang et al. (2024) highlight a marked contrast between regions rich in fossil resources, which are facing significant job losses, and those with strong renewable energy potential, such as wind and solar power, which are enjoying significant job growth. These studies converge on the need to adopt proactive policies adapted to local contexts, both to diversify vulnerable economies and to exploit the potential of regions advantaged by the transition, in order to maximize the opportunities offered by the energy transition while limiting its negative effects on the most vulnerable communities.

Furthermore, the transition to renewable energies is bringing about profound transformations in the energy sector, with major implications for employment and the skills required for such employment. While some jobs in the fossil fuel sector can be transposed relatively easily to renewable energies, due in particular to similar technical skills—for example, the chemical engineering skills used in oil refineries can be adapted to the production of clean fuels and hydrogen (IEA, 2022a)—this adaptability remains partial. A recent analysis based on 300 million job transitions shows that while transitions from fossil fuel-related jobs to green jobs have increased sharply in recent years, they are still rare: less than 1% of workers leaving fossil fuel-related jobs move directly to renewable energy-related jobs (Curtis et al., 2023). Many workers, especially those in extraction sectors such as oil drilling and mining, will have to undergo extensive retraining to meet the specific needs of clean technologies.

This challenge is even more critical given the existing shortage of skilled workers in the energy sector. A survey conducted by the IEA (2024) of over 190 employers in 27 countries on several continents, reveals that most companies are planning to hire more staff, but are finding it difficult to find qualified candidates for almost all job categories. These obstacles reflect the growing pressure for specialized skills in a rapidly changing sector. A study conducted by Arcelay et al. (2021) also points out that the new jobs linked to renewable energies require a combination of technical, digital, and project management skills, areas where training gaps are still significant, increasing the pressure on education systems and retraining programs.

This challenge is further complicated by a structural imbalance: public and policy discussions often present jobs in renewable energy as less stable or lower paying than those in fossil fuels, which may slow the transition for the workers most affected. This perception is partly supported by data: for example, salaries in the wind, solar, and hydrogen sectors are on average 15% to 30% lower than those in the nuclear, oil, and gas sectors, which are among the best paid in the entire economy (IEA, 2023). This gap is partly explained by the recent history of renewable energies, which limits the presence of powerful unions capable of negotiating favorable conditions, and by the fact that occupational risks are often lower than in fossil fuels, which reduces the danger premium associated with these jobs (Associated Press, 2022). To meet these challenges, it is essential to implement ambitious, coordinated policies combining training, retraining, and economic support.

Thus, the energy transition is profoundly transforming the job market, with the creation of significant opportunities in green sectors and the need to support professional reconversions in declining industries. These changes are taking place in a more global context, where the investments and technological innovation associated with this transition are also redefining the dynamics of economic growth.

Economic Growth

The relationship between electrification and economic growth has been the subject of numerous studies, revealing complex and sometimes divergent links depending on the context and methodologies employed.

Historically, economists and development scholars have regarded electrification as a major catalyst for economic growth. During the industrial revolution in Western countries, the introduction of electricity transformed production processes, improved energy efficiency, and stimulated technological innovation, thereby contributing to a significant increase in gross domestic product (GDP). Today, many human activities, such as industry, transport, services, and domestic use, depend on electricity to function efficiently. The various oil shocks in history, leading to sharp rises in electricity prices, have often been followed by recessions (Bretschger, 2009). These historical facts underline the close link between electricity and economic growth. We can also observe this link through descriptive statistics. Figure 5.1 illustrates, for the year 2022, a strong positive correlation between electricity consumption per capita and GDP per capita, with a high correlation coefficient of around 0.62.

Although the correlation between electricity consumption and economic growth is well established, it is essential to distinguish between correlation and causality. Correlation indicates an association between two variables, while causality implies that one is the cause of the other. The causal nature of this relationship remains a subject of research and debate within the economic community.

The economic literature presents a variety of results concerning the causality between energy consumption and economic growth. According to the classification proposed by Aqeel and Butt (2001), four main

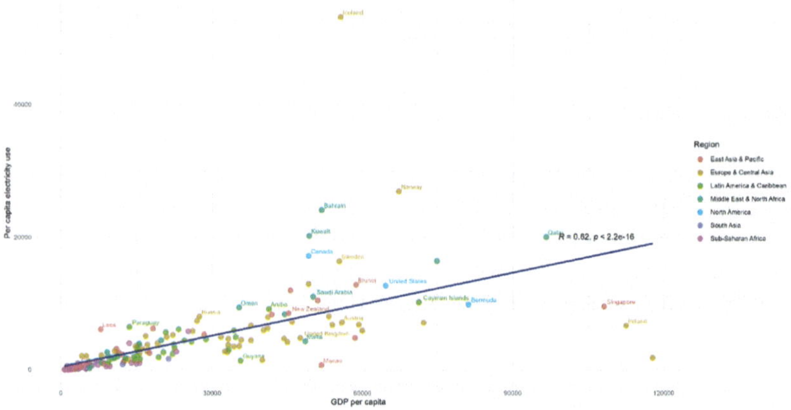

Fig. 5.1 Correlation between electricity generation and GDP in 2022, *Source* Author's own elaboration based on data from OurWorldinData.org

hypotheses emerge: the growth hypothesis, where energy consumption stimulates economic growth; the conservation hypothesis, suggesting that economic growth leads to an increase in energy demand; the feedback hypothesis, indicating a bidirectional relationship of interdependence between the two variables; and finally, the neutrality hypothesis, where no significant causal relationship is observed between energy consumption and economic growth. These discrepancies can be explained by methodological differences, the periods studied, and specific geographical contexts.

The energy transition, which involves a gradual shift from fossil fuels to renewable resources for electricity generation, further complicates the debate on the relationship between energy consumption and economic growth. This structural change modifies traditional dynamics by introducing new variables, such as investment in renewable technologies, dependence on specific infrastructures, and the impact of public policies favoring decarbonization.

On the one hand, proponents of renewable energies argue that this transition could stimulate economic growth through technological innovation, large investments, and the creation of green jobs. For example, in the wake of the COVID-19 health crisis, IRENA (2020) estimates that investment in renewable energies and the resulting job creation could boost global GDP by 1.3% by 2030. In addition, a study by Al-mulali et al. (2014) on Latin American countries showed that electricity consumption from renewable sources has a more significant impact on economic growth than that from non-renewable sources, suggesting that renewable energies are more effective in promoting economic development.

On the other hand, some studies point to short-term economic challenges, including high initial infrastructure costs and potential disruption to fossil fuel-dependent economies (Bhattacharya et al., 2016; Mohamad Taghvaee et al., 2017). In addition, studies also express concern about the long-term economic impact of the energy transition. Menegaki (2011) found evidence to support the neutrality hypothesis, suggesting that economic growth and renewable energy consumption in Europe are not significantly linked, which could indicate that investment in renewable energies will not have the expected economic impact. Some also question the feasibility of an energy system based entirely on renewable energies, due to technological, institutional, and economic constraints (Dupont et al., 2021; Moriarty & Honnery, 2020). Finally, commitments to a low-carbon energy transition, centered on the development of renewable energies, imply structural constraints which, according to Sers and Victor (2018) and Nieto et al. (2020), will inevitably reduce economic growth trajectories. These structural constraints stem from the very nature of renewable energies, whose energy efficiency ratio (EROI) is generally lower than that of fossil fuels, requiring massive capital and energy investments to maintain production levels. They also include demand saturation, the difficulty of decoupling growth and emissions, the intermittency of renewable sources, and the profound reorientation of existing infrastructures and economic systems.

These elements add a new dimension to the debate surrounding the classic hypotheses proposed by Aqeel and Butt (2001), by questioning whether the growth hypothesis or the conservation hypothesis remains valid in a world increasingly focused on renewable energy. In this way, the energy transition is not just a response to climate challenges, it is also redefining the way energy and economic growth interact, fueling discussions on the economic implications of this global mutation.

While the energy transition is profoundly transforming major national and international economic dynamics, it is also bringing about significant changes at the level of businesses, households and local communities.

Microeconomic Implication: Households and Businesses

At the microeconomic level, electrification via sustainable energy sources is having a transformative impact on households and businesses in both developed and developing countries. However, the effects of this transition to sustainable energy vary widely depending on the level of development and the challenges specific to each energy system: while developed countries are optimizing the efficiency and management of their consumption, developing countries are still focusing on basic access to energy.

Developed Countries: Energy Transition and the Fight against Fuel Poverty

In high-income countries, where access to electricity is almost universal, the energy transition remains crucial for households and businesses for a number of reasons. The recent energy crisis, exacerbated by geopolitical tensions, has pushed up energy costs. According to IEA (2024b), worldwide household energy costs rose by an average of 20% between 2021 and 2022, despite numerous measures taken by certain governments. At the same time, rising energy costs have led to a general rise in prices. Whether electricity, fuel, or gas, these energy sources are essential for production, transport, and essential services. As a result, rising energy prices impact the entire economy, increasing the cost of basic necessities and services. This underlines the need to diversify energy sources to stabilize costs. Many states have therefore introduced policies to support renewable energies and promote energy efficiency, in order to reduce greenhouse gas emissions while strengthening resilience to international market fluctuations.

The adoption of renewable energies by households and businesses is a sustainable solution for reducing bills and meeting government targets. Renewable energies, such as solar and wind power, enable consumers to generate their own electricity, reducing their dependence on traditional suppliers and fluctuating energy prices. A study in several US states (Fikru, 2019) found that households equipped with residential solar panels reduced their summer electricity bills by an average of 61%.

However, installation costs remain high, posing a challenge for many households and small businesses. Despite sharply declining costs (IEA,

2022b), a solar panel installation costs on average around $12,700 before subsidies for a 6 kW solar panel system (Ogletree et al., 2022). For some households, this remains an investment that is often difficult to access. Businesses, for their part, have to mobilize substantial resources to upgrade their infrastructure; an especially complex task for microenterprises or small firms with limited technical capacity, low cash flow, or no dedicated energy or facilities management team.

Energy crises and the energy transition disproportionately affect low-income households. These households spend a significant proportion of their budget on energy and often have limited access to energy efficiency technologies. Low-income households are thus the hardest hit by energy price rises. Battistini et al. (2022) show that rising energy prices affect low-income households almost four times more than those in the highest income quintile. This increased dependence leads to sacrifices in other essential areas, such as food and healthcare, creating a "fuel poverty" effect. In 2022, around 42 million people in the European Union, or 9.3% of the population, were unable to heat their homes adequately (European Economic and Social Committee, 2023). This figure is up from previous years, when fuel poverty affected 8% of the population in 2020 and 6.9% in 2021. According to the European Economic and Social Committee, this upward trend results from soaring energy prices and energy market instability, which have worsened the situation of vulnerable households.

In the context of the energy transition, these difficulties are further amplified: the high upfront costs of energy efficiency technologies, such as solar panels or efficient heating systems, remain out of reach for these households. At the same time, policies such as carbon taxes, designed to encourage emissions reductions, impose an additional burden on the bills of low-income households without offering them affordable alternatives (Wang et al., 2016). To make this transition more equitable, it is essential to provide specific financial support—in the form of subsidies, zero-interest loans and preferential tariffs—to enable low-income households to access sustainable technologies and reduce their vulnerability to fluctuations in energy prices.

Developing Countries: Reducing Energy Poverty through Sustainable Electrification

In developing countries, the energy transition poses challenges that go far beyond fuel poverty, touching on deeper realities of energy poverty. Here, the major problem is not just the cost of energy, but the very lack of access to modern energy services for a vast proportion of the population. By 2022, some 685 million people were living without electricity, and 2.1 billion still relied on harmful fuels for cooking (IEA et al., 2024). This situation is particularly acute in sub-Saharan Africa, where high population growth exacerbates energy needs, but where the necessary infrastructure is often non-existent or inadequate.

This situation profoundly hampers economic opportunities and degrades quality of life by limiting access to basic needs such as lighting, heating, cooking, and other services essential to the well-being of populations and the development of local businesses (Cecelski, 2003; Sule et al., 2022). The lack of modern energy hampers productivity, reduces access to education and healthcare, and increases health risks, particularly in households that rely on polluting fuels for cooking (Adom & Nsabimana, 2022; Sovacool, 2012).

A simple comparison of survey data from the Afrobarometer conducted in Angola in 2022 between populations with and without access to electricity reveals significant differences. This nationwide representative sample survey reveals that households with access to electricity report better living conditions (see Fig. 5.2). If we go into a little more detail, we see that individuals with access to electricity are less exposed to difficult situations in terms of food, health, and money (see Fig. 5.3). They are also more likely to have a job (see Fig. 5.4), a bank account, or also a motor vehicle (see Fig. 5.5).

In addition, several academic studies have demonstrated the positive effects of electrification on various human development indicators such as income, health, and education (da Silveira Bezerra et al., 2017). Indeed, electricity enables children to study for longer in conditions more conducive to learning, thereby increasing their academic performance. For example, Kanagawa and Nakata (2008), in a study carried out in India, show that a 1-point increase in the electrification rate leads to a 0.17-point improvement in the literacy rate.

In terms of health, electricity also plays a key role, reducing malnutrition, improving access to healthcare, and reducing respiratory diseases,

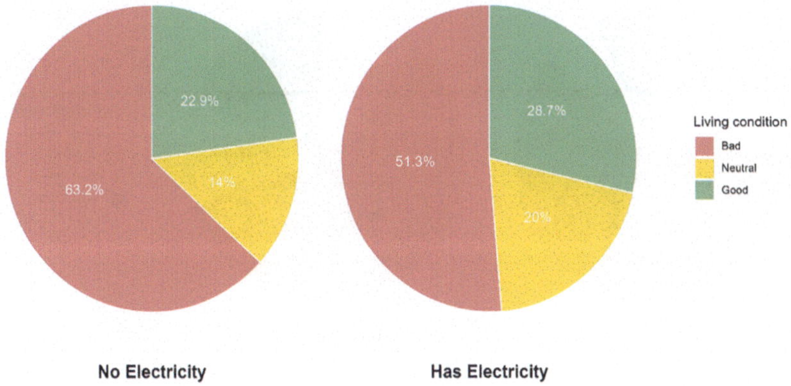

Fig. 5.2 Perceived living conditions according to access to electricity

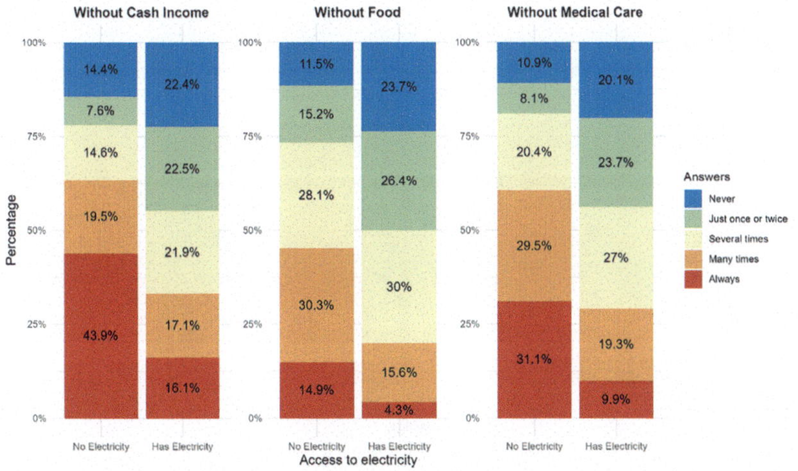

Fig. 5.3 Access to electricity and difficulties experienced

which are exacerbated by the use of traditional fuels. In a study covering 54 developing countries, Candelise et al. (2021) demonstrate that each additional point in the percentage of the population with access to electricity is associated with a 0.12-point drop in the rate of undernutrition. Similarly, Barron and Torero (2017) show that in El Salvador facilitated

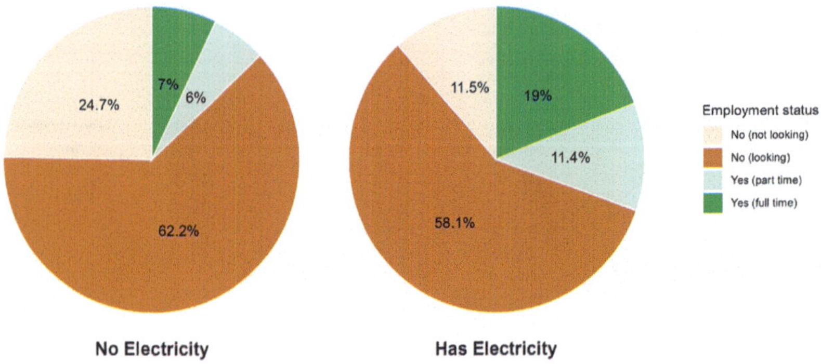

Fig. 5.4 Employment status according to access to electricity

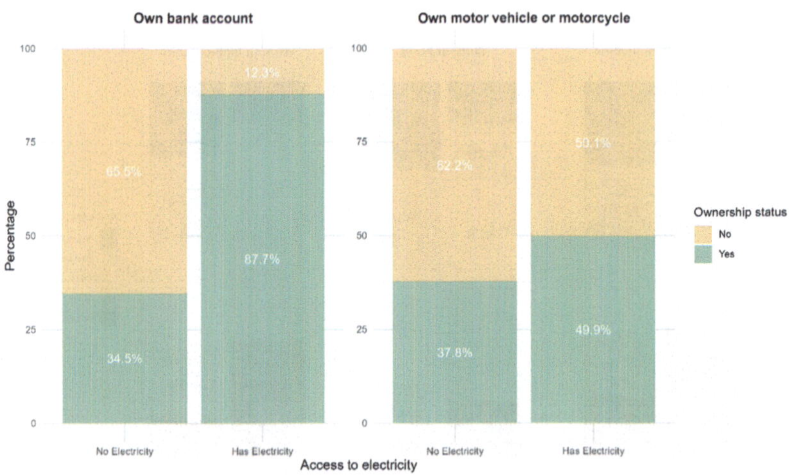

Fig. 5.5 Comparison of motor vehicle or motorcycle and bank account ownership by electricity access

access to electricity, through vouchers, leads to a 66% reduction in indoor air concentrations of fine particles (PM2.5) and an 8–14% drop in acute respiratory infections in children under six.

In terms of living standards, electricity enhances economic opportunities by enabling the use of appliances and power tools that boost productivity, promote local entrepreneurship, and free up time for income-generating activities. This, in turn, contributes to greater household economic stability. Adom and Nsabimana (2022), for example, show that in Rwanda, access to electricity improves equivalent consumption and increases labor force participation by 3.1% and 22 additional working days per year. Electrification also facilitates the integration of women into the labor market. For instance, Grogan and Sadanand (2013) demonstrate through a study in Nicaragua that access to electricity increases the probability of rural women working outside the home by 23%.

Electrification does not only benefit individual well-being; it also plays an essential role in local economic development, strengthening the entrepreneurial fabric and supporting the growth of industrial, craft, and agricultural activities in electrified areas. Indeed, electrification improves business performance by increasing productivity, reducing operating costs, and creating new investment opportunities. Access to reliable energy enables businesses to operate more efficiently, extends their operating hours, and supports the use of modern equipment essential for production.

Several studies confirm this positive impact. For example, Kirubi et al. (2009) found that access to electricity through a community electricity microgrid in Kenya enabled small and microenterprises to use electrical equipment, increasing productivity per worker by 50–200%, depending on the task performed. More specifically, they observed productivity gains of around 50–100% in tailoring tasks such as sewing and ironing clothes, and up to 200% in carpentry tasks like making coffee tables, where electric tools significantly reduced production time and increased daily output. In addition, the same study highlights how electrification in this region has enabled the cold storage of perishable agricultural products, such as meat and milk, as well as other consumer goods, thus improving preservation conditions and reducing losses for farmers.

The energy transition to renewable sources offers a significant opportunity to improve access to electricity in developing countries. Decentralized solutions, such as solar microgrids and stand-alone photovoltaic systems, make it possible to supply energy to remote areas where extending the national grid is economically or technically difficult. In recent work, L'Her et al. (2023) estimate that the use of photovoltaic systems coupled with batteries could provide reliable electricity to around 90% of unserved populations, with a production cost of less than $0.20 per kWh.

However, despite these advantages, the high initial cost of installation remains a major obstacle to widespread adoption. The investment required to acquire and install these systems can be prohibitive for local populations and governments in developing countries. To overcome these obstacles, innovative financing mechanisms, government subsidies, and public-private partnerships are essential to make these technologies accessible and sustainable in the long term. For example, Chapel (2022) has shown that renewable energy projects financed by World Bank and Chinese official development assistance between 2000 and 2014 have increased household electrification by 6% to 11% in 19 Sub-Saharan African countries.

The transformations brought about by the energy transition do not stop at the impact on households and businesses. They also affect all strategic economic sectors—that is, sectors whose functioning is essential to the overall stability and performance of the economy, such as energy, real estate, and finance. These sectors must adapt not only to changes in demand and regulation but also to deep structural shifts driven by decarbonization policies. The next section focuses on two of them: real estate and finance, and the strategic transformations they are undergoing in the face of the energy transition.

Real Estate and Finance Sector: Strategic Transformations in the Face of the Energy Transition

Impacts on the Real Estate Sector

The real estate sector plays a central role in the energy transition, due to its massive energy consumption and significant carbon footprint. In Europe, for example, buildings account for 40% of total energy consumption (European Commission, 2020), of which a significant proportion relies on electricity for lighting, appliances, heating, air conditioning, and connected technologies. At the same time, this sector is responsible for 36% of Europe's greenhouse gas emissions. Faced with these challenges, the energy transition is redefining the priorities of the real estate sector, driven by stricter regulations, growing demand for sustainable buildings, and the adoption of new technologies.

The transformation of the real estate sector begins with improving the energy efficiency of buildings, an approach that reduces both their ecological footprint and their operating costs. This means adopting technologies with high-energy standards. These improvements reduce energy consumption and associated costs. For example, the adoption of Light Emitting Diode (LED) lighting represents a major advance in energy efficiency compared with incandescent lamps and compact fluorescent lamps (CFLs). A study in Malaysia (Khorasanizadeh et al., 2015) shows that LEDs cut electricity costs by 88% compared with incandescent and 36% compared with CFLs, thanks to their low consumption and long life. Although they have a higher initial cost, their cost-effectiveness and durability make them an essential choice for efficient, sustainable buildings.

The proliferation of technological innovations in the building sector is not confined to lighting. Researchers and manufacturers are also driving significant advances in thermal insulation, double-glazed windows, and high-efficiency heating systems. These innovations play a key role in reducing the overall energy requirements of buildings. They are also accompanied by a revolution in energy management, made possible by connected technologies (Halhoul Merabet et al., 2021). Tools such as smart thermostats, connected meters, and automation systems make it possible to optimize electricity consumption in real time, according to the specific needs of occupants. By reducing energy wastage, these innovations contribute to both improved user comfort and more sustainable resource management. While these solutions naturally find their place in new-build projects, they are also essential in renovation projects, which are essential for modernizing the existing building stock, which is often outdated and energy-hungry.

Beyond the operational and environmental benefits, energy efficiency improvements increasingly influence property values, with growing empirical evidence highlighting heterogeneous effects depending on property type and tenure. For instance, Cerin et al. (2014) show that in the Swedish housing market, properties with strong energy performance certifications (EPCs) are sold at a premium, particularly for buildings constructed before 1960 and for those in the lower price-per-square-meter segment. This suggests that energy efficiency upgrades may be especially valued in older and more affordable housing, where buyers are more sensitive to potential long-term savings. Similarly, Kholodilin et al. (2017), using data from Berlin, demonstrate that energy-efficient dwellings are associated with higher sales prices and rents, but the magnitude of this premium depends on the mode of tenure: owner-occupied properties show stronger capitalization effects than rented units. This is likely due to the "landlord–tenant dilemma," where tenants benefit from lower energy bills while landlords bear the cost of retrofitting. These findings underscore the importance of considering property-specific and market-segment factors when assessing the return on investment of energy efficiency measures in real estate.

However, these technologies often have a higher initial cost and a long payback period, which can be an obstacle for owners and investors. While some simple upgrades such as improved air sealing, LED lighting, or double-glazed windows can have relatively short payback periods—ranging from 2 to 10 years, and even under 2 years in some cases—more advanced solutions like high-performance insulation, green roofs, or phase change materials can exceed 20 years or more, depending on the context and climate (Elaouzy & El Fadar, 2022; Sadineni et al., 2011).

To overcome these barriers, governments and public authorities around the world have introduced numerous public policies to encourage the adoption of energy efficiency measures. These include direct subsidies, tax credits, and low-interest loans, all of which help to reduce the entry cost of energy retrofits. In addition, some jurisdictions are adopting more coercive measures, such as prohibiting the letting of energy-inefficient buildings, thus encouraging landlords to undertake retrofits.

At the same time, the integration of renewable energies into buildings is revolutionizing their operation. Photovoltaic roofs, geothermal heat pumps, and energy storage systems not only reduce greenhouse gas emissions, but also improve buildings' energy autonomy. For example, Tsalikis and Martinopoulos (2015) show that a standard detached house equipped with solar photovoltaic panels and solar thermal systems can significantly reduce its primary energy consumption, achieving energy coverage rates of over 76%. This energy autonomy, combined with long-term cost reduction, is becoming a major asset in a context where fluctuating energy prices are of increasing concern to consumers.

In this context, environmental standards and certifications play a structuring role in guiding real estate industry practices. Labels such as Leadership in Energy and Environmental Design (LEED) or Building Research Establishment Environmental Assessment Method (BREEAM) have become essential references for assessing the sustainability of buildings. These certifications not only enhance the value of developers' and owners' efforts, but they also increase the attractiveness of real estate assets. Fuerst and McAllister (2011) show, for example, that the presence of a green label such as LEED or Energy Star increases US commercial office rents by 3 to 5% and sales prices by 18 to 25%.

These developments are profoundly redefining the dynamics of the real estate market. The energy transition marks a major turning point for the real estate sector. Sustainable electrification, the integration of renewable energies, and connected technologies are transforming not only buildings, but also consumer expectations and investor strategies. This transformation, while driven by climate challenges, also offers opportunities to build a more sustainable, competitive real estate market, adapted to the challenges of the twenty-first century.

Effects on the Banking Industry

The energy transition is also profoundly transforming the banking sector, redefining financial risks and creating new investment opportunities. Recently implemented policies and regulations—such as climate-related disclosure requirements, green taxonomy frameworks, and stricter lending criteria for carbon-intensive sectors—are impacting banking institutions both directly and indirectly.

On the one hand, as private players, banks are directly subject to these new regulations. These regulations, while demanding, are pushing banks to become key players in the energy transition by making their activities more transparent and encouraging them to actively support environmentally friendly projects. They must respond to growing demands for transparency and sustainability, notably through frameworks such as the European Green Taxonomy or extra-financial reporting obligations. The EU Taxonomy is particularly notable for being the first comprehensive, science-based classification system for sustainable economic activities, covering sectors responsible for up to 80% of the EU's greenhouse gas emissions, and designed to channel investments toward low-carbon technologies (Schütze & Stede, 2024). Its legal force, technical precision, and binding integration into EU regulations—such as the such as the Sustainable Finance Disclosure Regulation (SFDR), the Corporate Sustainability Reporting Directive (CSRD), and the EU Green Bond Standard—set it apart from other voluntary or principle-based frameworks. It has also become an international benchmark, inspiring convergence efforts via the International Platform on Sustainable Finance (IPSF). By requiring banks to demonstrate their contribution to the energy transition, these schemes push them to align their financing activities with climate objectives.

Moreover, as energy-consuming entities, they must themselves reduce their carbon footprint. For example, the Net-Zero Banking Alliance commits banking institutions to achieving carbon neutrality by 2050, not only for their internal operations but also for their investment portfolios. Launched in 2021 by the United Nations, this initiative brings together 138 banks from 44 countries in 2023, representing over 40% of the world's banking assets (Sastry et al., 2024). These commitments require significant efforts to decarbonize their processes, integrate environmental criteria into their risk models and adjust their governance. For instance, Oversea-Chinese Banking Corporation (OCBC) in Singapore launched the world's first sustainability-linked loan explicitly aligned with a 1.5 °C scenario. This innovative product, offered to a leading green real estate firm, adjusts its interest rate based on the company's actual progress toward decarbonization goals. It exemplifies how banks are reengineering financial instruments to embed climate objectives directly into lending conditions, moving beyond generic environmental, social, and governance (ESG) assessments (United Nations Environment Programme, 2024).

On the other hand, regulations indirectly affect banks through their asset portfolios and customer relationships, particularly when these involve carbon-intensive sectors. This exposure gives rise to what is known as transition risk—the financial risk associated with the shift to a low-carbon economy, driven by policy changes, technological innovation, or evolving market preferences. For instance, fossil fuel assets are likely to become "stranded assets" as a result of falling demand and restrictive climate policies. In addition, regulations targeting energy efficiency, such as the ban on renting out excessively energy-intensive housing in some European countries, are affecting landlords' solvency and increasing the risk of default. Thus, Alessi and Battiston (2022) estimate that Eurozone banks have an exposure to transition risk of about 5%, while their green alignment is only 0.8%. This indicates that banks are more exposed to energy transition risks than they are aligned with sustainable activities. These findings underline the need for these financial institutions to reassess their asset portfolios and strengthen their commitment to sustainable investments to mitigate the risks associated with the energy transition.

Furthermore, the energy transition offers banks significant opportunities to develop new activities and create innovative products and services, such as sustainability-linked lending, green project finance, and ESG investment advisory. The financing of sustainable projects, such as solar and wind power infrastructures and energy storage systems, is becoming a major strategic focus. In this context, green finance, and in particular green bonds, plays a crucial role. These financial instruments make it possible to mobilize funds specifically earmarked for environmentally friendly projects, responding to growing investor demand for products aligned with ESG criteria. Global issuance of green bonds has grown exponentially, from $800 million in 2007 to $545 billion in 2021, reflecting growing interest in this type of financial product (IRENA & CPI, 2023). They thus play a major role in investing in and promoting clean energy (Taghizadeh-Hesary et al., 2023; Ye & Rasoulinezhad, 2023).

Moreover, banks are innovating by offering services specifically adapted to the energy transition (Kim, 2018). They are developing loans dedicated to home energy renovation to finance work such as installing solar panels or improving insulation. Attractive-rate loans are also available for the purchase of electric vehicles, supporting the transition to more sustainable modes of transport. Additionally, green savings, via products such as green savings accounts, enable customers to invest directly in renewable energy projects, offering a positive environmental impact while meeting their return expectations. These initiatives reinforce the role of banks in decarbonizing the economy.

In addition to these various traditional green financial products, the energy transition brings unique opportunities to developing countries. First, banks are developing innovative products, such as microcredits, to enable rural and isolated populations to acquire renewable energy equipment, such as solar panels, and gain access to electricity (Boutabba et al., 2020; Pode, 2013). This financial support acts as a catalyst for energy access in areas often lacking traditional infrastructure. Moreover, by promoting electrification in these communities, the energy transition facilitates digitalization, particularly through the use of mobile phones and the Internet. This enables access to a wide range of digital services, including online banking, revolutionizing financial inclusion by making transactions, savings, and credit accessible even in the most remote regions (Houngbonon et al., 2021; Owolabi et al., 2021; Tryphone et al., 2023).

In conclusion, while the energy transition poses major challenges for banks, particularly in terms of regulation and transition risks, it also offers opportunities for transformation and innovation.

Implications for the Insurance Industry

Sustainable electrification is also reshaping the insurance industry, which finds itself at the forefront of new risks, but also of opportunities linked to energy transformation and the resilience of societies.

The energy transition is accompanied by the massive deployment of clean infrastructures and technologies: wind farms, solar panels, batteries, smart grids, and so on. While these assets are necessary to reduce carbon emissions, they also bring with them a host of new risks. Insurers must now cover specific claims, such as technical defects in these still-developing technologies, damage caused by extreme weather events, and cyberattacks on increasingly connected energy networks. These new risks require in-depth technical expertise and innovative claims management models (Kirillova et al., 2021; Mills, 2003). They represent both a challenge for traditional insurers and an opportunity for players capable of adapting quickly. Wu and Chiu (2023) illustrate this through carbon insurance, which compensates companies for the failure of emission reduction technologies, thereby reducing their exposure to financial risks. This mechanism encourages the adoption of green technologies, promotes investment, and contributes to an overall drop in emissions, while creating opportunities for insurers capable of adapting to these new challenges.

As major investors, insurers are also directly and indirectly concerned by the growing demands of sustainability. Regulatory and societal pressure is driving insurance companies to integrate ESG criteria into the management of their asset portfolios, which means gradually divesting from sectors deemed incompatible with the energy transition, such as fossil fuels, and directing funds toward green, sustainable projects. This repositioning is not without financial risk, particularly in terms of the volatility of green assets and regulatory uncertainty. However, it also offers long-term growth opportunities, by actively participating in the financing of the transition. In addition, it allows investors to diversify their asset portfolios, which could otherwise be heavily affected by new environmental regulations targeting the companies they currently finance. Alessi and Battiston (2022) assessed the alignment of financial portfolios with the European Union's environmental objectives and exposure to climate transition risks. Their results show that, for eurozone insurance companies, around 4.8% of portfolios are aligned with these objectives, while exposure to transition risks reaches 15.1%. These figures suggest that, although insurers have begun to invest in sustainable activities, a significant proportion of their investments remain exposed to the risks associated with the transition to a low-carbon economy.

Finally, insurers can also contribute to collective action by actively promoting sustainability through financial incentives for policyholders who adopt eco-responsible practices. These incentives include premium reductions for buildings that are certified green or meet high-energy efficiency standards, as well as for the installation of renewable energy systems such as solar panels or domestic wind turbines. For example, in the 1980s, Hanover Insurance Company offered a 10% discount on home insurance premiums in certain regions for solar, underground, or energy-efficient dwellings, based on the argument that these less-stressed heating systems presented a reduced risk of fire (Mills, 2003). In addition, insurers can offer financial benefits to owners of electric or hybrid vehicles, reflecting their lower environmental impact. By aligning policyholder behavior with sustainable development objectives, insurers are actively contributing to the transition to a greener economy. However, the insurance market, marked by strong competition and standardization of offerings, makes it difficult to grant premium reductions to encourage customers to adopt risk management measures (Mills, 2003).

In short, the energy transition is redefining the fundamentals of the insurance industry. Between increasing climate risks, the emergence of new technological threats and the need to finance green projects, insurers must combine adaptability and responsibility. This transition offers not only challenges, but also a unique opportunity for insurance to assert itself as a key player in building a sustainable and resilient future.

Conclusion

The energy transition is more than a simple technological switch to renewable sources: it is profoundly transforming economic and social dynamics. Its macroeconomic impacts reveal tensions between opportunities and challenges: while it promises to strengthen energy security by reducing dependence on fossil fuels, it also creates new vulnerabilities. These include structural unemployment in carbon-intensive sectors, unequal regional impacts, rising costs for consumers and firms, and investment uncertainty—all of which may hinder the inclusiveness and pace of the transition. It is also transforming the labor market, encouraging job creation in green sectors while requiring retraining efforts for workers in traditional industries. Finally, although significant investment in innovation and infrastructure is seen as a potential driver of growth, the meaning of this relationship remains debated. The energy transition is redefining the interactions between energy and economic growth, raising the question of whether the challenges of regional disparities and costs will really enable this potential to be realized. Countries that know how to integrate these transitions coherently into their economic strategies could make them a driver of competitiveness and resilience.

At the microeconomic level, disparities between the contexts of developed and developing countries remain marked. While the former must resolve issues of energy insecurity and support companies in costly adaptation processes, the latter see sustainable electrification as a lever for reducing energy poverty and accelerating their development. These differences underline the need for differentiated, targeted approaches.

Finally, strategic sectors such as real estate, banking, and insurance are not simply spectators of this transition: they are key players. Their ability to innovate, absorb new risks, and finance energy projects will determine the feasibility and effectiveness of energy policies. However, these sectoral transformations are not without friction, particularly in terms of integrating new standards and the associated costs.

In conclusion, the energy transition is a major economic opportunity, but also a complex challenge, requiring coordinated responses adapted to each scale. The challenge is not just to transform energy systems, but to do so in an inclusive and sustainable way, so that the benefits are not monopolized by a few, but shared equitably between countries, regions, sectors, businesses, and households.

References

Adom, P. K., & Nsabimana, A. (2022). Rural access to electricity and welfare outcomes in Rwanda: Addressing issues of transitional heterogeneities and between and within gender disparities. *Resource and Energy Economics, 70*, 101333. https://doi.org/10.1016/j.reseneeco.2022.101333

Alessi, L., & Battiston, S. (2022). Two sides of the same coin: Green taxonomy alignment versus transition risk in financial portfolios. *International Review of Financial Analysis, 84*, 102319. https://doi.org/10.1016/j.irfa.2022.102319

Al-mulali, U., Fereidouni, H. G., & Lee, J. Y. M. (2014). Electricity consumption from renewable and non-renewable sources and economic growth:

Evidence from Latin American countries. *Renewable and Sustainable Energy Reviews, 30*, 290–298. https://doi.org/10.1016/j.rser.2013.10.006

Aqeel, A., & Butt, M. S. (2001). The relationship between energy consumption and economic growth in Pakistan. *Asia-Pacific Development Journal, 8*(2), 101–110.

Arcelay, I., Goti, A., Oyarbide-Zubillaga, A., Akyazi, T., Alberdi, E., & Garcia-Bringas, P. (2021). Definition of the future skills needs of job profiles in the renewable energy sector. *Energies, 14*(9) Article 9. https://doi.org/10.3390/en14092609

Associated Press. (2022, September 8). *Clean energy jobs overtake fossil fuel sector but wages lag behind*. Climate Home News. https://www.climatechangenews.com/2022/09/08/clean-energy-jobs-overtake-fossil-fuel-sector-but-wages-lag-behind/

Barron, M., & Torero, M. (2017). Household electrification and indoor air pollution. *Journal of Environmental Economics and Management, 86*, 81–92. https://doi.org/10.1016/j.jeem.2017.07.007

Battistini, N., Di Nino, V., Dossche, M., & Kolndrekaj, A. (2022). Energy prices and private consumption: What are the channels? *Economic Bulletin Articles, 3*. https://ideas.repec.org//a/ecb/ecbart/202200031.html

Bhattacharya, M., Paramati, S. R., Ozturk, I., & Bhattacharya, S. (2016). The effect of renewable energy consumption on economic growth: Evidence from top 38 countries. *Applied Energy, 162*, 733–741. https://doi.org/10.1016/j.apenergy.2015.10.104

Boutabba, M. A., Diaw, D., Laré, A., & Lessoua, A. (2020). The impact of microfinance on energy access: A case study from peripheral districts of Lomé. *Togo. Applied Economics, 52*(45), 4927–4951. https://doi.org/10.1080/00036846.2020.1751800

Bretschger. (2009). Energie et croissance. *La lettre de l'AFSE, 74*, 15–18.

Candelise, C., Saccone, D., & Vallino, E. (2021). An empirical assessment of the effects of electricity access on food security. *World Development, 141*, 105390. https://doi.org/10.1016/j.worlddev.2021.105390

Cecelski, E. (2003). *Energy, Poverty, and Gender*. World Bank. https://documents1.worldbank.org/curated/en/850681468328564938/pdf/345310Equitable0electrification0access.pdf

Cerin, P., Hassel, L. G., & Semenova, N. (2014). Energy performance and housing prices. *Sustainable Development, 22*(6), 404–419. https://doi.org/10.1002/sd.1566

Chapel, C. (2022). Impact of official development assistance projects for renewable energy on electrification in sub-Saharan Africa. *World Development, 152*, 105784. https://doi.org/10.1016/j.worlddev.2021.105784

Curtis, E. M., O'Kane, L., & Park, R. J. (2023). *Workers and the green-energy transition: Evidence from 300 million job transitions (working paper no.*

31539). National Bureau of Economic Research. https://doi.org/10.3386/w31539

da Silveira Bezerra, P. B., Callegari, C. L., Ribas, A., Lucena, A. F. P., Portugal-Pereira, J., Koberle, A., Szklo, A., & Schaeffer, R. (2017). The power of light: Socio-economic and environmental implications of a rural electrification program in Brazil. *Environmental Research Letters, 12*(9), 095004. https://doi.org/10.1088/1748-9326/aa7bdd

Dupont, E., Germain, M., & Jeanmart, H. (2021). Feasibility and economic impacts of the energy transition. *Sustainability, 13*(10) Article 10. https://doi.org/10.3390/su13105345

Elaouzy, Y., & El Fadar, A. (2022). Energy, economic and environmental benefits of integrating passive design strategies into buildings: A review. *Renewable and Sustainable Energy Reviews, 167*, 112828. https://doi.org/10.1016/j.rser.2022.112828

European Commission. (2020). *In focus: Energy efficiency in buildings*. https://commission.europa.eu/news/focus-energy-efficiency-buildings-2020-02-17_en

European Commission. (2022). *Communication from the Commission to the European Parliament, the European Council, the Council, the European Economic and Social Committee and the Committee of the Regions: REPowerEU: Joint European Action for more affordable, secure and sustainable energy*. European Commission. https://eur-lex.europa.eu/legal-content/EN/TXT/?uri=COM:2022:108:FIN

European Economic and Social Committee. (2023). *Energy poverty—42 million people in the EU cannot afford to heat their homes adequately*. European Economic and Social Committee. https://www.eesc.europa.eu/en/news-media/press-releases/energy-poverty-42-million-people-eu-cannot-afford-heat-their-homes-adequately

Fattouh, B., & Economou, A. (2023). Oil relations and the balance of power between the big-3 oil producers: Transformations and impacts. *OIES Energy Comment, Oxford Institute for Energy Studies.*

Fikru, M. G. (2019). Electricity bill savings and the role of energy efficiency improvements: A case study of residential solar adopters in the USA. *Renewable and Sustainable Energy Reviews, 106*, 124–132. https://doi.org/10.1016/j.rser.2019.02.028

Fragkos, P., & Paroussos, L. (2018). Employment creation in EU related to renewables expansion. *Applied Energy, 230*, 935–945. https://doi.org/10.1016/j.apenergy.2018.09.032

Fuerst, F., & McAllister, P. (2011). Eco-labeling in commercial office markets: Do LEED and energy star offices obtain multiple premiums? *Ecological Economics, 70*(6), 1220–1230. https://doi.org/10.1016/j.ecolecon.2011.01.026

Gökgöz, F., & Güvercin, M. T. (2018). Energy security and renewable energy efficiency in EU. *Renewable and Sustainable Energy Reviews, 96*, 226–239. https://doi.org/10.1016/j.rser.2018.07.046

Grogan, L., & Sadanand, A. (2013). Rural electrification and employment in poor countries: Evidence from Nicaragua. *World Development, 43*, 252–265. https://doi.org/10.1016/j.worlddev.2012.09.002

Halhoul Merabet, G., Essaaidi, M., Ben Haddou, M., Qolomany, B., Qadir, J., Anan, M., Al-Fuqaha, A., Abid, M. R., & Benhaddou, D. (2021). Intelligent building control systems for thermal comfort and energy-efficiency: A systematic review of artificial intelligence-assisted techniques. *Renewable and Sustainable Energy Reviews, 144*, 110969. https://doi.org/10.1016/j.rser.2021.110969

Houngbonon, G. V., Le Quentrec, E., & Rubrichi, S. (2021). Access to electricity and digital inclusion: Evidence from mobile call detail records. *Humanities and Social Sciences Communications, 8*(1), 1–11. https://doi.org/10.1057/s41599-021-00848-0

IEA. (2021). *The Role of Critical Minerals in Clean Energy Transitions*. International Energy Agency. https://www.iea.org/reports/the-role-of-critical-minerals-in-clean-energy-transitions

IEA. (2022a). *Skills Development and Inclusivity for Clean Energy Transitions*. International Energy Agency. https://www.iea.org/reports/skills-development-and-inclusivity-for-clean-energy-transitions

IEA. (2022b). *Special Report on Solar PV Global Supply Chains*. International Energy Agency. https://iea.blob.core.windows.net/assets/d2ee601d-6b1a-4cd2-a0e8-db02dc64332c/SpecialReportonSolarPVGlobalSupplyChains.pdf

IEA. (2023). *World Energy Employment 2023*. International Energy Agency.

IEA. (2024a). *World Energy Employment 2024*. International Energy Agency. https://www.iea.org/reports/world-energy-employment-2024

IEA. (2024b). *World Energy Outlook 2024*. International Energy Agency. https://www.iea.org/reports/world-energy-outlook-2024

IEA, IRENA, UNSD, World Bank, & WHO. (2024). *Tracking SDG 7: The Energy Progress Report*. World Bank.

IRENA. (2020). *The post-COVID recovery: An agenda for resilience, development and equality*. International Renewable Energy Agency.

IRENA. (2023). *Geopolitics of the energy transition: Critical materials*. International Renewable Energy Agency.

IRENA, & CPI. (2023). *Global landscape of renewable energy finance, 2023*. International Renewable Energy Agency.

Jensen, L. (2023). *Global Decarbonization in Fossil Fuel Export-Dependent Economies: Fiscal and Economic Transition Costs*. United Nations Development Programme (UNDP). https://www.undp.org/sites/g/files/zskgke326/files/2023-05/Global%20Decarbonization%20in%20Fossil%20Fuel%20Export-Dependent%20Economies.pdf?utm_source=chatgpt.com

Kanagawa, M., & Nakata, T. (2008). Assessment of access to electricity and the socio-economic impacts in rural areas of developing countries. *Energy Policy, 36*(6), 2016–2029. https://doi.org/10.1016/j.enpol.2008.01.041

Kaushik, K., & Shastri, S. (2024). Oil prices, renewable energy consumption and trade balance nexus: Empirical evidence from Indian economy. *Sustainability Accounting, Management and Policy Journal, 15*(3), 731–751. https://doi.org/10.1108/SAMPJ-03-2023-0115

Kholodilin, K. A., Mense, A., & Michelsen, C. (2017). The market value of energy efficiency in buildings and the mode of tenure. *Urban Studies, 54*(14), 3218–3238. https://doi.org/10.1177/0042098016669464

Khorasanizadeh, H., Parkkinen, J., Parthiban, R., & David Moore, J. (2015). Energy and economic benefits of LED adoption in Malaysia. *Renewable and Sustainable Energy Reviews, 49*, 629–637. https://doi.org/10.1016/j.rser.2015.04.112

Kim, J. (2018). Leverage the financing role of banks for low-carbon energy transition. In V. Anbumozhi, K. Kalirajan, & F. Kimura (Eds.), *Financing for low-carbon energy transition: Unlocking the potential of private capital* (pp. 189–210). Springer. https://doi.org/10.1007/978-981-10-8582-6_8

Kirillova, N., Pukala, R., & Janowicz-Lomott, M. (2021). Insurance programs in the renewable energy sources projects. *Energies, 14*(20) Article 20. https://doi.org/10.3390/en14206802

Kirubi, C., Jacobson, A., Kammen, D. M., & Mills, A. (2009). Community-based electric micro-grids can contribute to rural development: Evidence from Kenya. *World Development, 37*(7), 1208–1221. https://doi.org/10.1016/j.worlddev.2008.11.005

L'Her, G. F., Osborne, A. G., Schweikert, A. E., Ramstein, C. S., Stoll, B. L., & Deinert, M. R. (2023). *Potential of Photovoltaics and Energy Storage to Address Lack of Electricity Access* (No. arXiv:2310.11615). arXiv. https://doi.org/10.48550/arXiv.2310.11615

McDowall, W., Reinauer, T., Fragkos, P., Miedzinski, M., & Cronin, J. (2023). Mapping regional vulnerability in Europe's energy transition: Development and application of an indicator to assess declining employment in four carbon-intensive industries. *Climatic Change, 176*(2), 1–23. https://doi.org/10.1007/s10584-022-03478-w

Menegaki, A. N. (2011). Growth and renewable energy in Europe: A random effect model with evidence for neutrality hypothesis. *Energy Economics, 33*(2), 257–263. https://doi.org/10.1016/j.eneco.2010.10.004

Mills, E. (2003). The insurance and risk management industries: New players in the delivery of energy-efficient and renewable energy products and services. *Energy Policy, 31*(12), 1257–1272. https://doi.org/10.1016/S0301-4215(02)00186-6

Mohamad Taghvaee, V., Khodaparast Shirazi, J., Boutabba, M. A., & Seifi Aloo, A. (2017). Economic growth and renewable energy in Iran. *Iranian Economic Review, 21*(4), 789–808. https://doi.org/10.22059/ier.2017.64081

Moriarty, P., & Honnery, D. (2020). Feasibility of a 100% global renewable energy system. *Energies, 13, 21*), Article 21. https://doi.org/10.3390/en13215543

Nieto, J., Carpintero, Ó., Miguel, L. J., & de Blas, I. (2020). Macroeconomic modelling under energy constraints: Global low carbon transition scenarios. *Energy Policy, 137*, 111090. https://doi.org/10.1016/j.enpol.2019.111090

Ogletree, A., Cellucci, N., & Akinwonmi, R. (2022, juillet 25). *How Much Do Solar Panels Cost? A Full-Breakdown of Associated Costs and Factors in 2024.* Forbes Home. https://www.forbes.com/home-improvement/solar/cost-of-solar-panels/

Owolabi, O. A., Oku, A.-R. O., Alejo, A., Ogunbiyi, T., & Ubah, J. I. (2021). Access to electricity, information and communications technology (ICT), and financial development: Evidence from West Africa. *International Journal of Energy Economics and Policy, 11*(2) Article 2. https://doi.org/10.32479/ijeep.10344

Ozkan, C., & Okay, N. (2024). Does renewable energy convey information to current account deficit?: Evidence from OECD countries. *Sustainability, 16*(18) Article 18. https://doi.org/10.3390/su16188241

Pode, R. (2013). Financing LED solar home systems in developing countries. *Renewable and Sustainable Energy Reviews, 25*, 596–629. https://doi.org/10.1016/j.rser.2013.04.004

Sadineni, S. B., France, T. M., & Boehm, R. F. (2011). Economic feasibility of energy efficiency measures in residential buildings. *Renewable Energy, 36*(11), 2925–2931. https://doi.org/10.1016/j.renene.2011.04.006

Sastry, P., Verner, E., & Marqués Ibáñez, D. (2024). *Business as usual: Bank climate commitments, lending, and engagement* (ECB Working Paper No. 2921).

Schütze, F., & Stede, J. (2024). The EU sustainable finance taxonomy and its contribution to climate neutrality. *Journal of Sustainable Finance & Investment, 14*(1), 128–160. https://doi.org/10.1080/20430795.2021.2006129

Sers, M. R., & Victor, P. A. (2018). The energy-emissions trap. *Ecological Economics, 151*, 10–21. https://doi.org/10.1016/j.ecolecon.2018.04.004

Sovacool, B. K. (2012). The political economy of energy poverty: A review of key challenges. *Energy for Sustainable Development, 16*(3), 272–282. https://doi.org/10.1016/j.esd.2012.05.006

Sule, I. K., Yusuf, A. M., & Salihu, M.-K. (2022). Impact of energy poverty on education inequality and infant mortality in some selected African countries. *Energy Nexus, 5*, 100034. https://doi.org/10.1016/j.nexus.2021.100034

Taghizadeh-Hesary, F., Phoumin, H., & Rasoulinezhad, E. (2023). Assessment of role of green bond in renewable energy resource development in Japan. *Resources Policy, 80*, 103272. https://doi.org/10.1016/j.resourpol.2022.103272

Tryphone, K., Joseph, C., & Ndanshau, M. O. A. (2023). Determinants of digital transformation in sub-Saharan Africa: Some fiscal policy implications. *African Journal of Economic Review, 11*(4) Article 4.

Tsalikis, G., & Martinopoulos, G. (2015). Solar energy systems potential for nearly net zero energy residential buildings. *Solar Energy, 115*, 743–756. https://doi.org/10.1016/j.solener.2015.03.037

United Nations Environment Programme. (2024). *Net-Zero Banking Alliance 2024 Progress Report*. United Nations Environment Programme. https://www.unepfi.org/wordpress/wp-content/uploads/2024/10/NZBA-2024-Progress-Report.pdf

Wang, Q., Hubacek, K., Feng, K., Wei, Y.-M., & Liang, Q.-M. (2016). Distributional effects of carbon taxation. *Applied Energy, 184*, 1123–1131. https://doi.org/10.1016/j.apenergy.2016.06.083

World Economic Forum. (2024). *Energy Transition and Geopolitics: Are Critical Minerals the New Oil?* https://www3.weforum.org/docs/WEF_Energy_Transition_and_Geopolitics_2024.pdf

Wu, Q., & Chiu, C.-H. (2023). The impacts of carbon insurance on supply chain and environment considering technology risk under cap-and-trade mechanism. *Transportation Research Part E: Logistics and Transportation Review, 180*, 103334. https://doi.org/10.1016/j.tre.2023.103334

Ye, X., & Rasoulinezhad, E. (2023). Assessment of impacts of green bonds on renewable energy utilization efficiency. *Renewable Energy, 202*, 626–633. https://doi.org/10.1016/j.renene.2022.11.124

Zhang, R., Li, W., Li, Y., & Li, H. (2024). Job losses or gains? The impact of supply-side energy transition on employment in China. *Energy, 308*, 132804. https://doi.org/10.1016/j.energy.2024.132804

PART II

Governance, Policy and Stakeholders

CHAPTER 6

The Role of Stakeholders in Sustainable Electrification

Capucine Chapel

Introduction

With global energy demand set to grow by 50% by 2050 (IEA, 2023b), renewable energy electrification is a strategic priority for governments, international organizations, and energy sector stakeholders seeking to meet these needs while reducing carbon footprints. According to the International Energy Agency (IEA, 2021), meeting the objectives of the Paris Agreement will require 70% of the world's electricity production to come from renewable sources by 2050. However, this transition cannot be undertaken unilaterally by any single government or actor: it relies on a constellation of stakeholders, each playing a decisive role. But what are the specific roles of these stakeholders in promoting sustainable electrification? How do they interact to make this transition accessible, inclusive, and effective?

C. Chapel (✉)
Université Clermont Auvergne, INRAE, VetAgro Sup, UMR Herbivores, Saint-Genès-Champanelle, France
e-mail: capucine.chapel@inrae.fr

T. Walker et al. (eds.), *Wiring the Future*, Palgrave Studies in Emerging Risk Management and Sustainable Finance,
https://doi.org/10.1007/978-3-032-14717-2_6

This chapter examines the contributions of the key players involved in the transition to cleaner energy. Section "Governments and Policymakers: Leading the Charge" explores the role of public decision-makers who create regulatory frameworks and mobilize resources to facilitate the transition. Section "Private Sector Engagement: Corporate Strategies and Responsibilities" analyzes the private sector as a driver of innovation and financing, capable of making renewable solutions more accessible and efficient. Similarly, Sect. "Community Involvement and Public Opinion" of this chapter emphasizes the importance of engaging local communities, whose active participation is a prerequisite for project ownership and sustainability. Finally, Sect. "Non-Governmental Organizations and Advocacy Groups" presented non-governmental organizations (NGOs) as agents of awareness and support, capable of filling the gaps in resources and expertise in the most vulnerable regions. Section "Conclusion" concludes by emphasizing that successful energy transition requires coordinated, multi-level governance, where states, private actors, and civil society must align their efforts to ensure both equity and effectiveness.

Through concrete examples and recent studies, this chapter analyzes the strategies implemented by each stakeholder to overcome the obstacles to energy transition, as well as the challenges specific to each stakeholder in this complex process of sustainable electrification. By exploring the interactions and synergies between these stakeholders, we aim to provide insight into the factors that contribute to—or limit—shared progress in the transition toward renewable, affordable, and accessible energy for all.

Governments and Policymakers: Leading the Charge

Governments and policymakers play a central role in the transition to sustainable electrification (Aklin & Urpelainen, 2018; Daszkiewicz, 2020; Meadowcroft & Rosenbloom, 2023). They are at the forefront of the electrification movement, using a variety of tools to guide the energy sector towards sustainability. They act as national planners, architects of legislative reforms, and coordinators of international cooperation. Their leadership is essential, both in providing the necessary regulatory frameworks and financial resources, and in fostering global cooperation to ensure that electrification is equitable and efficient. This section highlights their key responsibilities, detailing the political, financial, and infrastructural measures that are essential to accelerating the transition to sustainable energy sources.

Regulatory and Policy Frameworks

One of the main functions of governments is to define and implement long-term energy policies, in response to the economic, environmental, and social challenges posed by energy production and consumption. In recent decades, energy transition has become an increasingly important part of these policies, for several reasons. Firstly, in the face of the climate emergency, it has become essential to reduce greenhouse gas emissions, prompting governments to promote renewable energies and energy efficiency. Secondly, it is crucial to strengthen energy security by reducing dependence on imported fossil fuels, which are often exposed to market fluctuations. By mobilizing various tools such as tax incentives, subsidies, or strict regulations, governments are working to build a more sustainable and resilient energy system, while anticipating the future needs of their populations. This approach is essential for structuring a framework in which every player—companies, communities, and citizens—can be integrated and contribute effectively to the transition. It also aims to ensure a fair distribution of responsibilities and resources, guaranteeing that the energy transition benefits all groups in society.

As a starting point, political decision-makers are setting ambitious targets to steer the economy towards sustainability. These targets, such as carbon neutrality, provide a clear framework that encourages economic players to adopt ecological solutions. By structuring collective action and sending signals to markets, these commitments aim to gradually guide society toward a sustainable model. On a global scale, renewable energy targets are enjoying significant growth, with a diversity of approaches and a growing commitment to meeting climate challenges. As of the end of 2023, the Renewable Energy Policy Network for the 21st Century (REN21, 2024a) identifies a total of 398 renewable energy targets in 182 countries. In this group, 90 countries have set economy-wide targets, while others focus on specific sectors such as agriculture, transport, or construction.

Energy transition targets differ widely between countries, depending on national contexts, available resources, and sectoral priorities. As a result, the targets set, their scope, the sectors concerned, and the deadlines vary considerably (IRENA, 2022a). For example, Niger's National Action Plan for Renewable Energies (PANER) focuses primarily on electricity generation, with targets of 28% renewable installed capacity and 57% share in the generation mix by 2030. In a country where electrification rates remain low and infrastructure is limited, this strategy highlights a strong political commitment to expanding access and building up production capacity, despite tight constraints. In contrast, Bangladesh's Mujib Climate Prosperity Plan adopts a broader approach, setting targets of 30% renewable energy consumption and 30% electrification of the transport sector by 2030. These goals reflect the country's rapidly growing energy demand and its intention to reduce emissions in key sectors, especially urban transport. While electrification rates are already high, the system remains heavily reliant on fossil fuels, making the shift to renewables a major challenge. Meanwhile, Croatia's National Energy and Climate Plan (NECP) defines 2030 targets across several sectors, including a 36.4% share of renewable energy in final consumption, with 63.8% for electricity, 13.2% for transport, and 36.6% for heating and cooling. This integrated strategy reflects the country's intent to decarbonize its energy mix more broadly, building on an already significant contribution of hydropower, and aligns closely with the European Union's climate ambitions.

Beyond the differences between production and consumption targets for renewables, countries' ambitions are guided by their natural strengths, leading them to prioritize different renewable resources depending on their geographic and environmental conditions. In Brazil, the New Industry Brazil policy launched in 2024 emphasizes biofuels, setting a target of 50% biofuels in the transport energy mix by 2033. This ambition builds on the existing RenovaBio program, established in 2017 to reduce the carbon intensity of fuels through decarbonization credits. This strategic focus is motivated by the country's abundant biomass resources and aims to leverage them to reduce emissions from the transport sector. Germany, on the other hand, is focusing its efforts on the electricity sector, with strong targets for wind and solar power. It plans to achieve an 80% share of renewable energies in the electricity mix by 2030, targeting a capacity of 110 GW for onshore wind power, 30 GW for offshore wind power, and 200 GW for solar power. This reflects the country's commitment to accelerating the energy transition by relying on well-developed wind and solar potential, backed by strong infrastructure, consistent regulatory support, and major public investment.

Finally, timelines and ambitions also vary. For example, Papua New Guinea is adopting a longer-term approach compared to the previous examples, aiming for 100% renewable electricity by 2050. This extended timeframe reflects the country's limited grid infrastructure, dispersed population, and logistical constraints, which require a long-term vision to make the transition successful.

To achieve the targets set, governments deploy a variety of public policy instruments, encouraging economic players to adopt practices in line with these ambitions. Firstly, some governments are introducing regulations and technical standards to force economic agents to comply with set targets. For example, in France, the Climate and Resilience Act imposes progressive restrictions on the rental of energy-inefficient housing. Thus, from January 1, 2025, homes rated G on the Diagnostic de Performance Energétique (DPE), a mandatory energy performance certificate that rates dwellings from A (most efficient) to G (least efficient) based on their energy consumption and greenhouse gas emissions, will be banned from rental. This ban will extend to F-rated homes in 2028 and E-rated homes in 2034. Through these measures, the French government aims to improve the energy efficiency of the national housing stock, reduce energy consumption in the real estate sector, and advance progress toward national climate target (Ministère de la Transition Écologique, 2018). Another example comes from Kenya, where since 2012, authorities have required the installation of solar water heaters in any building where occupants use more than 100 liters of hot water per day. This policy promotes access to clean energy and reduces dependence on fossil fuels (VOA Afrique, 2017).

Governments also often encourage the private sector and individual citizens to invest in renewable energies through subsidies and tax incentives. These grants and tax incentives aim to reduce installation costs and make renewable technologies more accessible. For example, the Renewable Heat Incentive (RHI) in the UK, in place from 2011 to 2022, offered payments to encourage the production of heat from renewable sources in the commercial and residential sectors (Ofgem, 2022). In the US, the Federal Investment Tax Credit (ITC) allows people to deduct 30% of the cost of installing solar systems (U.S. Department of Energy, 2020).

One of the best-known measures, however, is the feed-in tariff, a widely implemented financial incentive that allows renewable energy producers to sell the electricity they generate at a fixed rate set by the government, usually for a period of 15 to 20 years. This approach has been adopted by numerous countries—particularly in Europe and Asia—to foster long-term investments in renewable energy. While the United States has not adopted feed-in tariffs at the federal level, some states (such as California and Vermont) have implemented similar schemes at the state level. As of the end of 2022, 83 countries had in place feed-in tariff or premium payment policies (REN21, 2023). This incentive mechanism has the advantage of providing financial security for producers, guaranteeing them a stable income, which encourages investment in renewable energy projects.

Another important instrument in advancing renewable energy adoption is the use of taxes and quotas, which play a crucial role in encouraging businesses and consumers to shift toward more sustainable practices. In recent years, carbon pricing has been at the heart of debates on energy transition and climate change, as it represents a potentially effective and flexible approach to reducing emissions and promoting renewable energies. Carbon pricing is an economic mechanism aimed at reducing carbon dioxide (CO_2) emissions by assigning a cost to these emissions, thus encouraging businesses and consumers to adopt more sustainable practices. It mainly takes the form of carbon taxes, which impose a direct cost on CO_2 emissions, and emissions trading schemes, or "carbon markets," which set a limit on emissions and allow companies to buy or sell carbon credits according to their needs. Both approaches aim to integrate the cost of emissions into economic decisions, encouraging investment in low-carbon technologies and practices. Figure 6.1 illustrates the global distribution of carbon pricing instruments in 2024: by that year, 47 countries had implemented carbon-pricing policies, supporting the transition to renewable energies. Among them, Sweden introduced a carbon tax in 1991, initially set at 22 euros per ton, and gradually increased to around 122 euros per ton in 2023, making it one of the highest in the world (Government of Sweden, 2023).

The implementation of these public policy instruments can be complex and sometimes controversial. Measures such as fuel taxes or restrictions on certain industrial activities often lead to opposition from citizens, who are frequently concerned about the impact on their purchasing power and lifestyle. A notable example of this resistance is the "yellow vest" (gilets

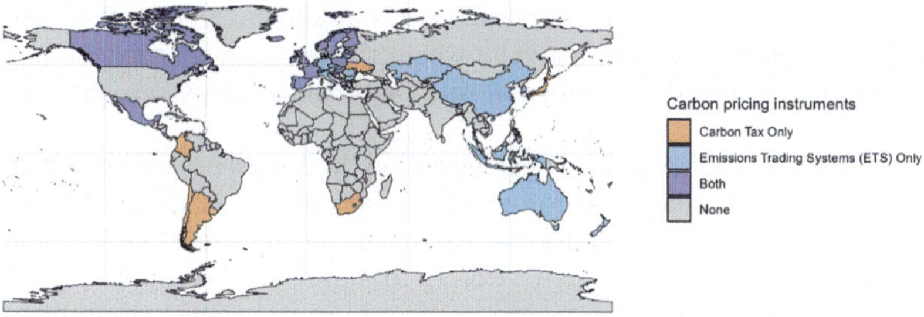

Fig. 6.1 Countries with carbon pricing instruments (*Source* Author's own elaboration based on data from the World Bank's carbon pricing dashboard)

jaunes) protests in France, which began in late 2018 in response to a planned increase in the carbon tax on transport fuels. The movement—named after the high-visibility vests that all drivers are legally required to keep in their vehicles—quickly spread across the country, expressed broader frustration with the rising cost of living and perceived inequalities in climate policy, particularly among rural and working-class populations who rely heavily on private vehicles. At its height, the protests mobilized hundreds of thousands of people and lasted for several months, forcing the government to suspend the planned tax increase and reconsider the social dimension of its energy transition policies.

Additionally, the private sector also exerts significant pressure, aiming to protect profit margins and minimize the impact of new regulations on its operations. For instance, the American Petroleum Institute (API), funded by major oil companies, has supported biased economic studies to exaggerate the costs of a carbon tax. As noted by Franta (2022), these analyses not only influenced policymakers and public opinion but also helped to halt the implementation of this measure in the Senate, thus protecting the profits of the oil industry. These events highlighted the political and social challenges associated with introducing ecological taxes and underscored the need for a balanced approach that considers both citizens' concerns and the economic imperatives of the private sector.

Despite various tools implemented to encourage and compel economic players to commit to the energy transition, achieving this shift still incurs significant costs. Indeed, the transition requires infrastructures and technologies that are often costly. In this context, political decision-makers play an essential role in mobilizing financial resources. Governments invest directly in the energy sector to accelerate the integration of renewable energies. They often intervene where the private sector is reluctant to invest, particularly in projects with high initial costs or uncertain profitability.

Because of the technical challenges and high costs associated with new technologies, the transition to renewable energies requires constant innovation. Consequently, governments can devote considerable funds to Research and Development (R&D) to stimulate this innovation. A notable example is Finland, which stands out for its strong commitment to R&D in the renewable energy and energy efficiency sectors. In 2020, the country devoted almost 15% of its public R&D expenditure to these fields, one of the highest percentages in the world (OECD, 2021).

Governments are also financing critical infrastructure such as smart grids, which modernize energy distribution, or large-scale solar and wind farm projects, which are needed to meet ambitious renewable energy targets. A case in point is Morocco, which has allocated significant public funds and acquired the land needed to develop the Noor solar power plant, one of the world's largest solar projects, illustrating the importance of public investment in promoting renewable energy projects (Toga et al., 2019).

These different public investments are essential to creating a favorable environment for energy transition and to attracting the private sector to these fields. Indeed, public resources remain limited and today 75% of investments in renewable energies are made by the private sector (IRENA, 2023). Governments therefore need to preserve these investments. Thus, in addition to creating a climate of trust and financial incentives, policymakers are also setting up public–private partnerships (PPPs) aimed at aligning the interests of public and private players around common objectives. These collaborations make it possible to pool risks, accelerate innovation and strengthen the technological capabilities needed to achieve climate objectives. An illustrative example of these partnerships is the commissioning, in 2023, of Europe's largest industrial solar thermal power plant, located in Seville, Spain. With a capacity of 30 megawatts-thermal, this project is the result of cooperation between Heineken and Engie, via a 20-year heat purchase agreement. Partly financed by a 14.3 million USD grant from the European Regional Development Fund, this PPP demonstrates how such collaborations can share financial risks, while stimulating innovation and the adoption of clean technologies to accelerate the energy transition (REN21, 2024b).

International Cooperation and Leadership

Energy transitions do not take place in a vacuum. Sustainable electrification often transcends national borders, requiring international coordination to achieve global climate goals. Governments therefore need to cooperate internationally to harmonize standards, share best practices, and coordinate responses to global climate challenges. Governments play a crucial role in promoting regional and international cooperation.

International agreements, such as the United Nations Sustainable Development Goals (SDGs), and international forums such as Conferences of the Parties (COPs), are key to structuring this cooperation. The SDGs aim to eradicate poverty, protect the planet, and ensure prosperity for all, while incorporating specific targets related to sustainable energy and climate change. Among these SDGs, goal number 7, titled *"Ensure access to affordable, reliable, sustainable and modern energy for all,"* is of particular importance in terms of the energy transition, as it is entirely dedicated to ensuring access to affordable, reliable, sustainable, and modern energy for all, by promoting the adoption of renewable energies and improved energy efficiency. Similarly, the COPs, which bring together the signatory countries of the United Nations Framework Convention on Climate Change, enable governments to commit to concrete actions to reduce greenhouse gas emissions and support the transition to more sustainable energy systems. At COP28, held in 2023, 130 countries pledged to triple the world's renewable energy capacity and double the annual rate of improvement in energy efficiency by 2030 (REN21, 2024a).

Beyond formal commitments, concrete initiatives illustrate the benefits of international cooperation. For example, the West Africa Clean Energy Corridor (WACEC), launched in 2016, supports the integration of renewable energies across the region by promoting joint planning, regulatory harmonization, and the development of cross-border infrastructure (IRENA Coalition for Action, 2023). It also assists in identifying high-potential renewable energy zones to guide investment and planning. Coordinated by International Renewable Energy Agency (IRENA), ECOWAS Centre for Renewable Energy and Energy Efficiency (ECREEE), and the West African Power Pool (WAPP), WACEC aims to integrate up to 10 GW of solar capacity into the regional mix by 2030. By facilitating cooperation around shared energy resources, it contributes to energy security and sustainable development goals.

To provide a more nuanced perspective, it is essential to note that, despite the necessity for international cooperation to address climate challenges, not all governments are equally committed to these initiatives. While international agreements and concrete initiatives such as WACEC demonstrate progress, some nations take a more hesitant stance or withdraw from their commitments. A notable example is the United States, which withdrew from the Paris Agreement in 2020 under President Trump, rejoined in 2021 under President Biden, and initiated a second withdrawal in 2025 following Trump's return to office. As one of the world's largest greenhouse gas emitters, the U.S. plays a critical role in achieving global climate targets; its renewed withdrawal was accompanied by additional actions, such as ending international climate finance contributions, suspending clean energy subsidies, and declaring a national energy emergency to support fossil fuel expansion. This sequence of reversals highlights the fragility of international cooperation: despite shared agreements, national priorities and political shifts can significantly undermine the collective fight against climate change.

These challenges make international solidarity all the more essential in supporting the diverse paths countries must take in their energy transitions. While each nation encounters unique obstacles, the impacts of these challenges ripple across the global community. In this context, developed countries have committed to supporting the transition through substantial international aid. This support takes various forms, including financing renewable energy projects, transferring clean technologies, and providing expertise in energy policy. Notably, a number of developed countries have pledged to allocate 0.7% of their gross domestic product (GDP) to official development assistance (ODA), a commitment formalized at the United Nations Conference on Development in 1970 (UNCTAD, 2022). ODA is instrumental in advancing renewable energy, particularly in developing nations where access to energy remains a critical issue. Through this commitment, governments and international organizations invest in initiatives to strengthen local capacity and introduce renewable technologies such as solar, wind, and hydropower. Figure 6.2 shows the significant increase in ODA allocated to the energy sector, with an ever-increasing share devoted to renewable energies over the years.

However, this commitment to international assistance has faced challenges, with signs of dwindling support and, at times, misallocation of funds. In 1970, United Nations member states—led by developed countries—set the target of allocating 0.7% of gross national income (GNI)

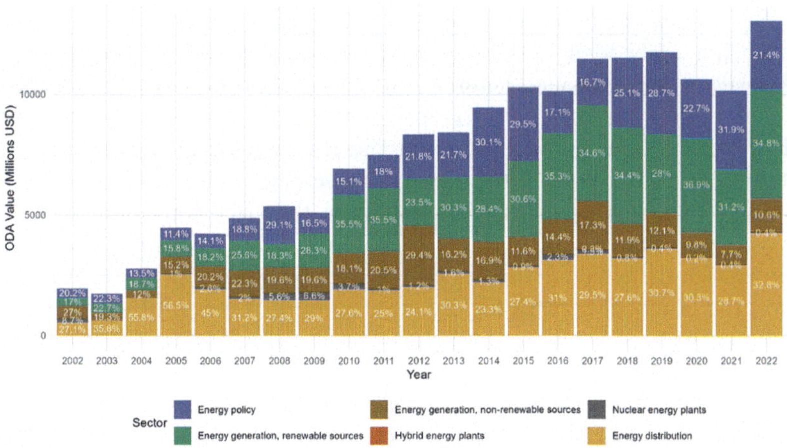

Fig. 6.2 Part of each sector in total ODA for energy by year (*Source* Author's own elaboration based on data from the OECD Creditor Reporting System (CRS))

to official development assistance (ODA), a commitment reaffirmed in subsequent international agreements. However, many developed countries have consistently fallen short of this goal. According to Sumner et al. (2020), since its adoption, only a handful of countries—namely Sweden, the Netherlands, Norway, Denmark, the UK, Germany, and Luxembourg—have met the 0.7% target at least once, and most Organisation for Economic Co-operation and Development (OECD) countries have failed to meet it in any year since it was introduced. More recently, support appears to be weakening even further, due to a combination of budgetary constraints, shifting political priorities, and growing doubts about aid effectiveness. For example, in 2024, France announced a historic reduction of almost €2 billion in its ODA budget for 2025, a decrease of 34.5% compared to the previous year (Focus 2030, 2024). Additionally, even when donor countries or agencies provide aid, they sometimes misallocate funds by supporting projects that overlook local needs, or, in some cases, local actors divert the funds through corruption (Briggs, 2018; Chong et al., 2009; Williamson, 2010). For example, in South Africa, the repurposing of the Komati coal-fired power station into a renewable energy site—financed in part through a World Bank loan under the Just Energy Transition Investment Plan—has drawn criticism not only for its limited socioeconomic benefits and modest job creation, but also for being embedded in a broader system affected by past corruption scandals and governance failures. Moreover, a portion of the international funding has been used to repay debts linked to earlier controversial coal projects, raising concerns about the effective use of climate finance (Reboredo & Carmody, 2024). In some cases, aid is diverted into administrative expenses or short-term initiatives that lack long-term impact (Easterly & Pfutze, 2008; Knack & Rahman, 2007). This misallocation can reduce the effectiveness of ODA in addressing energy challenges and limit the potential benefits for the populations it is meant to serve, highlighting the need for more strategic, transparent, and impactful allocation of international aid.

The role of governments as regulators and policymakers is fundamental, but only part of the solution. The private sector, as a catalyst for innovation and growth, is a key player in accelerating this transformation. Section "Private Sector Engagement: Corporate Strategies and Responsibilities" examines how companies are adjusting their strategies and fulfilling their responsibilities in response to current challenges.

Private Sector Engagement: Corporate Strategies and Responsibilities

The private sector is an essential player in the global energy transition, representing a source of both technological innovation and financing. To meet global targets for reducing greenhouse gas emissions, companies need to review their strategies, adapt to new regulations and meet growing expectations in terms of environmental and social responsibility.

Adaptation: Rethinking Business Models and Strategies

Faced with the urgency of climate change and pressure from society and government, more and more companies are setting ambitious targets to contribute to the energy transition. These commitments are often in line with international agreements. For example, giants such as TotalEnergies, Microsoft, and ArcelorMittal have pledged to achieve carbon neutrality by 2050, in line with the goals of the Paris Agreement (2015), which calls for net-zero emissions in the second half of the century to limit global warming to 1.5 °C. These decarbonization objectives are not only a response to environmental imperatives, but also to the growing expectations of consumers, regulators, and investors.

To achieve these objectives, companies have a number of technological and organizational levers at their disposal. Their decarbonization efforts include integrating renewable energy sources, improving energy efficiency, and adopting clean and innovative technologies. For example, the steel industry, a sector with high emissions, is beginning to replace coal with hydrogen in steel production through pilot projects in countries such as China, Sweden, and Germany (Karakaya et al., 2018; Zhang et al., 2021). Another example is Amazon, which delivered more than 20 million parcels to customers in North America and Europe using electric vehicles in 2020 (Yu et al., 2022).

Digital transformation is also playing a key role in process energy optimization, with the growing use of artificial intelligence (AI) and the Internet of Things (IoT). While these technologies contribute to a certain increase in emissions, they can also promote greater energy efficiency in many sectors. In practice, companies use IoT sensors to monitor energy use in real time, while AI systems analyze the data to adjust equipment performance, predict energy needs, and detect anomalies—leading to smarter energy management. A recent study of commercial buildings found that the adoption of AI could reduce energy consumption and carbon emissions by 8% to 19% by 2050 (Ding et al., 2024). By modernizing their production lines and using more sophisticated energy management tools, companies are boosting their environmental performance while reducing their operating costs.

These concrete actions are often part of a broader framework of corporate social responsibility (CSR) and environmental, social, and governance (ESG) criteria. CSR encourages companies to integrate more sustainable practices into their business model, by reducing their environmental impact and adopting more ethical approaches. For instance, as part of its commitment to CSR, Nestlé has made significant strides in the use of renewable electricity at its manufacturing sites. By the end of 2023, according to a recent report published by the company, 91.9% of their electricity consumption will come from renewable sources, up from 78.4% in 2022 and 63.7% in 2021 (Nestlé, 2024). This target is in line with Nestlé's strategy of guaranteeing 100% renewable electricity by 2025. However, while these improvements are notable, it is important to assess them in the context of Nestlé's overall environmental footprint. Like other major multinationals, Nestlé has been criticized for its contributions to plastic pollution, water usage controversies, and carbon emissions throughout its supply chain. Therefore, these renewable energy gains must be considered alongside the company's broader environmental impact to evaluate whether such claims reflect genuine sustainability efforts or are part of a greenwashing narrative.

ESG criteria, for their part, provide a measurable framework for evaluating these efforts, making it possible to monitor and improve companies' sustainable performance. These criteria are increasingly taken into account by investors, who favor companies aligned with ambitious climate and social objectives. Major asset managers such as BlackRock and Goldman Sachs are integrating ESG criteria into their investment decisions, putting positive pressure on companies to improve their sustainability and social performance (Archer, 2022). These companies have outlined, in their own institutional publications, how ESG criteria are incorporated into their investment strategies. BlackRock, for example, has committed to embedding ESG considerations across all its active strategies, treating them as material financial risks and using tools such as Aladdin Climate to evaluate climate-related exposures (BlackRock, 2020). Goldman Sachs similarly approaches ESG as a long-term value driver, developing thematic funds focused on environmental sustainability and incorporating ESG metrics into its financial analysis frameworks (Goldman Sachs, 2020). In both cases, ESG factors are primarily understood as risk indicators—whether environmental, social, or governance-related—that can impact long-term financial performance and are therefore essential to sound investment decision-making.

In this way, combining CSR and ESG criteria not only enhances a company's brand image but also encourages concrete actions towards decarbonization objectives, building stakeholder confidence—from investors to consumers—through a transparent and measurable approach to sustainable commitment. However, this approach faces criticism. While companies often showcase their sustainability progress, some practices are more about image than impact, a phenomenon known as greenwashing. For instance, a 2021 investigation by the European Commission reviewed 344 websites and found that 42% of "green" claims were exaggerated, false, or deceptive (European Commission, 2021). Additionally, 59% of cases lacked easily accessible evidence to substantiate their environmental claims, and 37% used vague terms like "eco-friendly" or "sustainable" without clear definitions.

Although commitments and measures may appear to be aligned with climate objectives, they sometimes lack transparency and follow-through, calling into question the authenticity of the actions undertaken by companies. The *Corporate Climate Responsibility Monitor 2023* (Day et al., 2023) revealed that many major companies rely on ambitious climate pledges without significant follow-through, often excluding critical emissions sources. In fact, the report found that, on average, these companies' commitments would only achieve 15% emissions reductions by 2030, far below the necessary 43% to align with the 1.5 °C target. These findings highlight the need for stronger regulatory standards and heightened oversight to ensure that CSR and ESG initiatives translate into genuine climate action.

Financing: Mobilizing the Necessary Financial Resources

Financing is a central pillar of a successful energy transition, and the private sector, with its financial and innovation capabilities, plays a decisive role in providing the capital needed to develop low-carbon infrastructures and technologies. The private sector is investing massively in renewable energies through direct financing, strategic partnerships, and innovative initiatives. According to IRENA and CPI (2023), between 2013 and 2020, private-sector investment amounted to 1.825 billion USD, three times as much as public-sector investment. However, this ratio varies according to technology and context, with the private sector generally more selective in its investment choices than the public sector. Thus, while 83% of investment in solar energy in 2020 will come from the private sector, this percentage drops to just 32% for geothermal energy (IRENA & CPI, 2023).

Direct investments in renewable energies include not only the purchase of green technologies, but also the financing of infrastructure projects, whether internal or external. Between 2010 and 2013, for example, Google invested over 1 billion USD in solar and wind energy projects, largely to power its own data centers (Carus, 2013). In 2017, Google thus became the first major company to cover 100% of its annual electricity consumption for its global operations through renewable energy purchases. By the end of 2022, the company had achieved this target for the sixth consecutive year (Google, 2023). These investments enable companies to become energy self-sufficient while reducing their carbon footprint.

Foreign direct investment (FDI) represents a crucial subset of direct investment, referring to investments made by a company in a country other than its own, with the aim of acquiring assets, building facilities, or creating local operations. For example, companies such as TotalEnergies, Shell, and Enel Green Power are investing in solar and wind power projects in Africa, Asia, and Latin America. These FDIs are helping to accelerate the energy transition in these regions by providing financing, appropriate technologies, and training, while improving access to sustainable energy for local communities. Of all renewable energy projects in 2019, nearly 40% were financed by foreign companies, underlining the growing importance of FDI in this sector. According to the World Bank, this type of investment in the renewable energy sector has grown rapidly in recent years, from just under 10 billion euros in 2003 to just under 120 billion euros today (World Bank & Energy Charter Secretariat, 2023).

At the same time, indirect investments represent an alternative strategy for supporting renewable energies. These investments enable companies to finance renewable energy projects without developing the projects themselves, often through investments in renewable energy funds or purchases of green bonds. By 2021, the green bond market had reached 545 billion USD, with a large majority dedicated to renewable energies, underlining the growing interest in these sustainable financing tools (IRENA & CPI, 2023). Unlike direct investments, where a company takes full control of the asset, indirect investments offer companies an opportunity to diversify their portfolio while participating in the energy transition. Companies can also issue green bonds to finance specific projects, enabling them to access new capital while reinforcing their commitment to sustainable development. In short, this collaborative approach between companies and banks in financing renewable energies is essential to achieving global climate objectives.

In addition to more traditional direct and indirect investments, other forms of financing are emerging from the private sector to support the energy transition. Among these alternatives, PPPs enable companies to commit to renewable energy projects while sharing risks and responsibilities with the public sector. This collaboration offers an effective model for the development of sustainable energy infrastructure by pooling financial resources, technical expertise, and management capacity of both sectors, as exemplified by Gujarat's rooftop solar program (International Finance Corporation, 2013), launched in 2010 by the Gujarat state government in India with support from the World Bank Group through its International

Finance Corporation (IFC). This program aims to develop solar energy in the capital, Gandhinagar, by involving private companies selected through a tender process, such as Azure Power and SunEdison, to install and manage solar panels on public and private buildings. This PPP model maximizes the use of existing resources, reduces obstacles associated with land acquisition, and promotes rapid development of solar infrastructure. The 5 MW capacity installed in Gandhinagar provides power for around 10,000 people while reducing CO_2 emissions by over 7000 tons a year, illustrating how this collaboration optimizes the financial and technical resources of both sectors to accelerate the sustainable energy transition. Thanks to this success, the Gandhinagar model has been replicated in several other cities, such as Vadodara, reinforcing the state's energy transition (International Finance Corporation, 2015).

Moreover, philanthropic funding or giving brings a unique dimension to supporting the energy transition. Unlike traditional investments, philanthropic donations are not intended to generate a financial return, but to maximize social and environmental impact. By funding innovative or high-risk projects that are often overlooked by traditional investors, philanthropists help catalyze initiatives that address climate issues and promote access to sustainable energy solutions for communities. In 2020, the OECD estimated private philanthropic giving flows for sustainable development at 8.7 billion US dollars, underlining their growing significance in the energy transition (OECD, 2023).

A striking example of philanthropic funding is the joint initiative launched in 2021 by the Rockefeller Foundation and the IKEA Foundation to finance renewable energy projects in Africa and Asia. With a pledged investment of 1 billion USD, the project—carried out under the Global Energy Alliance for People and Planet—aims to provide clean electricity to over one billion people by 2030, particularly in rural areas where energy infrastructure is lacking, while reducing greenhouse gas emissions through the deployment of solar microgrids and other renewable technologies (United Nations, 2021). As of 2023, the initiative had enabled access to renewable energy for approximately 1.2 million people, contributed to avoiding 147 million tons of CO_2 emissions, and supported the creation of over 600,000 jobs in countries such as Nigeria, India, and Sierra Leone (Rockefeller Foundation, 2023). While these figures indicate promising early progress, they remain modest compared to the project's long-term objective. Reaching over a billion people will require sustained and scaled-up efforts in the coming years. Nonetheless, this initiative illustrates the potential of philanthropic capital to unlock investment in high-impact, socially driven projects that often lack appeal for private investors, and contributes meaningfully to the global transition toward a green economy.

Innovation: Accelerating the Transition Through Technology

Private-sector-led innovation is a key driver for accelerating the sustainable energy transition. Thanks to substantial investment in R&D, private companies can introduce cutting-edge technologies, addressing challenges that the public sector alone could not solve. R&D spending by companies reached 160 billion USD in 2023, marking a 7% increase on the previous year and representing more than three times public sector R&D spending (IEA, 2024).

This investment dynamic is also reflected in a significant increase in patent filings in the energy technology sector. Between 2000 and 2022, patents for renewable energies rose substantially, driven largely by the solar energy sector (see Fig. 6.3). Among these patents is Tesla's Powerwall battery, launched in 2015. This energy storage system enables users to store electricity generated by solar panels for later use. What sets the Powerwall apart is its integration of advanced energy management technologies, optimizing the use of renewable resources in the home. This innovation not only improves energy efficiency; it also makes solar energy more accessible to consumers, supporting the transition to sustainable solutions (Wang et al., 2022).

Technological innovation in the renewable energy sector is essential, as it not only enables the development of more efficient and accessible technologies, but is also crucial to achieving global climate goals. According to IRENA, this innovation is a key factor in reducing renewable energy costs, boosting productivity, and increasing the market share of sustainable solutions. For example, between 2010 and 2020, significant advances in the design, manufacture, and materials of solar photovoltaic panels have driven PV module prices down by 93% and reduced total installation costs by 81%. At the same time, module efficiency has risen by 24% (IRENA, 2022b). These advancements make solar energy an increasingly competitive option in the energy market, boosting its adoption and fostering public support for renewable energy initiatives.

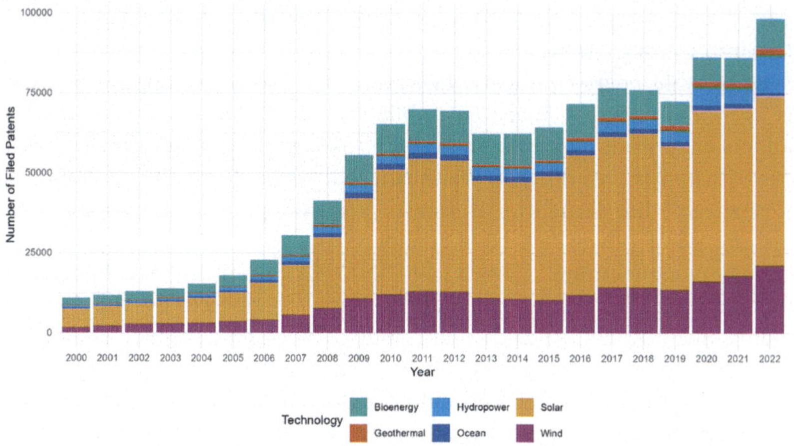

Fig. 6.3 Renewable energies patents by year and technology (*Source* Author's own elaboration based on data from IRENA INSPIRE)

Community Involvement and Public Opinion

The involvement of local communities and the incorporation of public opinion play central roles in the success of energy transition projects. The construction of new electricity infrastructures, whether solar power plants, wind farms, or distribution networks, has a direct impact on the people living in the regions concerned. A successful energy transition therefore depends not only on technological advances and political decisions, but also on the support and participation of citizens and local communities. This section explores the mobilization of communities through energy projects, the importance of public awareness, and the role of public opinion in defining energy priorities.

The Impact of Public Opinion on the Energy Transition

Public opinion has a decisive impact on the political viability of energy projects and the direction of public policy. In several countries, public pressure has accelerated government commitments to carbon neutrality and pushed companies to adopt greener strategies. In the United States, for example, a study in 2020 shows that 79% of American adults support initiatives to increase the share of renewable energies in the country's energy mix, despite political polarization on the climate issue (Pew Research Center, 2020).

Renewable energy projects, however, can also spark opposition that hampers the progress driven by public policies. Communities and stakeholders may raise concerns about land use, environmental impact, and local disruption, resulting in resistance that slows down the deployment of renewable infrastructure. In France, for example, the development of onshore wind power has been hampered by the rise of an opposition movement, sometimes dubbed the "NIMBY" (Not In My Backyard) phenomenon. According to a 2021 study conducted by the French polling institute IFOP (2021), although 77% of French people support renewable energies, only 53% say they are in favor of installing wind turbines near their homes. This kind of reluctance highlights the need for better communication and transparency on the local benefits of energy projects to overcome resistance.

Conversely, pressure from public opinion has led to significant changes in the energy transition in both the public and private sectors. Indeed, citizens have the power to influence both politicians and the private sector. A study of energy policies in Europe between 1974 and 2015 has shown that when public opinion focuses on environmental issues, this leads to an increase in policies in favor of renewable energies (Anderson et al., 2017). There are several reasons for this dynamic: on the one hand, political leaders seek to maintain their popularity and electoral support by conforming to public expectations; on the other, social movements mobilize public opinion, thus exerting pressure on decision-makers. In short, this interplay between public opinion and political choices plays a key role in promoting sustainable energy solutions. Public opinion can also influence the private sector, in particular the behavior of companies, which depend heavily on their reputation with consumers. Companies often respond to customer pressure by adopting more sustainable practices and innovating in the development of eco-friendly solutions. This dynamic is key to catalyzing positive change, as consumers increasingly favor brands that demonstrate a commitment to the environment (Lestari et al., 2021; Onjewu et al., 2023).

Citizen Awareness, Training, and Involvement

Awareness-raising and education are essential to engaging citizens in the energy transition. Indeed, numerous studies have shown that greater environmental awareness and concern translates into more environmentally-friendly behavior (Bang et al., 2000; Lillemo, 2014). Informing the public about climate issues and sustainable solutions therefore encourages eco-responsible behavior. A meta-analysis of 156 studies showed that information strategies led on average to a 7.4% reduction in people's electricity consumption (Delmas et al., 2013). Better information can also break down preconceived ideas and reduce reluctance. For example, Bidwell's (2016) experimental study shows that residents who took part in in-depth information sessions on wind power showed a greater change in their general support for wind energy.

The participatory aspect also strengthens citizen involvement in the energy transition. Citizen participation in the energy transition can take many forms, such as community workshops, public forums, energy cooperatives, and online consultation platforms. These approaches enable citizens to express their opinions, contribute to the planning of local projects, and become directly involved in the production of renewable energy. As Schroeter et al. (2016) show through an empirical case study of the BEKO project in Germany, citizen participation plays a central role in the social acceptance of energy policies by promoting transparency and legitimacy. Using a mixed-methods approach—combining surveys, interviews, and participatory process evaluation—they demonstrate how inclusive consultation enhances trust in institutions and increases the perceived legitimacy of policy outcomes. It is from this participatory dynamic that concepts such as energy democracy and energy citizenship are born. These concepts emphasize the importance of citizen involvement in decision-making on energy policies and sustainable development projects. Energy democracy promotes a model in which citizens not only have the right to express themselves, but also the opportunity to co-determine actions relating to the energy transition, thereby reinforcing their power to act and their responsibility towards the environment.

Renewable Energy Source cooperatives (REScoops) offer concrete and significant examples of citizen involvement in the energy transition. These citizen-owned renewable energy cooperatives enable individuals to organize themselves collectively to finance, manage, and consume green energy projects. REScoops are based on democratic governance principles, with each member having one vote, regardless of their financial investment. This structure enables citizens to become both producers and consumers of renewable energy, reinforcing their direct involvement in local energy decisions.

By 2023, REScoop.eu—the European federation of energy cooperatives—lists 2,450 renewable energy cooperatives involving over 1.5 million citizens across Europe (REScoop.eu, 2023). Initiatives such as Ecopower in Belgium and Som Energia in Spain have become emblematic of this model due to their pioneering role, large membership bases, and influence as national references. According to Koltunov et al. (2023), Ecopower counts over 64,000 members and Som Energia more than 83,000, making them outliers in terms of average membership compared to other European energy communities. Both operate as retail cooperatives, supplying renewable electricity directly to their members at preferential rates, and exemplify successful citizen-driven models that combine democratic governance, local reinvestment, and large-scale participation. According to a study by Wierling et al. (2018), energy cooperatives improve local acceptance of energy projects by integrating residents' concerns early on in the decision-making process, thereby reducing the opposition often associated with energy infrastructure (Wierling et al., 2018). In addition, these initiatives help to conserve energy and reduce gas emissions, with reductions of up to 20% according to a study by Sifakis et al. (2019). To further strengthen community involvement and meet public expectations, NGOs also play a central role in raising awareness and supporting energy transition initiatives, as the following section reveals.

Non-Governmental Organizations and Advocacy Groups

NGOs and advocacy groups are key players in the energy transition, both in developing and developed countries. They act as essential intermediaries between governments, private companies, and local communities, playing multiple roles ranging from raising public awareness to implementing energy projects. They raise public awareness, lobby governments and the private sector to accelerate the adoption of clean energy, and implement concrete projects.

Pushing for Transition: Campaigns and Influence

NGOs and advocacy groups play a central role in influencing the energy transition, using mobilization strategies to raise public awareness and push policymakers and corporations to change course. Organizations such as Greenpeace, 350.org, and Extinction Rebellion orchestrate large-scale campaigns, calling on governments and companies to reduce their dependence on fossil fuels. Through petitions, viral publications on social networks, shock videos, and demonstrations, these organizations harness public awareness to create powerful leverage.

A representative example is the "Break Free from Fossil Fuels" campaign launched in 2016 by 350.org in collaboration with several other NGOs, which brought together thousands of people around the world to demand a transition to renewable energy sources (Hopke & Hestres, 2017). These actions aim not only to raise awareness of the dangers of climate change, but also to exert direct pressure on governments to adopt environmentally friendly policies. Social networks have also been essential in these mobilizations, as witnessed by the "Fridays for Future" movement, initiated by Greta Thunberg, which succeeded in transforming an online campaign into a global movement advocating urgent climate action (Hopke & Paris, 2021; Rizzoli et al., 2024; Shim, 2024). These initiatives have led to unprecedented levels of youth mobilization worldwide, with millions participating in climate strikes across more than 150 countries. They have helped bring climate concerns to the forefront of public debate, influenced media narratives, and increased political pressure on decision-makers at both national and international levels.

Alongside awareness-raising efforts, NGOs work closely with governments, providing active support and, at times, lobbying for stronger environmental policies. These organizations use scientific reports and case studies to influence legislation and energy policies. For example, Friends of the Earth and Climate Action Network actively lobby governments to adopt more ambitious climate measures (Parks et al., 2023). Their sustained advocacy has played a key role in pushing the European Union to strengthen its climate targets, particularly in the context of the Green Deal and recent emissions reduction frameworks (WWF, 2025). The Rocky Mountain Institute (RMI) is another non-profit organization that plays a crucial role in the energy transition through collaborating with governments and businesses, to help design innovative solutions to promote renewable energies. A notable example of their commitment is the Partnership for Just Energy Transition in Indonesia, launched in 2022. This partnership aims to support Indonesia in its transition to renewable energy, while ensuring that communities dependent on fossil fuels benefit from this transformation. RMI is contributing its expertise to develop strategies that promote the integration of renewable energies, working on projects that support both the country's climate goals and the well-being of workers affected by this change (Fauzianto, 2023; IEA, 2023a).

Some NGOs go even further by taking direct action against governments and polluting industries, including through legal proceedings. For instance, in the United States, the Sierra Club has been a key player in the energy transition through its "Beyond Coal" program, launched in 2010. This initiative targeted the coal industry, judicially challenging operating permits for coal-fired power plants, considered a threat to public health and the environment (Drake & York, 2021). The Sierra Club reports that, through these efforts, it has helped shut down more than 380 coal plants nationwide, contributing to a broader shift toward renewable energy (Sierra Club, n.d.). Supporting this claim, an independent study by Drake and York (2021) shows that the campaign significantly shortened the operating life of targeted coal units by an average of more than two years, demonstrating the campaign's substantial influence beyond what market dynamics or federal regulations alone could explain.

In developing countries, local and international NGOs are also putting pressure on international financial institutions to redirect their funding toward clean energy. For example, the NGO Oil Change International has revealed that billions of dollars are still being allocated to fossil fuel projects (Oil Change International, 2024). Between 2020 and 2022, international public funding for fossil fuels, from G20 countries and multilateral development banks, averaged 47 billion USD a year, almost 1.4 times the support given to renewable energies, which amounted to around 35 billion USD annually. This imbalance highlights the need to readjust investments towards clean energies if we are to meet our climate commitments. NGOs such as Oil Change International are actively working to secure commitments from international donors, and this advocacy has helped some development banks, like the World Bank, pledge to stop financing fossil fuel projects (World Bank, 2017).

Building the Energy Transition: Support and Implementation

In addition to advocacy, many NGOs focus on field programs, working directly with communities to implement sustainable energy solutions. They act as catalysts, raising communities' awareness of environmental issues and providing them with the tools they need to adopt renewable energy solutions. An excellent example of this type of commitment is the Eco-Schools program, run by the Foundation for Environmental Education (FEE), an international NGO. Since its inception in 1994, the program has expanded to over 74 countries, involving more than 49,600 schools and reaching millions of students worldwide (Foundation for Environmental Education, 2023). Eco-Schools offers schools a framework for developing environment-related projects, such as waste management and energy conservation. By actively involving students in concrete initiatives, such as the installation of solar panels, Eco-Schools is shaping a generation aware of the challenges of the energy transition. This approach not only raises students' awareness, but also creates a culture of sustainability that can ripple out into their communities (Andreou, 2020).

NGOs are also implementing practical solutions to help those often left behind in the energy transition. They target vulnerable populations, particularly in rural areas, who find it difficult to access renewable energies. By building local capacity, these organizations offer training in the use and maintenance of sustainable technologies, enabling beneficiaries to adopt and maintain these solutions independently. This approach helps overcome barriers to energy access while creating local employment opportunities. By promoting an inclusive energy transition, NGOs ensure that the benefits of renewable energies reach as many people as possible, contributing to a sustainable energy future for all.

As an example, SolarAid is an NGO that Solarcentury, a UK solar company, founded in 2006. Primarily active in Africa, it works to reduce energy poverty by providing affordable solar lamps to rural communities, thereby reducing reliance on costly and harmful kerosene lamps (Munro et al., 2023). In addition to selling solar lamps at low prices, SolarAid trains local entrepreneurs and repair technicians, creating jobs and essential skills for sustainable autonomy. Through this approach, the organization not only ensures continuous access to clean energy but also enables communities to maintain and repair equipment locally, reducing costs and increasing energy resilience. With millions of families already impacted, SolarAid actively contributes to the economic and sustainable development of the most remote regions, while promoting an inclusive energy transition. Since its founding, SolarAid has installed more than 2.3 million solar systems (SolarAid, 2023), providing a sustainable solution for those often excluded from the benefits of the energy transition.

However, NGO activities remain largely dependent on external funding (Vincent, 2006). For example, in 2022, over 70% of SolarAid's income came from donations and grants, making the organization vulnerable to economic fluctuations and changing donor priorities (SolarAid, 2023). This dependence limits their ability to maintain long-term actions and invest in large-scale initiatives. In addition, environmental NGOs often suffer from fragmentation and a lack of inter-organization coordination, leading to redundancies and a dilution of efforts. Cooley and Ron (2002) describe how competition for external funding drives NGOs to secure their resources rather than collaborate on common goals. This dynamic leads to the fragmentation of efforts and the duplication of projects, diluting their overall impact. In the absence of concerted action, NGOs find themselves competing for the attention of donors, encouraging individual actions and a lack of shared strategic vision, to the detriment of their collective mission of environmental protection and energy transition.

Conclusion

In conclusion, the transition to sustainable electrification relies on close collaboration between multiple stakeholders, each of whom makes essential and complementary contributions. Public decision-makers, by laying the regulatory foundations and facilitating international cooperation, set the framework for this transformation. The private sector, as the driving force behind innovation and financing, makes possible the technological advances crucial to the adoption of renewable solutions. Local communities, through their support and involvement, ensure that these projects are anchored in everyday practices and needs, reinforcing their sustainability and impact. Finally, NGOs, as catalysts for change, raise awareness and support electrification projects in the most fragile areas, playing a key role in bridging gaps in resources and expertise.

However, despite these combined efforts, the figures show that the road to achieving the Paris Agreement targets is still a long one. According to the International Energy Agency, to achieve 1.5 °C target by 2050, global investment in clean energy would have to quadruple, reaching around 5 trillion USD a year (IRENA, 2023). In reality, although global investment in all energy transition technologies reached a record 1.3 trillion USD in 2022, current levels remain largely insufficient, and billions of people still lack access to modern electricity.

These figures highlight the urgent need to intensify financial and technological efforts and to expand international cooperation to bridge these gaps. Despite the obstacles, examining the roles and strategies of each actor shows that it is precisely through their interaction and cooperation that the key to successful and inclusive electrification can be found. These stakeholders' abilities to work together, share resources, and adapt to local contexts will determine the scale and effectiveness of sustainable electrification. A coordinated and inclusive approach will thus be able to turn the ambitions of the energy transition into concrete realities, offering everyone a cleaner, safer and more equitable energy future. The successes of local energy cooperatives and multi-actor alliances show the potential of decentralized models rooted in community participation and adapted to territorial realities. At the same time, the recurring obstacles—fragmentation of governance, lack of long-term financial commitment, and limited integration of citizen voices—highlight the need for mechanisms that foster transparency, accountability, and inclusive planning. Strengthening regulatory frameworks to support civil society initiatives, facilitating cross-sectoral dialogue platforms, and promoting flexible financing tools that align private and public interests may thus represent meaningful steps. In this sense, coordination should not be understood as top-down control, but as the capacity to align diverse actors around shared goals, while maintaining responsiveness to local contexts and social dynamics.

References

Aklin, M., & Urpelainen, J. (2018). *Renewables: The politics of a global energy transition*. MIT Press.

Anderson, B., Böhmelt, T., & Ward, H. (2017). Public opinion and environmental policy output: A cross-national analysis of energy policies in Europe. *Environmental Research Letters, 12*(11), Article 114011. https://doi.org/10.1088/1748-9326/aa8f80

Andreou, N. (2020). Towards a generation of sustainability leaders: Eco-schools as a global green schools movement for transformative education. In A. Gough, J. C. K. Lee, & E. P. K. Tsang (Eds.), *Green schools globally: International explorations in outdoor and environmental education* (pp. 31–45). Springer. https://doi.org/10.1007/978-3-030-46820-0_3

Archer, M. (2022). The ethics of ESG: Sustainable finance and the emergence of the market as an ethical subject. *Focaal, 2022*(93), 18–31. https://doi.org/10.3167/fcl.2022.930102

Bang, H.-K., Ellinger, A. E., Hadjimarcou, J., & Traichal, P. A. (2000). Consumer concern, knowledge, belief, and attitude toward renewable energy: An application of the reasoned action theory. *Psychology & Marketing, 17*(6), 449–468.

Bidwell, D. (2016). The effects of information on public attitudes toward renewable energy. *Environment and Behavior, 48*(6), 743–768. https://doi.org/10.1177/0013916514554696

BlackRock. (2020). *Sustainability as BlackRock's new standard for investing*. BlackRock. Retrieved April 14, 2025, from https://www.blackrock.com/corporate/investor-relations/2020-blackrock-client-letter

Briggs, R. C. (2018). Poor targeting: A gridded spatial analysis of the degree to which aid reaches the poor in Africa. *World Development, 103*, 133–148. https://doi.org/10.1016/j.worlddev.2017.10.020

Carus, F. (2013). Google—Leading the way on renewable energy. *The Guardian*. https://www.theguardian.com/sustainable-business/google-renewable-green-energy

Chong, A., Gradstein, M., & Calderon, C. (2009). Can foreign aid reduce income inequality and poverty? *Public Choice, 140*(1/2), 59–84.

Cooley, A., & Ron, J. (2002). The NGO scramble: Organizational insecurity and the political economy of transnational action. *International Security, 27*(1), 5–39.

Daszkiewicz, K. (2020). Policy and regulation of energy transition. *The Geopolitics of the Global Energy Transition, 203–226*. https://doi.org/10.1007/978-3-030-39066-2_9

Day, T., Mooldijk, S., Hans, F., Smit, S., Posada, E., Skribbe, R., Woollands, S., Fearnehough, H., Kuramochi, T., Warnecke, C., Kachi, A., & Niklas, N. (2023). *Corporate Climate Responsibility Monitor 2023*. https://newclimate.org/sites/default/files/2023-02/NewClimate_CorporateClimateResponsibilityMonitor2023_Feb23.pdf

Delmas, M. A., Fischlein, M., & Asensio, O. I. (2013). Information strategies and energy conservation behavior: A meta-analysis of experimental studies from 1975 to 2012. *Energy Policy, 61*, 729–739. https://doi.org/10.1016/j.enpol.2013.05.109

Ding, C., Ke, J., Levine, M., & Zhou, N. (2024). Potential of artificial intelligence in reducing energy and carbon emissions of commercial buildings at scale. *Nature Communications, 15*(1), 5916. https://doi.org/10.1038/s41467-024-50088-4

Drake, D. F., & York, J. G. (2021). Kicking ash: Who (or what) is winning the "war on coal"? *Production and Operations Management, 30*(7), 2162–2187. https://doi.org/10.1111/poms.13360

Easterly, W., & Pfutze, T. (2008). Where does the money go? Best and worst practices in foreign aid. *Journal of Economic Perspectives, 22*(2), 29–52. https://doi.org/10.1257/jep.22.2.29

European Commission. (2021). *Screening of websites for 'greenwashing': Half of green claims lack evidence*. European Commission. Retrieved November 11, 2024, from https://ec.europa.eu/commission/presscorner/detail/en/ip_21_269

Fauzianto, R. (2023). *Insight from Indonesia: Just energy transitions hinge on well-equipped energy leaders*. RMI. Retrieved November 5, 2024, from https://rmi.org/insight-from-indonesia-just-energy-transitions-hinge-on-well-equipped-energy-leaders/

Focus 2030. (2024). *France reneges on its official development assistance commitments*. Focus 2030. Retrieved November 12, 2024, from https://focus2030.org/france-reneges-on-its-official-development-assistance-commitments

Foundation for Environmental Education. (2023). *Annual Report 2023*. Foundation for Environmental Education. https://static1.squarespace.com/static/550aa28ae4b0f34c8f787b74/t/6672c711ebabf17adb52e837/1718798104766/FEE+Annual+Report+2023_Final+%28compressed%29.pdf

Franta, B. (2022). Weaponizing economics: Big oil, economic consultants, and climate policy delay. *Environmental Politics, 31*(4), 555–575. https://doi.org/10.1080/09644016.2021.1947636

Goldman Sachs. (2020). *Sustainable ESG investing: Turning promises into performance*. Goldman Sachs. https://www.goldmansachs.com/pdfs/insights/pages/gs-research/sustainable-esg-investing-f/report.pdf

Google. (2023). *Google Environmental Report 2023*. Google. https://www.gstatic.com/gumdrop/sustainability/google-2023-environmental-report.pdf

Government of Sweden. (2023). *Carbon tax in Sweden*. Carbon Tax in Sweden—General Information.

Hopke, J. E., & Paris, L. (2021). Environmental social movements and social media. In *The Handbook of international trends in environmental communication*. Routledge.

Hopke, J., & Hestres, L. (2017). Fossil fuel divestment and climate change communication. *Oxford Research Encyclopedia of Climate Science, 1–35*. https://doi.org/10.1093/acrefore/9780190228620.013.566

IEA. (2021). *Net zero by 2050*. International Energy Agency. https://www.iea.org/reports/net-zero-by-2050

IEA. (2023a). *Navigating Indonesia's power system decarbonisation with the Indonesia just energy transition partnership*. International Energy Agency.

https://iea.blob.core.windows.net/assets/c5427919-6f48-48e4-aae3-e2d8e874d385/NavigatingIndonesiasPowerSystemDecarbonisationwiththeIndonesiaJustEnergyTransitionPartnership.pdf

IEA. (2023b). *World Energy Outlook 2023.* International Energy Agency. https://iea.blob.core.windows.net/assets/86ede39e-4436-42d7-ba2a-edf61467e070/WorldEnergyOutlook2023.pdf

IEA. (2024). *World Energy Investment 2024.* https://iea.blob.core.windows.net/assets/60fcd1dd-d112-469b-87de-20d39227df3d/WorldEnergyInvestment2024.pdf

IFOP. (2021). *Les connaissances des Français en matière d'énergie éolienne.* IFOP. https://www.ifop.com/wp-content/uploads/2021/02/rapport-enquete-ifop-connaissance-francais-energie-eolienne-2021.pdf

International Finance Corporation. (2013). *Public-private partnership stories: India: Gujarat Solar*. World Bank Group. https://www.ifc.org/content/dam/ifc/doc/2010/2013-india-gujarat-solar-ppp-brief.pdf

International Finance Corporation. (2015). *Public-private partnership stories: India—Vadodara Solar*. World Bank Group. https://documents1.worldbank.org/curated/pt/641011467986288773/pdf/96898-BRI-PUBLIC-PPPStories-India-VadodaraSolar.pdf

IRENA. (n.d.). *West Africa clean energy corridor.* Retrieved November 3, 2024, from https://www.irena.org/Energy-Transition/Country-engagement/Regional-Initiatives/West-Africa-Clean-Energy-Corridor

IRENA. (2022a). *Renewable energy targets in 2022: A guide to design.* International Renewable Energy Agency. https://www.irena.org/-/media/Files/IRENA/Agency/Publication/2022/Nov/IRENA_RE_targets_2022.pdf

IRENA. (2022b). *Renewable technology innovation indicators: Mapping progress in costs, patents and standards.* International Renewable Energy Agency. https://www.irena.org/-/media/Files/IRENA/Agency/Publication/2022/Mar/IRENA_Tech_Innovation_Indicators_2022_.pdf

IRENA. (2023). *World energy transitions outlook 2023: 1.5°C pathway.* International Renewable Energy Agency. https://www.irena.org/-/media/Files/IRENA/Agency/Publication/2023/Jun/IRENA_World_energy_transitions_outlook_2023.pdf

IRENA Coalition for Action. (2023). *Scaling up renewable energy investments in West Africa.* International Renewable Energy Agency.

IRENA, & CPI. (2023). *Global landscape of renewable energy finance, 2023.* International Renewable Energy Agency.

Karakaya, E., Nuur, C., & Assbring, L. (2018). Potential transitions in the iron and steel industry in Sweden: Towards a hydrogen-based future? *Journal of Cleaner Production, 195*, 651–663. https://doi.org/10.1016/j.jclepro.2018.05.142

Knack, S., & Rahman, A. (2007). Donor fragmentation and bureaucratic quality in aid recipients. *Journal of Development Economics, 83*(1), 176–197. https://doi.org/10.1016/j.jdeveco.2006.02.002

Koltunov, M., Pezzutto, S., Bisello, A., Lettner, G., Hiesl, A., van Sark, W., Louwen, A., & Wilczynski, E. (2023). Mapping of energy communities in Europe: Status quo and review of existing classifications. *Sustainability, 15*(10), 8201. https://doi.org/10.3390/su15108201

Lestari, E. R., Dania, W. A. P., Indriani, C., & Firdausyi, I. A. (2021). The impact of customer pressure and the environmental regulation on green innovation performance. *IOP Conference Series: Earth and Environmental Science, 733*(1), Article 012048. https://doi.org/10.1088/1755-1315/733/1/012048

Lillemo, S. C. (2014). Measuring the effect of procrastination and environmental awareness on households' energy-saving behaviours: An empirical approach. *Energy Policy, 66*, 249–256. https://doi.org/10.1016/j.enpol.2013.10.077

Meadowcroft, J., & Rosenbloom, D. (2023). Governing the net-zero transition: Strategy, policy, and politics. *Proceedings of the National Academy of Sciences, 120*(47), Article e2207727120. https://doi.org/10.1073/pnas.2207727120

Ministère de la Transition Écologique. (2018). *Diagnostic de performance énergétique—DPE*. Retrieved November 3, 2024, from https://www.ecologie.gouv.fr/politiques-publiques/diagnostic-performance-energetique-dpe

Munro, P. G., Samarakoon, S., Kearnes, M., & Paisley, C. (2023). The right to repairable energy: A political ecology of off-grid solar repair in Zambia. *Political Geography, 106*, Article 102962. https://doi.org/10.1016/j.polgeo.2023.102962

Nestlé. (2024). *Creating Shared Value and Sustainability Report 2023*. https://www.nestle.com/sites/default/files/2024-02/creating-shared-value-sustainability-report-2023-en.pdf

OECD. (2021). OECD environmental performance reviews: Finland 2021. *OECD Publishing*. https://doi.org/10.1787/d73547b7-en

OECD. (2023). *Private philanthropy for sustainable development, 2018–20: Data and analysis* (No. 32; OECD Development Perspectives). OECD. https://doi.org/10.1787/0164439d-en

Ofgem. (2022). *Domestic renewable heat incentive (RHI) essential guide*. https://www.ofgem.gov.uk/sites/default/files/2022-03/Domestic%20RHI_Essential%20guide.pdf.

Oil Change International. (2024). *Public enemies: Assessing MDB and G20 international finance institutions' energy finance*. Oil Change International. https://www.oilchange.org/wp-content/uploads/2024/04/G20-Public-Enemies-April-2024.pdf

Onjewu, A.-K.E., Jafari-Sadeghi, V., Kock, N., Haddoud, M. Y., & Sakka, G. (2023). The catalyzing role of customer pressure on environmental initiatives and export intensity: A study of family firms. *Journal of Business Research, 166*, Article 114134. https://doi.org/10.1016/j.jbusres.2023.114134

Parks, L., Porta, D. della, & Portos, M. (2023). Environmental and climate activism and advocacy in the EU. In *Handbook on European Union climate change policy and politics* (pp. 98–112). Edward Elgar Publishing. https://www.elgaronline.com/edcollchap-oa/book/9781789906981/book-part-9781789906981-18.xml

Pew Research Center. (2020). *Two-thirds of Americans think government should do more on climate*. https://www.pewresearch.org/wp-content/uploads/sites/20/2020/06/PS_2020.06.23_government-climate_REPORT.pdf

Reboredo, R., & Carmody, P. (2024). Between energy and politics: Ruin, renewal, and the contours of state power in post-apartheid South Africa. *Environment and Planning D*, 02637758241242213. https://doi.org/10.1177/02637758241242213

REN21. (2023). *Renewables 2023 global status report collection.*

REN21. (2024a). *Renewables 2024 Global Status Report*. Renewable Energy Policy Network for the 21st Century (REN21). https://www.ren21.net/reports/renewables-2024-global-status-report/

REN21. (2024b). *Renewables 2024 Global Status Report Collection, Energy Demand*. REN21 Secretariat.

REScoop.eu. (2023). *Annual report 2023*. REScoop.eu. https://www.rescoop.eu/toolbox/annual-report-2023

Rizzoli, V., Casara, B. G. S., & Sarrica, M. (2024). Tweeting about a revolution? A cross-national analysis of tweets on climate change during the rise of "Fridays for future." *Social Psychological Bulletin, 19*, 1–26. https://doi.org/10.32872/spb.12383

Rockefeller Foundation. (2023). *Impact Report 2023—Energy chapter: Speeding the energy transition*. Rockefeller Foundation. https://www.rockefellerfoundation.org/wp-content/uploads/2024/07/RF-2023-Impact-Report-Energy-Chapter-Speeding-the-Energy-Transition.pdf

Schroeter, R., Scheel, O., Renn, O., & Schweizer, P.-J. (2016). Testing the value of public participation in Germany: Theory, operationalization and a case study on the evaluation of participation. *Energy Research & Social Science, 13*, 116–125. https://doi.org/10.1016/j.erss.2015.12.013

Shim, D. (2024). Personalising climate change—How activists from Fridays for Future visualise climate action on Instagram. *Humanities and Social Sciences Communications, 11*(1), 1–9. https://doi.org/10.1057/s41599-024-03556-7

Sierra Club. (n.d.). *We're moving beyond coal and gas | Beyond coal*. Retrieved November 5, 2024, from https://coal.sierraclub.org/campaign

Sifakis, N., Savvakis, N., Daras, T., & Tsoutsos, T. (2019). Analysis of the energy consumption behavior of European RES cooperative members. *Energies, 12*(6), 970. https://doi.org/10.3390/en12060970

SolarAid. (2023). *Annual Review 2022–2023*. SolarAid. https://solar-aid.org/wp-content/uploads/2024/01/SolarAid-Annual-Review-22-23.pdf

Sumner, A., Gulrajani, N., Wickstead, M., & Glennie, J. (2020). A proposal for a new universal development commitment. *Global Policy, 11*(4), 478–485. https://doi.org/10.1111/1758-5899.12844

Toga, M., Monyl, N., & Dobrotkova, Z. (2019). *The role of the public sector in mobilizing commercial finance for grid-connected solar projects: Lessons learned and case studies*. World Bank Group.

UNCTAD. (2022). *Global crisis response group: Official Development Assistance (ODA) Report*. United Nations Conference on Trade and Development (UNCTAD). https://unctad.org/system/files/official-document/un-gcrg-oda-report_en.pdf

United Nations. (2021). *Foundations announce $1 billion fund for renewables in emerging economies as over 45 ministers review plans to achieve clean, affordable energy for all*. https://www.un.org/sites/un2.un.org/files/press-release-foundations-announce-fund-to-achieve-clean-affordable-energy.pdf

U.S. Department of Energy. (2020). *Guide to the federal investment tax credit for commercial solar PV*. https://www.energy.gov/sites/prod/files/2020/01/f70/Guide%20to%20the%20Federal%20Investment%20Tax%20Credit%20for%20Commercial%20Solar%20PV.pdf

Vincent, F. (2006). NGOs, social movements, external funding and dependency. *Development, 49*(2), 22–28. https://doi.org/10.1057/palgrave.development.1100253

VOA Afrique. (2017). *Le Kenya table sur le solaire pour étendre son électrification*. Retrieved November 3, 2024, from https://www.voaafrique.com/a/le-kenya-mise-sur-le-solaire-pour-etendre-son-electrification/3853239.html

Wang, J., Duan, Y., & Liu, G. (2022). A study of specific open innovation issues from perspectives of open source and resources—The series cases of Tesla. *Sustainability, 14*(1), Article 1. https://doi.org/10.3390/su14010142

Wierling, A., Schwanitz, V. J., Zeiß, J. P., Bout, C., Candelise, C., Gilcrease, W., & Gregg, J. S. (2018). Statistical evidence on the role of energy cooperatives for the energy transition in European countries. *Sustainability, 10*(9), 142. https://doi.org/10.3390/su10093339

Williamson, C. R. (2010). Exploring the failure of foreign aid: The role of incentives and information. *The Review of Austrian Economics, 23*(1), 17–33. https://doi.org/10.1007/s11138-009-0091-7

World Bank. (2017). *World Bank Group announcements at one planet summit*. World Bank. Retrieved November 5, 2024, from https://www.worldbank.org/en/news/press-release/2017/12/12/world-bank-group-announcements-at-one-planet-summit

World Bank, & Energy Charter Secretariat. (2023). *Enabling foreign direct investment in renewable energy sector: Reducing regulatory risks and preventing investor-state conflicts*. World Bank. https://www.energycharter.org/fileadmin/DocumentsMedia/Occasional/Renewable_Energy_FDI_Final__032923.pdf

WWF. (2025). *NGOs call for strong EU policies to beat the 2030 renewable energy target*. Retrieved November 5, 2024, from https://www.wwfmmi.org/?247930/NGOs-call-for-strong-EU-policies-to-beat-the-2030-renewable-energy-target

Yu, W., Hassan, A., & Adhikariparajuli, M. (2022). How did Amazon achieve CSR and some Sustainable Development Goals (SDGs)—Climate change, circular economy, water resources and employee rights during COVID-19? *Journal of Risk and Financial Management, 15*(8), 368. https://doi.org/10.3390/jrfm15080364

Zhang, X., Jiao, K., Zhang, J., & Guo, Z. (2021). A review on low carbon emissions projects of steel industry in the world. *Journal of Cleaner Production, 306*, Article 127259. https://doi.org/10.1016/j.jclepro.2021.127259

CHAPTER 7

Empowering the Grid: Stakeholder Engagement in Advancing Electrification

Fadi Sayegh

Introduction: Background Information on Electrification

Electrification is at the center of achieving carbon neutrality, since the generation of electricity using renewable and clean energy sources like solar and wind can significantly lower carbon emissions. The main sectors rapidly transitioning to electricity include transportation using electric vehicles (EVs), real estate (construction), and economic and agricultural sectors. The International Energy Agency (IEA, 2023) has predicted that electrification will increase the Net Zero Emissions (NZE) Scenario to 27 percent in 2030. IEA (2023) has also documented that heat pumps outsold fossil fuel-based heating systems in the US and France. In 2022, China's EV sales reached 29 percent, and approximately half of India's three-wheeler registrations were electric. This shows that electrification is not only becoming more popular in the countries mentioned but also has

F. Sayegh (✉)
Rochester Institute of Technology (RIT), Rochester, NY, USA
e-mail: academic@mail.rit.edu; fadi_sayegh@hotmail.com

T. Walker et al. (eds.), *Wiring the Future*, Palgrave Studies in Emerging Risk Management and Sustainable Finance,
https://doi.org/10.1007/978-3-032-14717-2_7

significant potential for reducing final energy demand. Electric technologies are inherently more efficient than fossil fuel-based alternatives and go hand-in-hand with increasing renewable energy.

The influence of governments, the private sector, local communities, non-governmental organizations (NGOs), advocacy groups, and policymakers is paramount in spearheading electrification since they are at the forefront of crafting strategic policies, establishing regulations, and funding infrastructure developments. Their leadership stimulates innovation, guides public and private investments, mitigates risks associated with large-scale projects, and ensures equal benefits from electrification on every level of society (Meckling et al., 2015; Sovacool & Hess, 2017). However, there are few studies that examine the diverse roles of key stakeholders who directly influence sustainability development. Abiddin et al. (2022) evaluate the contributions of NGOs in fostering sustainable community development and find that NGOs play a significant role in advocating for sustainability. The study also concludes that further engagement of other stakeholders, like the government, is crucial to ensure the effectiveness of implemented efforts. Batidzirai et al. (2021) also evaluate the existing people-private-public partnerships aiming at spearheading sustainability. Existing studies often overlook the specific roles and interactions of different stakeholders in driving sustainable electrification. The study focuses on electrification and establishes that the traditional top-down programs and approaches are inadequate for capturing respective community needs. The present study fills this gap by evaluating the role of governments, the private sector, communities, NGOs, advocacy groups, and policymakers in the transition to electrification using real-world examples and scholarly analyses and research findings from peer-reviewed literature. It provides a comprehensive understanding of how collaborative approaches can overcome challenges and accelerate the shift toward a low-carbon, electrified future.

The Strategic Role of Governments in Electrification

Governments play a pivotal role in creating policies and regulations that drive the electrification transition by fostering a stable, investment-friendly, and innovation-driven environment.. Governments set national energy goals, which provide direction and clarity on priorities and actions so that all stakeholders may work together toward common objectives. The interplay of policy, regulation, and technological innovation in shaping the future of the energy system is complex and dynamic. For example, Denmark's decision to achieve 100% renewable energy before 2050 demonstrates the feasibility of such goals, encouraging other countries to follow suit and sending (Danish Energy Agency, 2019). The Danish wind energy strategy has been among the leaders in this field globally. Such success can only be achieved when there are clear national energy objectives, aligning efforts among different contributors, and outlining a clear path forward. Denmark's dedication to reaching zero carbon emissions by 2050 is evident in its substantial investments in near Zero Energy Buildings (nZEBs) (Drysdale et al., 2019). Policies such as the investment into nZEBs are also supported by strong political will, commitment to development and research, and investment in major renewable projects like harnessing solar and wind energy to promote electrification (Pinson et al., 2025).

Governments also play a role in funding and implementing public works projects that provide infrastructure to support the increasing share of renewable energy within the grid. For example, programs like California's Self-Generation Incentive Program (SGIP) exemplify the proactive policies governments have set to foster sustainable infrastructural development. By offering financial incentives, the program increases the adoption of distributed energy technologies, rooftop solar panels, and battery storage systems and allows consumers and businesses to more easily participate in and further develop a decentralized and resilient energy network (California Public Utilities Commission, 2021). As of 2022, the program had facilitated the allocation of approximately $1.2 billion in incentives for supporting the development of at least 750 MW of distributed generation and 620 MWh of energy storage capacity (Fero & Ferguson, 2025). This helped reduce greenhouse emissions by 16 kg/kWh. Each program feature is an integrated part of the general objective of transition toward a more sustainable and resilient energy system.

Finally, governments must ensure that all the benefits of electrification are accessible to everyone by implementing programs that address historical disparities in energy access. Addressing the social impact of energy transition can counteract negative sentiments toward electrificationin communities and work toward ensuring equitable availability of renewable energy and electrification resources. For example, the South African Integrated National Electrification Programme has connected millions of people to the grid, particularly in the historically underserved rural and peri-urban areas (Department of Energy, 2019). This has improved access to education, healthcare, and economic opportunities and has reduced reliance on hazardous energy sources like kerosene and firewood.

Challenges in Government-Led Electrification

Governments face the challenge of balancing the need for economic growth with the imperative to reduce carbon emissions, especially in emerging markets with a rapid rise in energy demand. This challenge reflects the significant economic management difficulties faced by countries heavily reliant on fossil fuel exports as they transition to electrification. To ease these difficulties, some governments are pioneering programs aimed at facilitating a smoother transition. For example, Brazil's large-scale ethanol program, Proálcool, significantly reduced the country's reliance on imported fossil fuels, positioning it as a global leader in biofuels. Goldemberg et al., 2008) establish that tapping into biofuels specifically in Brazil reduces the consumption of fossil fuels, although this project caused serious debates on land use, food production, and nature preservation. Biofuel production raises concerns over deforestation, competition with food crops, and land-use conflicts, illustrating the complex trade-offs governments must navigate when implementing sustainable energy policies.

In addition, the global nature of energy transition necessitates international cooperation, mandating the government to ensure that national policies align with global goals provided by international agreements, such as the Paris Agreement. The Paris Agreement, adopted in December 2015 and entered into force in November 2016, initially had 175 parties, including 174 countries and the European Union. As of February 2023, 195 members of the United Nations Framework Convention on Climate Change (UNFCCC) have become parties to the agreement and aim to limit global warming to well below 2 °C by reducing greenhouse gas emissions. Aligning national goals with global goals is challenging, and governments must simultaneously drive economic growth, protect jobs, and maintain energy security while committing to sustainable development goals.

Private Sector Engagement: Corporate Strategies and Responsibilities in Driving Electrification

The integration of sustainability into the core of corporate strategy is a signature attribute of industry leadership. Companies are increasingly voicing an understanding of the imperative for sustainability and embedding it within their operations. For example, in 2018, Google committed to employing a sustainability strategy to ensure carbon-free energy by 2030 (Chenic et al., 2022; Google Sustainability, 2018; Koningstein & Fork, 2014). This goal is being pursued by matching electricity consumption with renewable energy credits, whereby Google shows environmental responsibility and secures its economic stability against fluctuating energy costs. Google has invested in long-term power purchase agreements (PPAs), AI-driven energy optimization, and investments in wind, solar, and battery storage, ensuring environmental benefits and resilience against volatile energy markets.

The private sector is also increasingly held accountable for investing in the clean energy revolution. Through strategic investments in research and development, companies are expanding the frontiers of clean energy technology and securing a more sustainable future. For instance, Ørsted, a Danish multinational energy company, has successfully reduced its coal consumption by 73% since 2006 and aims to become carbon–neutral by 2025 through large-scale investments in wind and solar power (Ørsted, 2020). In addition, it has heavily invested in offshore wind farms, including Hornsea One, the world's largest operational offshore wind farm, which generates enough electricity to power over one million homes in the UK (ØrstedUK, 2018). The company's transformation from one of Europe's most fossil-fuel-intensive energy companies to a global leader in renewable energy is a prime example of corporate commitment to sustainability.

As companies become major drivers of electrification, they are accountable for the social and environmental impacts their operations create. This includes ensuring fair distribution of benefits from electrification to ensure that already existing inequalities are not further cemented. Corporations are the harbingers of innovation and ethical progress in the shifting landscape of global electrification. For instance, Apple has made significant efforts to ensure its suppliers uphold high responsibility standards, actively protecting labor rights within its supply chain. Apple enforces these standards through rigorous supplier audits, mandatory compliance with its Supplier Code of Conduct, and initiatives to eliminate exploitative labor practices, such as child labor and excessive working hours (Apple, 2021; Nicolas, 2018). This illustrates that corporations are encouraged to weave ethics into their operational principles, with growing regulations, industry standards, and societal expectations urging businesses to align their growth with broader objectives of social welfare and environmental integrity.

Lastly, corporations are increasingly held responsible for ensuring transparency and accountability, both to the public and regulatory bodies, which are stressed in their ethos and operations. These efforts are vital in building public trust and forging long-term relationships with all stakeholders. For this reason, companies use sustainability reporting and disclosure frameworks that guide their sustainability performance, such as the Global Reporting Initiative (GRI) and the Task Force on Climate-related Financial Disclosures (TCFD). For instance, Unilever, a leader in corporate sustainability, provides comprehensive sustainability reports that detail its progress toward reducing GHG emissions, improving energy efficiency, and transitioning to renewable energy (Unilever, 2024). These practices reveal the changing face of corporate sustainability; accountability, transparency, and stakeholder engagement are intrinsic elements in pursuing ecological and social objectives.

Challenges Facing Private Sector Engagement in Electrification

The electrification movement within the private sector is a dynamic balance between proactive strategy and reactive adaptation (Sayegh, 2023). Internal goals and the external forces of regulatory frameworks and market dynamics shape companies' electrification efforts. Companies' electrification efforts are shaped by both internal goals and external factors, such as regulatory frameworks and market dynamics (Vrikki, 2024). While many companies express commitments to sustainability and carbon neutrality, their actual actions and planned initiatives may not always align with these claims. This highlights the need for a critical evaluation of their sustainability narratives and transparency, ensuring that the intentions behind these commitments are matched by concrete progress. The ability to respond quickly to regulatory changes is crucial, as seen with British Petroleum's (BP) strategic shifts in anticipation of stricter emissions regulations and global carbon reduction targets (BP, 2024). BP has expanded its renewable energy portfolio, invested in offshore wind projects, and committed to reducing oil and gas production, aligning its business model with evolving policies aimed at limiting carbon emissions. This adaptability is not just about survival but about seizing opportunities that arise from new regulations, such as advancements in renewable energy incentives and carbon pricing. However, such actions warrant scrutiny to ensure they genuinely reflect sustainable practices rather than mere greenwashing.

Market risks of electrification, including volatile energy prices and the variable nature of consumer demand, need an equally strong risk management strategy. IKEA's forward-looking investment in renewable energy and efficiency technologies exemplifies how companies can protect themselves from market volatility while strengthening their sustainability credentials in at the same time (IKEA, 2020). IKEA owns and operates wind and solar farms, invests in energy-efficient store designs, and implements on-site renewable energy solutions, reducing reliance on fluctuating energy markets and cutting long-term operational costs However, as with all such efforts, it is essential that these initiatives be critically assessed to ensure they align with the company's stated sustainability goals and deliver genuine benefits, rather than being a mere marketing tool.

The private sector's journey toward electrification is thus characterized by foresight to plan and agility to pivot so that resilience is assured against uncertainty. However, this journey must also be accompanied by a commitment to transparency and accountability to validate the authenticity of sustainability claims (Wei et al., 2019).

The Role of Communities in the Energy Transition and Public Opinion

Community ownership in a renewable energy project drives local acceptance and creates a foundation for a sustainable development path. A practical example is the Danish model, in which almost half of the wind turbines in the country are owned by local cooperatives. Some studies suggest that the widespread acceptance and success of wind energy in Denmark could be related to this facilitating factor (Walker, 2008). A model like this has significant financial dividends for the community but also brings a sense of collective responsibility and achievement. It gives people a sense of empowerment regarding their energy future, reinforcing the connection between local action and national sustainability objectives

(Van Der Schoor & Scholtens, 2015). This synergy of community ownership and renewable energy initiatives provides a replicable model for other nations willing to galvanize support for sustainable practices while driving economic and social benefits at the grassroots level.

Another vivid example of what can be achieved through community action is the Eigg Electric cooperative on the Isle of Eigg, Scotland. Eigg Electric planned and installed hydroelectric, wind, and solar to provide a diversified mix of energy able to cope with the unpredictable weather on the island. This effort resulted in a sharp drop in carbon emissions, giving the community a steady, reliable power source at much less cost than diesel generators, which were costly, polluting, and provided only limited electricity access. The success of the renewable energy project on Eigg reinforces that even the smallest local initiatives can cause a tectonic shift globally toward sustainable energy and reduced carbon footprint (Chmiel & Bhattacharyya, 2015). This is a case study where a community, through a collective vision for a better tomorrow, has faced the challenges involved in the transition toward renewable energy and can thus serve as a pathfinder for others.

The media significantly influences public opinion on energy matters by framing the narrative around electrification and sustainable energy practices. The portrayal of renewable energy benefits or the environmental and economic impacts of fossil fuels can sway public support for clean energy initiatives. Whenever the media presents positive stories about renewable energy, such as reporting on the number of jobs it would create and how it would benefit the environment, support for these initiatives significantly increases. On the other hand, public opposition is generated by discouraging or critical stories about certain kinds of energy projects. One such example is media coverage of fracking in the UK. The media extensively covered the potential environmental and health risks of fracking, such as water contamination, seismic activity, and air pollution, which fueled public anxiety and opposition (Jaspal & Nerlich, 2013). This sustained negative media attention intensified grassroots activism, which successfully pressured policymakers into imposing strict regulations and, ultimately, implement a moratorium on fracking in 2019.

Social media has become a critical battleground for the public. Platforms like X (formerly Twitter) and Facebook enable information diffusion at incredible speeds and serve as an arena for opposing ideologies. Incidents such as the Dakota Access Pipeline controversy have taken this dynamic to a new level, illustrating how social media can act as double-edged sword, able to galvanize support but also sharpen divisions. Specifically, the Dakota Access Pipeline protests demonstrates how social media can amplify grassroots activism through allowing movements like #NoDAPL to gain global traction, mobilizing millions to oppose the project and pressure policymakers (Presley, 2019). However, the same platforms also became spaces for misinformation and polarization, with industry supporters and critics spreading conflicting narratives, making it challenging for the public to reach a consensus on energy infrastructure projects (Presley, 2019). Therefore, grasping the mechanisms of social media influence becomes critical for stakeholders in energy debates who are trying to navigate the labyrinthine landscape of public opinion.

Communities might be resistant and demonstrate "Not In My Backyard" (NIMBY)-ism toward projects geared toward sustainability and electrification. This is particularly true when developers construct infrastructure projects close to residential areas, due to perceived adverse impacts. For example, local residents voiced concerns about potential disruptions to marine life and fishing industries, highlighting the importance of addressing such adverse impacts during community consultations. Devine-Wright (2009) contends that this opposition is not simply mindless local resistance but rather often reflects genuine concerns about environmental, aesthetic, or health issues. This shows that community support is a key factor in large project development. For instance, the success of the Hornsea Project, developed by Ørsted in the North Sea, demonstrates how public consultations and the promise of economic gains can translate to regional support (Ørsted, 2020). Community engagement during the establishment of this project involved a series of public consultations, where local residents were invited to discuss concerns and provide input on the development. Ørsted facilitated meetings, information sessions, and surveys, ensuring that community members had a voice in the decision-making process. The engagement highlighted the importance of employment opportunities, regional investments, and the potential for community shares, which helped to secure a broad base of local support for the project.

Social acceptance plays a crucial role in the successful deployment of new energy technologies, especially when significant infrastructure changes are involved, such as Electric Vehicles (EVs), or even smarter grids, which necessitate shifts in how people interact with and manage energy use in their daily lives. A smart grid, for instance, requires consumers to adapt to new systems of energy monitoring, real-time consumption data, and possibly new pricing models that could alter daily routines and consumption habits. These shifts require public understanding and trust, which are critical for the smooth integration of such technologies. For example, Norway's EV success is attributed to strong governmental incentives, widespread public awareness campaigns, and visible EV infrastructure. Policies such as tax exemptions, reduced toll fees, free public parking, and access to bus lanes have made EVs financially attractive, and extensive charging networks and societal normalization of EVs have further reinforced their acceptance (Figenbaum & Kolbenstvedt, 2016). All these factors together form a basis for increasing perceived benefits and reducing perceived apprehensions, positively influencing public opinion, building trust in technology, promoting institutions, and ensuring public participation in the technological transition. This case illustrates the complex interplay of influences underlying new technologies' social reception.

Non-Governmental Organizations and Advocacy Groups

NGOs have been critical in key international climate change and energy policy agreements. In the lead-up to the Paris Agreement, influential organizations such as the Climate Action Network and Friends of the Earth placed strong pressure on state actors to make stringent environmental commitments. Due to the relentless advocacy of these organizations, this agreement included critical targets like the international commitment to keep global temperature rise well below 2 degrees Celsius (Giese, 2017). However, despite these efforts, recent reports indicate that global temperatures in 2024 reached unprecedented levels, with the annual average temperature exceeding pre-industrial levels by 1.55 °C, surpassing the previous record set in 2023 (Bevacqua et al., 2025). This alarming trend underscores the challenges in meeting the targets set by the Paris Agreement and exemplifies the urgent need for nations to intensify their climate actions to avoid further deviation from the established goals. The Paris Agreement established a collective goal to limit global warming to well below 2 °C, preferably to 1.5 °C, above pre-industrial levels. Nevertheless, he current trajectory suggests that existing commitments and actions may be insufficient to achieve these objectives.

NGOs also champion and monitor accountability in the realization of energy policies, ensuring governments and corporations meet their pledges. Independent research, report publishing, and public campaign running are some of the activities they engage in to ensure that all stakeholders remain committed to environmental pledges. Watchdog organizations are key players in the oversight process for electrification and climate change policies. Organizations, including the Environmental Defense Fund (EDF), carefully monitor corporations and governments for transparency and accountability (Environmental Defense Fund, 2024). They publish highly detailed reports on disparities between commitments versus reality in actions taken. A leading example is the work done by EDF on repeatedly tracking methane emissions in the oil and gas sector; indeed, this has set a new standard of regulatory scrutiny and enhanced corporate accountability.

Furthermore, NGOs have been influential in encouraging corporations to adopt and promote sustainable practices through targeted campaigns and movements for divestment. They have progressively succeeded in shifting business investment toward renewable energy sources. This shift has been driven by the divestment from fossil fuels movement, which has emerged as a potent force reshaping investment decisions across diverse institutions. Led by groups such as 350.org, the movement has forced many universities, pension funds, and other institutions to divest their investments in fossil fuels (350.org, 2020). The fossil fuel divestment movement has significantly influenced global investment patterns, with over 1500 institutions committing to divest from fossil fuels, representing approximately $40 trillion in assets as of 2021 (Stand.earth, 2022). This shift has redirected substantial capital toward renewable energy projects, decreasing carbon emissions by reducing fossil fuel exploration and extraction funding. As the movement has pointed out, the environmental and financial risks involved in such investment have brought more awareness and shifted massive funding to renewable energy projects, making it one of the most crucial turns toward sustainable development.

NGOs also promote partnership and collaboration, especially in the research and innovation sphere, furthering the sustainability and electrification process. For instance, the efforts of Rocky Mountain Institute (RMI), headquartered in Basalt, Colorado, provide examples of relationships that work with businesses to develop and implement new energy solutions. RMI is a nonprofit organization focused on advancing clean energy and sustainable practices. It collaborates with businesses, governments, and organizations to accelerate the transition to a low-carbon future, with a particular emphasis on energy efficiency, renewable energy, and climate policy. Most of RMI's collaborative works are pilot projects that can prove the viability and, at the same time, advantages of different new energy models, including microgrids and energy-efficient technologies (Rocky Mountain Institute, 2023). The organization actively partners with businesses and governments to scale clean energy solutions, such as microgrids and distributed renewable energy systems, which enhance energy access and resilience. The Forest Stewardship Council (FSC) is another organization that partners with companies, working to ensure that the biomass used for energy comes from well-managed forests that are responsibly harvested. This encourages good industry practices while contributing to wider goals of environmental conservation and social responsibility (Forest Stewardship Council, 2021). Notably, FSC is an international nonprofit NGO headquartered in Germany and operates through a network of national and regional offices globally. This global presence enables it to promote responsible forest management practices across diverse regions.

NGOs also take the lead in corporate responsibility campaigns. The basis of such campaigns is the need to hold corporations responsible, particularly those that trade in energy, for their environmental impact. In reality, most Rainforest Action Network (RAN) campaigns leave a mark, and many of the campaigns do indeed cause a shift in investing among the biggest banks and major financial institutions. RAN is a nonprofit environmental organization that focuses on protecting forests, climate, and human rights by challenging corporate practices that contribute to environmental degradation. RAN has successfully pressured major financial institutions like Citibank to adopt more stringent policies on financing fossil fuel projects, reducing funding for coal and other high-emission industries (Rainforest Action Network, 2021). Their campaigns have contributed to banks committing to align their portfolios with net-zero emissions goals, thereby accelerating the global shift toward renewable energy investments. Many are now introducing policies that limit coal project financing, thus hastening the transition toward alternative, cleaner energy solutions.

Additionally, NGOs are fundamental in developing industry standards, especially in electrification. They qualify the chain with certification schemes guaranteeing high environmental and social criteria for bioenergy goods.

Lastly, NGOs organize campaigns and public education to solicit public support for electrification. Through awareness raising and coalition building, they provide much-needed social and political momentum that helps bring forward clean energy solutions. An example is the "Global 100% Renewable Energy" campaign, which is driven by a group of NGOs, including the World Wildlife Fund (WWF) and Greenpeace, and advocates for the attainment of 100% renewable energy production globally. The campaign has received much public support and has influenced policy development, with many communities coming up with and working toward ambitious renewable energy targets (Global 100% Renewable Energy, 2024). The campaign's advocacy and support has influenced cities like Sydney, Australia, and Frankfurt, Germany, to accelerate local clean energy transitions and influenced national policy dialogues on sustainable energy, owing to the Global 100% Renewable Energy Platform advocacy and support. NGOs also play an important role in public awareness and political advocacy on the use of electrification to tackle climate change. For instance, the Union of Concerned Scientists (UCS), a nonprofit organization in the United States, works to advance science-based solutions to global challenges, including climate change and energy transition (Union of Concerned Scientists, 2020). UCS played a significant role in advocating for the Clean Power Plan in the U.S. and has been instrumental in pushing for stronger vehicle emissions standards, which have resulted in reductions in greenhouse gas emissions. Additionally, UCS has worked to secure the inclusion of clean energy provisions in federal and state-level policies, such as the California Clean Energy Jobs Act. It creates educational resources and public events promoting science-based policies on energy transition and climate change. UCS is active in the public debate around sustainable solutions for the energy transition.

Challenges Facing NGOs in Electrification

The various challenges facing NGOs include political resistance and financial constraints that prevent them from achieving their goals. NGOs encounter especially substantial political resistance to electrification advocacy in areas where the fossil fuel industry has more hold. If such resistance is to be effectively overcome, NGOs must have strong coalitions of various stakeholders, such as businesses, municipal authorities, and community organizations. Such strategic alliances tend to underline the pervasive benefits of electrification while mustering the necessary support to overcome the potential political impediments to such proposals (Betsill & Corell, 2007). An example of such a coalition is the Fossil

Fuel Non-Proliferation (FFNP) Treaty Initiative, which was launched by a group of NGOs in 2020 to advocate for phasing out fossil fuel production and facilitating a just transition to renewable energy sources. The coalition was comprised of over 2700 civil society organizations, including environmental NGOs, research institutions, and community groups from different continents (Newell et al., 2022; Treaty, 2022). This collaboration demonstrates the power of unified action in advancing global electrification efforts. One notable achievement is the successful inclusion of the Fossil Fuel Non-Proliferation Treaty Initiative in the UN's High-Level Dialogue on Energy in 2021, which highlighted the need for a just energy transition. This initiative also helped to influence the commitment of countries like Costa Rica, which has pledged to align its energy policies with the goal of phasing out fossil fuels and transitioning to renewable energy sources.

Financial sustainability remains one of the major challenges for NGOs, particularly those involved in long-term advocacy and community development processes. To meet this challenge, NGOs increasingly look beyond traditional grants and donations as sources of funds, seeking partnerships within the private sector and exploring novel financing options such as impact investing and crowdfunding. For example, the FFNP Treaty leverages crowdfunding campaigns and impact investments to fund its advocacy efforts (Newell et al., 2022), which helps sustain long-term policy engagement and grassroots mobilization. Such financial strategies allow NGOs to expand and find new venues for their important work (Moore et al., 2015). NGOs must be able to quickly adapt in this fast-changing energy landscape, necessitating fast response to new challenges and opportunities. Awareness of such changes in technology, policy, and markets is essential for NGOs to remain relevant and effective in achieving their goals (Lachapelle & Paterson, 2013). NGOs that can quickly adjust their strategies because of the mentioned dynamics develop an enhanced capability to achieve their objectives and remain relevant and effective in a changing world.

Synergies and Conflicts in Stakeholder Collaboration

The transition process toward sustainable energy systems, especially electrification, incorporates various stakeholders at different levels of government, private companies, communities, and NGOs. Each of these groups brings along peculiar interests, resources, and strategies that not only offer opportunities for synergy but also areas of probable conflict. Governments are central in setting the regulatory framework and offering incentives that favor private sector investment in electrification projects. Government policies on developing renewable energy through subsidies or tax incentives create favorable opportunities that encourage private companies to invest in sustainable infrastructure. For example, in Germany, the Renewable Energy Act (EEG) has been instrumental in bringing about public–private partnerships, which resulted in the rapid growth of renewable energy (Ahrens, 2017; Gründinger, 2017). Such synergy shows how coordinated governmental policy and corporate strategy have accelerated the deployment of sustainable energy technologies. Similarly, NGOs often play the role of linking agents and act to bridge the gaps between government goals and needs in the community for inclusive energy policies and grassroots-level participation.

However, conflict arises when the interests of such stakeholders clash. Private sector actors are usually driven by motives for profit, which often do not align with governments' or communities' social or environmental priorities. For instance, in India, large-scale solar projects led by corporations have sometimes displaced locals, leading to protests and community resistance to exclusion from the benefits of electrification (Stock, 2022). These protests arose in response to projects like Gujarat Solar Park and the Kurnool Solar Park because land acquisitions for these projects disrupted traditional agrarian systems and exacerbated existing social inequalities. Such issues have raised the need for more integrated approaches to planning involving communities in decision-making processes so as not to be trumped by corporate or government interests.

Lastly, regional disparities complicate stakeholder interaction. Strategies that are successful in one geographical and economic context may prove less effective for others. In sub-Saharan Africa, where so many rural areas remain unelectrified, NGOs have often stepped in to implement decentralized solar energy projects in cooperation with regional governments and private firms. Examples include Kenya, where collaboration between NGOs and local governments has resulted in community-owned mini-grid systems that can provide low-cost off-grid power supplies to dispersed villages (Ulsrud et al., 2015). Such projects demonstrate how local approaches to collaboration models can balance the interests of many different stakeholders and provide a template for other regions facing similar electrification challenges.

Conclusions

Overcoming the multi-faceted challenges of electrification in sustainable energy transitions requires strong collaboration between governments, the private sector, communities, and NGOs. While each stakeholder group has distinct roles and priorities, their successful coordination is essential to achieving broad and lasting progress in electrification efforts. Probably, the most efficient way to develop collaboration is through public–private partnerships (PPPs). These partnerships allow governments to capture private firms' financial and technical expertise while meeting social and environmental objectives. In this regard, PPPs have seen considerable success in Southeast Asia, especially where governments have partnered with private companies to develop renewable energy projects and expand grid infrastructure (Baucia, 2017). For example, in the Philippines, the public–private partnership between the government and Solar Philippines has successfully deployed solar farms and microgrids, ensuring electricity interconnectivity to remote islands previously reliant on costly diesel generators. In Vietnam, Le and Lam (2023) argue that PPP initiatives have driven the rapid expansion of wind and solar energy, with the government offering feed-in tariffs and private firms investing in large-scale projects. Consequently, with public oversight, large-scale investments that such partnerships can mobilize would ensure that the electrification projects align with national development goals and local needs.

Decentralized community models offer another promising pathway for overcoming some of the electrification challenges, at least in those rural or remote areas where centralized grids are unlikely to expand. These decentralizing models enhance ownership and long-term sustainability since they involve local communities in an active role in the management of energy projects. Community-managed solar mini-grids have successfully extended electricity access to sub-Saharan Africa, especially in regions with a slow access speed due to national grid expansion or impracticability (Utoh et al., 2024). For such projects to succeed, cooperation between local governments, NGOs, and private businesses must be strong, with each party playing a distinct yet complementary role. A good example of this approach is seen in Tanzania's Rural Energy Agency (REA) mini-grid projects, where local cooperatives manage solar-powered microgrids to supply electricity to off-grid villages (Ilskog et al., 2005; Smart Villages,

2025). NGOs provide technical training, private companies supply equipment, and the government offers regulatory support, ensuring long-term viability and community ownership of the energy infrastructure.

Another essential consideration in overcoming electrification challenges will be policy harmonization across sectors. More often than not, mismatched government regulations, corporate strategy, and community priorities give rise to conflicts that hinder progress. For example, the Nord Stream 2 pipeline project, a major natural gas pipeline running from Russia to Germany through the Baltic Sea, encountered conflicting regulations and priorities, significantly hindering its progress. Germany supported the project for energy security reasons, the EU pursued it for energy diversification, and the US opposed it, citing geopolitical concerns related to transatlantic relations and Russian influence (Goldthau, 2016). In order to avoid such conflict, governments and relevant stakeholders must develop multisectoral communication platforms for regular information exchange and stakeholder coordination. These platforms ensure that all parties are aware of each other's goals and constraints, and are thus able to predict and handle the emergence of potential conflicts before they become a problem. For example, the *Clean Energy for All Europeans* package provided policies that created coherence at the level of the European Union member states, hence allowing further cooperation and simplifying the transition toward renewable sources of energy across borders. The package created this coherence among EU member states by harmonizing electricity market regulations and setting a common renewable energy target of at least 32% by 2030 (Nieto et al., 2020). The package also facilitated cross-border cooperation through regional coordination centers, enabling seamless energy trading, and integration of renewables.

Moreover, long-term financing is crucial for making such electrification projects viable. Governments should engage multilateral development banks and private investors in formulating financing structures that reduce the risks for private sector groups to invest, while keeping the projects affordable for the end-users, especially in low-income areas. Blended finance combines public and private capital, and can be a key tool in financing large-scale renewable energy projects and ensuring long-term sustainability (Popovic et al., 2024). These financial mechanisms can help overcome one of the most substantial barriers to electrification: the high upfront costs associated with developing and deploying energy infrastructure.

Lastly, the decisions on electrification are to be made inclusively and transparently by governments, private companies, and local communities, considering the real needs from the perspective of all stakeholders, particularly the most marginalized (Sayegh, 2024). The orientation of governments and private companies should be toward local communities and local population representation within the planning and implementation phases to prevent and minimize conflict, build trust, and induce a sense of mutual responsibility, leading to more resilient and sustainable electrification outcomes. As outlined in Table 7.1, successful electrification hinges on collaborative efforts, including public–private partnerships, decentralized models, policy harmonization, innovative financing, and inclusive decision-making. Sustained cooperation across sectors is essential to ensure equity, resilience, and long-term sustainability in electrification initiatives.

Table 7.1 Summary of key takeaway and collaborative pathways

Key takeaway	*Why it matters*	*Success strategy*	*How*
Public–Private Partnerships (PPPs)	Governments can leverage private sector expertise and resources to achieve large-scale electrification goals while ensuring alignment with social and environmental objectives	Foster partnerships where private technical expertise is paired with public oversight for equitable outcomes	Develop frameworks for investment incentives, risk-sharing mechanisms, and regulatory support to attract private sector participation
Decentralized Community Energy Models	Centralized grids may not be viable in rural or remote areas. Localized, community-managed systems (e.g., solar mini-grids) promote ownership and sustainability	Strengthen cooperation among local governments, NGOs, and private entities to ensure sustainable implementation and management of projects	Conduct feasibility studies, provide capacity-building programs, and establish regulatory support for community energy governance
Policy Harmonization	Misaligned policies across sectors can hinder progress. Multisectoral platforms facilitate communication, prevent conflicts, and align goals	Develop communication platforms to synchronize regulations, corporate strategies, and community priorities	Organize policy dialogues, establish inter-ministerial task forces, and implement digital platforms for regulatory coordination
Innovative Financing Mechanisms	High upfront costs are a significant barrier to electrification projects, especially in low-income regions	Engage multilateral development banks and private investors in creating long-term, risk-reduced financing structures for energy projects	Utilize blended finance models, issue green bonds, and develop micro-financing options to lower financial barriers

(continued)

Table 7.1 (continued)

Key takeaway	*Why it matters*	*Success strategy*	*How*
Inclusive and Transparent Decision-Making	Including marginalized groups in planning ensures equity, builds trust, and minimizes conflicts during implementation	Governments and private entities should actively involve local populations in decision-making to foster mutual responsibility and resilience	Conduct stakeholder consultations, implement participatory planning approaches, and establish grievance redress mechanisms
Building Long-Term Collaboration Frameworks	Electrification requires sustained efforts across various sectors and regions	Establish frameworks for ongoing collaboration, including technical innovation, policy development, and community engagement	Develop joint action plans, create knowledge-sharing networks, and facilitate cross-sectoral partnerships

References

350.org. (2020). *350 Campaign Update: Divestment—350*. https://350.org/350-campaign-update-divestment/

Abiddin, N. Z., Ibrahim, I., & Aziz, S. A. A. (2022). Non-governmental organisations (NGOs) and their part towards sustainable community development. *Sustainability, 14*(8), 4386. https://doi.org/10.3390/SU14084386

Ahrens, C. D. (2017). Transition to very high share of renewables in Germany. *CSEE Journal of Power and Energy Systems*, *3*(1), 17–25. https://doi.org/10.17775/CSEEJPES.2017.0004

Apple. (2021). *People and Environment in Our Supply Chain*.

Batidzirai, B., Trotter, P. A., Brophy, A., Stritzke, S., Moyo, A., Twesigye, P., Puranasamriddhi, A., & Madhlopa, A. (2021). Towards people-private-public partnerships: An integrated community engagement model for capturing energy access needs. *Energy Research & Social Science, 74*, Article 101975. https://doi.org/10.1016/J.ERSS.2021.101975

Baucia, S. (2017). *Financing of renewables in developing countries: Financial models of microgrid development in the South East Asia*. https://www.politesi.polimi.it/retrieve/a81cb05d-06e1-616b-e053-1605fe0a889a/2018_12_Baucia.pdf

Betsill, M. M., & Corell, E. (2007). NGO Diplomacy: The influence of nongovernmental organizations in international environmental negotiations. *NGO Diplomacy*. https://doi.org/10.7551/MITPRESS/7512.001.0001

Bevacqua, E., Schleussner, C. F., & Zscheischler, J. (2025). A year above 1.5 °C signals that Earth is most probably within the 20-year period that will reach the Paris Agreement limit. *Nature Climate Change*, *15*(3), 262–265. https://doi.org/10.1038/s41558-025-02246-9

BP. (2024). *Net Zero Ambition: Progress Update*. https://www.bp.com/en/global/corporate/news-and-insights/press-releases/bernard-looney-announces-new-ambition-for-bp.html

California Public Utilities Commission. (2021). *Self-Generation Incentive Program*. https://www.cpuc.ca.gov/industries-and-topics/electrical-energy/demand-side-management/self-generation-incentive-program

Chenic, A. Ştefania, Cretu, A. I., Burlacu, A., Moroianu, N., Vîrjan, D., Huru, D., Stanef-Puica, M. R., & Enachescu, V. (2022). Logical analysis on the strategy for a sustainable transition of the world to green energy—2050. Smart cities and villages coupled to renewable energy sources with low carbon footprint. *Sustainability, 14*(14), 8622. https://doi.org/10.3390/SU14148622

Chmiel, Z., & Bhattacharyya, S. C. (2015). Analysis of off-grid electricity system at Isle of Eigg (Scotland): Lessons for developing countries. *Renewable Energy, 81*, 578–588. https://doi.org/10.1016/J.RENENE.2015.03.061

Danish Energy Agency. (2019). *Denmark's energy and climate outlook 2019*. http://www.ens.dk/

Department of Energy, S. A. (2019). *Integrated National Electrification Programme (INEP)*. https://www.dmre.gov.za/energy-resources/programmes-and-projects/integrated-national-electrification-programme/overview

Devine-Wright, P. (2009). Rethinking NIMBYism: The role of place attachment and place identity in explaining place-protective action. *Journal of Community & Applied Social Psychology, 19*(6), 426–441. https://doi.org/10.1002/CASP.1004

Drysdale, D., Mathiesen, B. V., & Paardekooper, S. (2019). Transitioning to a 100% renewable energy system in Denmark by 2050: Assessing the impact from expanding the building stock at the same time. *Energy Efficiency, 12*(1), 37–55. https://doi.org/10.1007/S12053-018-9649-1

Environmental Defense Fund. (2024). *Methane pollution from oil and gas—Environmental Defense Fund (EDF)*. https://www.edf.org/issue/methane

Fero, A., & Ferguson, P. (2025). *Evolution of the Self-Generation Incentive Program | Energy & Environmental Law Insights | Davis Wright Tremaine*. https://www.dwt.com/blogs/energy--environmental-law-blog/2020/06/california-sgip-wildfire-power-shutoffs

Figenbaum, E., & Kolbenstvedt, M. (2016). *Learning from Norwegian battery electric and plug-in hybrid vehicle users—Results from a survey of vehicle owners: Transportøkonomisk Institutt*. https://www.toi.no/publications/learning-from-norwegian-battery-electric-and-plug-in-hybrid-vehicle-users-results-from-a-survey-of-vehicle-owners-article33869-29.html

Forest Stewardship Council. (2021). *FSC Certification-Forest Stewardship Council (FSC)*. https://ca.fsc.org/ca-en/get-certified/certification

Giese, L. J. (2017). *The role of NGOs in International Climate Governance: A case study of Indian NGOs*. https://digitalcommons.csbsju.edu/honors_thesis

Global 100% Renewable Energy. (2024). *Making 100% the new normal*. https://global100re.org/

Goldemberg, J., Coelho, S. T., & Guardabassi, P. (2008). The sustainability of ethanol production from sugarcane. *Energy Policy, 36*(6), 2086–2097. https://doi.org/10.1016/J.ENPOL.2008.02.028

Goldthau, A. (2016). *Assessing Nord Stream 2: Regulation, geopolitics & energy security in the EU, Central Eastern Europe and the UK*. King's Russia Institute & Department of War Studies, King's College London. https://pure.royalholloway.ac.uk/en/publications/assessing-nord-stream-2-regulation-geopolitics-amp-energy-securit

Google Sustainability. (2018). *Google 2018 Environmental Report*.

Gründinger, W. (2017). The Renewable Energy Sources Act (EEG). *Drivers of Energy Transition, 257–419*,. https://doi.org/10.1007/978-3-658-17691-4_6

IEA. (2023). *World Energy Outlook 2023—Analysis: IEA*. https://www.iea.org/reports/world-energy-outlook-2023

IKEA. (2020). *IKEA Sustainability Report FY20*. https://www.inter.ikea.com/en/performance

Ilskog, E., Kjellström, B., Gullberg, M., Katyega, M., & Chambala, W. (2005). Electrification co-operatives bring new light to rural Tanzania. *Energy Policy, 33*(10), 1299–1307. https://doi.org/10.1016/J.ENPOL.2003.12.006

Jaspal, R., & Nerlich, B. (2013). Fracking in the UK press: Threat dynamics in an unfolding debate. *Public Understanding of Science, 23*(3), 348–363. https://doi.org/10.1177/0963662513498835

Koningstein, R., & Fork, D. (2014). Energy's creative destruction. *IEEE Spectrum, 51*(12), 30–35. https://doi.org/10.1109/MSPEC.2014.6964924

Lachapelle, E., & Paterson, M. (2013). Drivers of national climate policy. *Climate Policy, 13*(5), 547–571. https://doi.org/10.1080/14693062.2013.811333

Le, T.-H., & Lam, T. P. (2023). Dynamic principal component analysis for sustainable energy transition: A case study of Vietnam. https://doi.org/10.2139/SSRN.4680153

Meckling, J., Kelsey, N., Biber, E., & Zysman, J. (2015). Winning coalitions for climate policy. *Science, 349*, 1170–1171. https://doi.org/10.1126/science.aab1336

Moore, M.-L., Riddell, D., & Vocisano, D. (2015). Scaling out, scaling up, scaling deep: Strategies of non-profits in advancing systemic social innovation. *Journal of Corporate Citizenship, 2015*(58), 67–84. https://www.jstor.org/stable/jcorpciti.58.67

Newell, P., van Asselt, H., & Daley, F. (2022). Building a fossil fuel non-proliferation treaty: Key elements. *Earth System Governance, 14*, Article 100159. https://doi.org/10.1016/J.ESG.2022.100159

Nicolas, E. (2018). RSE, GRH et supply chain. La stratégie de réponse d'Apple à la demande sociale d'amélioration des conditions de travail chez les sous-traitants. *@GRH, 24(3), 55–76.* https://doi.org/10.3917/GRH.173.0055

Nieto, J., Carpintero, Ó., Lobejón, L. F., & Miguel, L. J. (2020). An ecological macroeconomics model: The energy transition in the EU. *Energy Policy, 145*, Article 111726. https://doi.org/10.1016/J.ENPOL.2020.111726

Ørsted. (2020). *Ørsted to become carbon neutral by 2025.* https://orsted.com/en/media/news/2020/01/373943464489368

ØrstedUK. (2018). *Ørsted Hornsea 4 Consultation Summary Report.*

Pinson, P., Mitridati, L. M.-J. M., Ordoudis, C., & Østergaard, J. (2025). Towards fully renewable energy systems-Experience and trends in Denmark. *CSEE Journal of Power and Energy Systems, 3*(1), 26–35. https://doi.org/10.17775/CSEEJPES.2017.0005

Popovic, T., Lygnerud, K., Denk, I., Fransson, N., & Unluturk, B. (2024). Blended finance as a catalyst for accelerating the European heat transition? *Smart Energy, 14*, Article 100136. https://doi.org/10.1016/J.SEGY.2024.100136

Presley, R. E. (2019). *Decolonizing Dissent: Mapping Indigenous Resistance onto Settler Colonial Land.*

Rainforest Action Network. (2021). *Banks and Climate Change: Campaigning for Responsible Financing.* RAN.

Rocky Mountain Institute (RMI). (2023). *Energy service agreements for deep efficiency and electrification retrofits of affordable multifamily housing in California.* https://rmi.org/

Sayegh, F. (2023). Proactive sustainable decision-making and climate change awareness: A Canadian study. *GeoJournal, 1–27*,. https://doi.org/10.1007/S10708-023-10976-6/TABLES/16

Sayegh, F. (2024). Leveraging emotional intelligence to foster proactive climate change adaptation: A study of engineering decision-making. *Journal of Environmental Management, 365*, Article 121669. https://doi.org/10.1016/J.JENVMAN.2024.121669

Smart Villages. (2025). *Energy policies for off-grid villages in Tanzania*. https://e4sv.org/energy-policies-off-grid-villages-tanzania/

Sovacool, B. K., & Hess, D. J. (2017). Ordering theories: Typologies and conceptual frameworks for sociotechnical change. *Social Studies of Science, 47*(5), 703–750. https://doi.org/10.1177/0306312717709363/ASSET/IMAGES/LARGE/10.1177_0306312717709363-FIG8.JPEG

Stand.earth. (2022). *Fossil fuel divestment movement hits $40 trillion in represented assets—Stand.earth*. https://stand.earth/press-releases/fossil-fuel-divestment-movement-hits-40-trillion-in-represented-assets/

Stock, R. (2022). Triggering resistance: Contesting the injustices of solar park development in India. *Energy Research & Social Science, 86*, Article 102464. https://doi.org/10.1016/J.ERSS.2021.102464

Treaty. (2022). *The Fossil Fuel Non-Proliferation Treaty Initiative*. https://fossilfueltreaty.org/

Ulsrud, K., Winther, T., Palit, D., & Rohracher, H. (2015). Village-level solar power in Africa: Accelerating access to electricity services through a socio-technical design in Kenya. *Energy Research & Social Science, 5*, 34–44. https://doi.org/10.1016/J.ERSS.2014.12.009

UNFCCC. (2015). *The Paris Agreement*. https://unfccc.int/process-and-meetings/the-paris-agreement

Unilever. (2024). *Unilever Climate Transition Action Plan updated 2024*.

Union of Concerned Scientists. (2020). *Science for a healthy planet and safer world*. UCS. https://www.ucs.org/

Utoh, I. O., Ekpotu, W., & Obialor, M. (2024). Assessing the viability and impact of off grid systems for sustainable electrification of rural communities in Sub-Saharan Africa. *Society of Petroleum Engineers - SPE Nigeria Annual International Conference and Exhibition, NAIC 2024*. https://doi.org/10.2118/221709-MS

Van Der Schoor, T., & Scholtens, B. (2015). Power to the people: Local community initiatives and the transition to sustainable energy. *Renewable and Sustainable Energy Reviews, 43*, 666–675. https://doi.org/10.1016/J.RSER.2014.10.089

Vrikki, P. (2024). Measuring up? The illusion of sustainability and the limits of big tech self-regulation. *Sustainability, 16*(23), 10197. https://doi.org/10.3390/SU162310197

Walker, G. (2008). What are the barriers and incentives for community-owned means of energy production and use? *Energy Policy, 36*(12), 4401–4405. https://doi.org/10.1016/J.ENPOL.2008.09.032

Wei, M., McMillan, C. A., & de la Rue du Can, S. (2019). Electrification of industry: Potential, challenges and outlook. *Current Sustainable/Renewable Energy Reports, 6*(4), 140–148. https://doi.org/10.1007/S40518-019-00136-1/METRICS

CHAPTER 8

Stakeholder Engagement and Governance in the Transition to Renewable Energy Systems

Kofi Nyarko and Tania Urmee

Introduction

Power systems are large technical systems that generate electricity for residential, industrial, commercial, and agricultural consumption. Most of these systems consist of individual power plants interconnected to form a centralized grid. These power plants are often distant from the load centers and require transmission lines to transfer power to the consumers. The emergence of distributed generated schemes at the turn of the twenty-first century paved the way for a generation of electricity close to the load in a decentralized setting. Electricity generated and consumed at the local, decentralized level is fundamentally different from the standard

K. Nyarko · T. Urmee (✉)
School of Engineering and Energy, Murdoch University, Perth, Australia
e-mail: t.urmee@murdoch.edu.au

K. Nyarko
e-mail: kofi.nyarko@murdoch.edu.au

T. Walker et al. (eds.), *Wiring the Future*, Palgrave Studies in Emerging Risk Management and Sustainable Finance,
https://doi.org/10.1007/978-3-032-14717-2_8

grid. It is usually small-scale and environmentally and socially benign, and prioritizes community benefits over profit-making (African Development Bank, 2016; Alvial-Palavicino et al., 2011). In conventional and decentralized power grids, electricity can be produced from both renewable and fossil-based energy sources.

Despite the growth of the renewable energy (RE) sector, electricity generated from fossil fuels (oil, natural gas, and coal) continues to dominate the global share. As of 2023, 60.65% of global electricity production used fossil fuels, while 30.24% and 9.11% came from renewable and nuclear sources, respectively (Ritchie & Rosado, 2020). These statistics indicate that significant action is required to decarbonize the electricity sector to meet net-zero ambitions. Decarbonization starts with progressively substituting fossil fuels with cleaner technologies by retrofitting existing plants to integrate renewable energy sources, thereby reducing emission levels. According to the International Renewable Energy Agency (IRENA), transitioning to local renewable energy sources for power generation and end-use services by 2050 could lower fossil fuel use by 90% in the power sector and by 65% in end-use sectors (IRENA, 2022). Replacing or supplementing conventional energy sources with RE will decrease the amount of carbon dioxide and other pollutants released into the atmosphere, directly contributing to a more sustainable environment (Haldar et al., 2023).

Moreover, modern RE technologies promote more inclusive economic and social development in small-scale communities than fossil-fuel-based options (IRENA, 2015). However, renewable energy electrification poses various challenges to the stability of both centralized and decentralized grids. Its intermittency challenges energy storage, demand response, and frequency control. Therefore, unwavering commitment and collaboration among multiple stakeholders is essential to achieve resilient, cleaner, and more sustainable RE-based power systems.

These stakeholders play crucial roles in electricity planning, financing, provision, regulation, and compliance. Governments play a starring role by providing an enabling environment for all other components to flourish. This is achieved through government enacting relevant legislation and regulations that state regulatory agencies then enforce. These agencies work closely with businesses, investors, utilities, and non-governmental organizations (NGOs). International development agencies also play a critical role in ensuring that developing economies have access to financing and the technologies required to achieve net-zero emissions on time. Technical factors, social and cultural roles, and the insights of beneficiaries influence distributed generation for rural electrification. Therefore, community involvement is crucial for the success of sustainability efforts, as public attitudes towards renewable energy electrification projects (REEPS) are often complex and multi-dimensional. Apart from the key stakeholders, other players such as suppliers, trade unions, and employees provide ancillary services to ensure smooth operation of systems. The role played by stakeholders during the implementation of REEPs is covered extensively by other researchers. However, this study advances our understanding of this role by investigating how stakeholders can collaborate to achieve sustainable development through REEPs. In doing so, the study will answer the following research questions:

- What are the roles of key stakeholders during the implementation of REEPs?
- How can the key stakeholders collaborate to overcome the challenges of REEPs to ensure their sustainability?

The outcome of the study is relevant to all stakeholders involved in activities that promote RE and universal electricity access. The remainder of the chapter is organized as follows. Section 2 delves into the methods the study uses to achieve its conclusions, Section 3 describes the key stakeholders' roles, and Section 4 examines the results of a stakeholder consultation in Ghana and a comparative analysis with Sierra Leone. Finally, Section 5 describes the key insights and action plan required to promote the sustainable development of future RE projects.

Methods

We employed a qualitative approach to achieve the study's objectives and draw our conclusions. We answered the first research question by conducting a comprehensive content analysis of relevant literature. We began by using search terms related to RE development to retrieve articles published in peer-reviewed journals and indexed with Scopus and Google Scholar. Additionally, we included reports published by international agencies such as IRENA, the International Energy Agency (IEA), and the World Bank) in the review. We then used the NVIVO software to analyze the documents to identify pertinent roles played by the key stakeholders in the sustainable development of REEPs.

To address the second research question, we interviewed seventeen expert stakeholders of five pilot mini-grids that served five island communities in Ghana. We asked them to share their experiences during the implementation of the mini grid. We randomly sampled these stakeholders, which included government agencies, academic, public utility, and RE consultants who were directly involved in the project, for the interview. The interviews focused on identifying the challenges of mini-grid implementation and how stakeholder collaborations could help mitigate those challenges. We sought the consent of the experts to record the interviews and then transcribed the recordings and returned them to the participants for their endorsement. Finally, we analyzed the transcripts using NVIVO and Microsoft Excel.

Role of Stakeholders in Addressing Pressing Interventions Required to Sustain Renewable Energy Electrification Projects

The various stakeholders that play different roles in the electrification process can be classified into two main categories, shown in Fig. 8.1 (Edomah et al., 2021). Decision makers put primary measures in place to ensure that a REEP is sustainable. These interventions are related to national energy frameworks and are spearheaded by public institutions such as the Ministry of Energy, public utilities, and regulatory agencies. Non-decision-making stakeholders engage in non-energy-specific responsibilities such as land rights, environmental protection, banking, capacity building, data collection, and logistics.

Decision-Making Stakeholders Mapping

Government

As aforementioned, governments play a crucial role in REEPs by creating a conducive environment that encourages clean energy adoption. A combination of policies and regulations, institutional frameworks, delivery models and financing options, technological advancements, capacity-building initiatives, and cross-sector linkages creates a level playing field that encourages investment in renewable energy (IRENA, 2018).

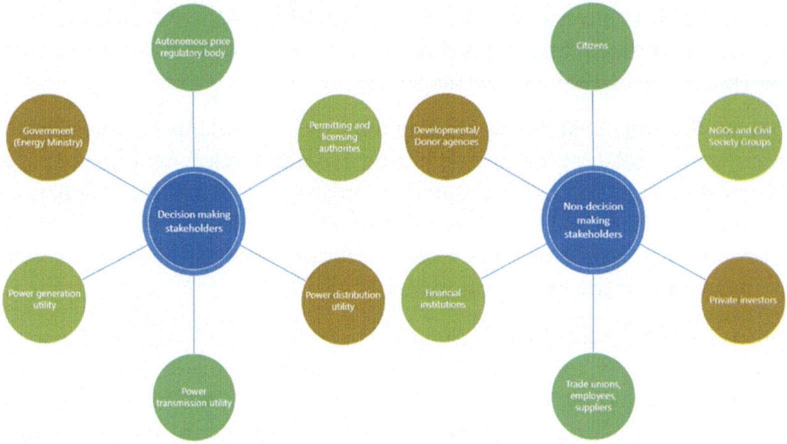

Fig. 8.1 Categories of stakeholders of REEP. *Source* Edomah et al. (2021, pp. 2–6)

Policies and Regulations: The government establishes national RE policies and targets which should be adaptable to changing technological and market conditions. Effective policies, including incentives, subsidies, and carbon pricing, can encourage investment by making RE projects more economically attractive. As noted by Ali et al. (2017), more than one policy is required to transform the clean energy market. Such a transformation often involves combining policies and amendments to existing national policies to promote, incentivize, utilize, research, and develop sustainable REEPs. Governments usually employ fiscal incentives to incentivize REEP developers. Tax reduction (such as custom and import-duty exemptions) is the most prevalent fiscal incentive, especially in developing countries. This option usually does not require budgetary allocation and involves less bureaucracy and regulatory supervision (IRENA, 2015). Though such fiscal policies effectively support standalone power systems, larger energy systems like mini-grids require more regulatory measures. Key factors include access to concession schemes, regulations that enable cost recovery, and measures to mitigate risks related to the arrival of the main grid (IRENA, 2015).

Regulations are crucial for aligning renewable energy projects with national and international goals. Effective regulations provide stability and predictability for investors, reducing financial risks associated with renewable energy projects. Streamlined regulations also reduce barriers to entry, allowing more investors to participate.

Create Strong Institutional Capacity: Governments can establish specialized regulatory bodies dedicated to RE development to manage project approvals, establish industry standards, and enforce compliance. These agencies create a structured framework ensuring projects adhere to environmental, safety, and operational requirements, promoting a stable environment for the expansion of RE.

Innovative Financing and Delivery Approaches: Novel delivery strategies, including collaborations between public and private entities, along with diverse funding options such as eco-friendly bonds and financial incentives, can alleviate monetary constraints and improve how public funds are managed. When governments streamline processes such as permits and licenses, it reduces costs and delivery times and attracts a range of investors, from individuals to institutions, enhancing the accessibility and scalability of RE initiatives.

Advancements in Technology: Government investment into research and development of RE technologies, particularly in areas like energy storage and grid integration, can reduce expenses and boost efficiency. This increased performance makes RE more economically viable than conventional energy sources.

Skill Development: Educational initiatives and knowledge exchange through the development of skills-centered curriculum by the state can equip local stakeholders with the necessary skills, enhancing expertise and workforce preparedness. This ensures the sustainable implementation of projects and promotes the long-term adoption of RE.

Public Utilities

As major stakeholders, public utilities are central to electricity production and use. They invest directly in RE by building and operating generation, transmission, and distribution infrastructures, such as solar farms, wind turbines, and energy storage facilities. Public utilities are responsible for maintaining grid stability, which is crucial as RE sources like solar and wind are intermittent. However, maintaining a stable grid requires sustainable financial and technical solutions for operations, maintenance, and management of REEP (Alliance for Rural Electrification, 2014). As a system operator, public utilities utilize advanced forecasting techniques to balance generation and consumption. The techniques used take into account business needs, population growth, and the resulting increases in load demand due to lifestyle changes (Azimoh et al., 2017). In most countries, power generation, transmission, and distribution are segregated and handled by public utilities, but it is important that they maintain cooperation among themselves to ensure system reliability. Establishing official targets and clear demarcation of on- and off-grid concessions provides clear guidance for the utilities, helping to align investment efforts and improve resource allocation (IRENA, 2018).

Public utilities typically partner with independent power producers (IPPs) through power purchase agreements (PPAs) to secure a reliable power supply. These agreements outline the terms under which the utility purchases electricity from an IPP, helping utilities meet energy demand while enabling IPPs to finance and operate RE projects. To enhance the bankability of PPAs, it is advisable for public utilities to explore credit enhancement tools from development finance institutions (DFIs). These instruments boost the creditworthiness or debt profile of a REEP, since when a reputable third party guarantees the fulfillment of the contractual obligations of a public utility under a PPA, lenders gain confidence in the utility's ability to meet its loan repayment commitments (African Development Bank, 2016).

Public utilities actively participate in policy discussions and advocate for favorable policies and incentives. In developing countries, most utilities usually require some form of funding from donor agencies to supply electricity to consumers efficiently. Funding is supported through capital subsidies, performance-based subsidies, and cross-subsidies. However, if they are not properly executed, subsidies may become a barrier rather than a driver to sustainable REEPs. For instance, in many Sub-Saharan African countries, social justice principles dictate that lifeline customers in decentralized areas should not pay higher tariffs than central grid users. Regulatory authorities adopt cross-subsidy approach to cater for the gap in tariff. However, the public utilities do not receive the subsidies on schedule impacting revenue mobilization. To operate sustainably, public utilities need revenue to exceed operation and maintenance costs. Although flat-rate tariffs with different tiers are commonly used for small-scale renewable energy projects, this approach becomes ineffective when consumption is deficient.

Regulatory Authorities

Government-established state regulators play a crucial role in promoting the adoption of RE. Energy regulators enforce technical regulations and standards designed to ensure that system operators adhere to best practices to enhance the safety and reliability of REEPs. Moreover, they are responsible for issuing licenses and permits to energy providers, including renewable energy developers, IPPs, and public utilities. By regulating entry into the market, regulators ensure that all energy projects meet technical, environmental, and operational standards.

Regulatory bodies overseeing electricity prices also play a crucial role in establishing and monitoring the rates that utility companies can impose on consumers. Regulators' primary objective is to strike a balance between ensuring fair pricing for customers and maintaining the financial viability of energy suppliers. These regulators examine power generation, transmission, and distribution expenses to guarantee that prices accurately reflect actual costs while safeguarding consumers from sudden price increases and market instability. By promoting transparency, fostering competition, and ensuring affordability, these regulatory entities contribute to creating a stable, equitable, and sustainable electricity market that serves the interests of consumers and energy providers. However, governing bodies often pressure energy price regulators to avoid implementing upward tariff adjustments, as such changes could potentially sway public opinion against the government.

Effective tariff regulations largely influence the success and long-term sustainability of mini-grids. In many instances, rural communities cannot afford tariffs that cover the total cost of electrification. As a result, regulators face the challenge of balancing two critical goals: maintaining affordable, high-quality service for rural consumers while ensuring financial viability and cost recovery for private mini-grid providers.

Non-Decision-Making Stakeholders

Donor/Developmental Agencies

International donor and development agencies and regional actors play an essential role in funding REEPs, particularly in underserved areas in developing and emerging economies where access to capital may be limited. Developed nations such as the United States of America, Germany, France, and Sweden partner with developmental banks like the World Bank and Africa Development Bank to initiate and scale RE projects

Table 8.1 Funds available for RE development

Name of Fund	*Description*
The Africa renewable energy fund	The goal is to improve the energy supply in sub-Saharan Africa, excluding South Africa, by investing in renewable energy projects that deploy mature technologies
Beyond the grid Fund for Africa	An initiative of the Swedish International Development Cooperation Agency that provides funds for companies into off-grid solutions and offers technical assistance and capacity building for local energy authorities
Clean technology fund	Part of the climate investment fund aimed to facilitate the scaleup of low-carbon technologies
Green climate fund	Established within the United Nations framework convention on climate change framework to help developing countries meet their nationally determined contributions. Financial support includes equity, loans, readiness grants, and guarantees
Sustainable energy fund	The African development Bank manages this fund to unlock private finance for small-to-medium-scale projects in RE and energy efficiency in Africa

Source IRENA and African Development Bank (2022, p. 112–113)

that might otherwise be financially unfeasible through grants, low-interest loans, and concessional financing. This support often reduces the financial risk for local stakeholders and attracts further investment. Examples of funds extended to RE development are shown in Table 8.1 (IRENA & Africa Development Bank, 2022).

Most development assistance is provided as financial support, such as loans or grants. However, technical support for feasibility studies and environmental and social impact assessments, which is often enabled by development funds, is essential for implementing renewable energy projects.

Financial Institutions and Investors
Access to finance at the various stages of REEP development is critical, and the role played by financial institutions is valuable. Worldwide, most RE financing has come from the private sector, with public funding making up only 14% of direct investment in RE projects, primarily

through DFIs (IRENA & Africa Development Bank, 2022). However, public financing in Africa has a more significant role because most projects need help to attract private capital from financial institutions due to political, legal, and economic risks.

According to IRENA (2015), loan guarantees and concessional financing can de-risk investments and leverage private capital in Africa. However, the unattractive local currency loans with high interest rates offered by commercial banks make it unpopular. For instance, interest rates for debt financing in Ghana remain over 25% per annum, leading to low uptake by private investors (African Development Bank, 2015). Once renewable energy technologies (RETs) have demonstrated success, the investment risks associated with RE projects decrease. Consequently, in countries with well-established RETs and research and development facilities more financial institutions are inclined to offer favorable loans for implementing RE projects (Abdmouleh et al., 2015).

To have a profound and lasting sustainability, financing programs for RE systems must address three critical levels described in Table 8.2: end-user finance, business finance, and project finance (Alliance for Rural Electrification, 2014).

Investors

Significant investment must be mobilized and efficiently deployed by implementing parties to speed up the global energy transition and foster growth in RE markets. Despite a positive outlook in global RE investments, as depicted in Fig. 8.2, more is needed to meet the benchmarks for sustainable development, climate objectives, and the broader benefits of a clean energy shift (IRENA & CPI, 2023).

This slower progress is due to lingering real and perceived risks that prevent many investors from confidently committing their capital to the market. These risks include a high administrative burden and inflexibility with project implementation. Commercial investors often need more risk tolerance, making it difficult to secure funding for innovative or unconventional projects (African Development Bank, 2016). The small ticket size of transactions further complicates matters and makes investing in RE projects unattractive to more prominent investors. The remaining risks are depicted in Fig. 8.3 (IRENA & African Development Bank, 2022).

Over the years, governments with support from developmental partners invest in REEPS to extend electricity access to serve the domestic needs of residents in remote communities. Nonetheless, a critical strategy

Table 8.2 Critical financing required to ensure the sustainability of REEP

Type of finance	*Description*	*Providers*
End-user finance	This financing level allows consumers to afford the cost of connecting to mini-grids or standalone systems and acquiring electric appliances. It ensures that energy access is within reach for rural households and small farmers, directly improving their quality of life and economic productivity.	Rural banking networks
Business finance	This level supports local small enterprises or entrepreneurs responsible for delivering and operating energy systems, including mini-grids, solar home systems, or appliance leasing. By providing financing tailored to these businesses, programs can ensure local providers have the resources to scale up, maintain equipment, and offer reliable services over the long term.	Microfinance organizations, agricultural cooperatives.
Project finance	Project finance provides the necessary capital for the development and installation of energy infrastructure, including large community-scale systems and mini-grids. Access to project finance enables sustainable growth of renewable energy systems and attracts private investment, which is crucial for long-term expansion.	Utilities

Source Alliance for Rural Electrification (2014, p. 33)

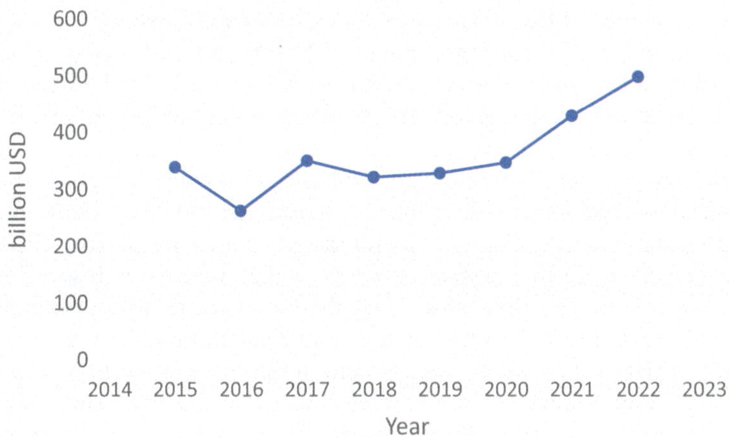

Fig. 8.2 Global trend of RE investment. *Source* IRENA & CPI (2023, p. 11)

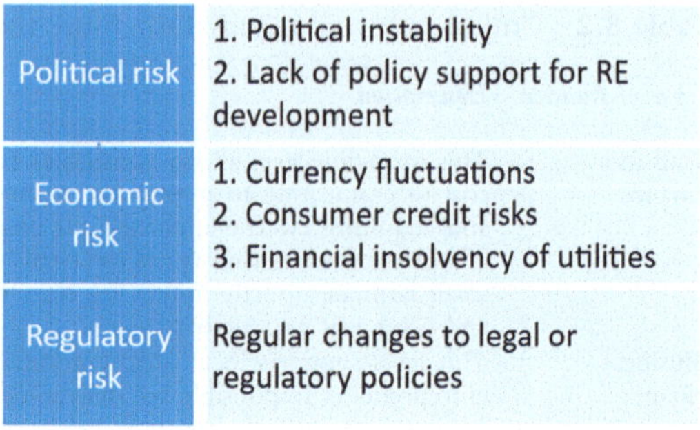

Fig. 8.3 Risks of RE investments. *Source* IRENA and African Development Bank (2022, p. 127)

that can reduce the capital cost of REEPs and make them more economically attractive to investors is to increase the market size. This goal can be realized by developing the system around existing or upcoming commercial ventures, which helps to build critical mass, enhance profit potential, and foster greater local participation (Alliance for Rural Electrification, 2014). The parallel development of productive services within the project boosts local industries—such as carpentry, agricultural irrigation, and telecommunications—and creates a multiplier effect, fostering economic growth and community development.

Non-Governmental Organizations/Civil Society Organizations
While the structure, scope, and range of NGOs and civil society organizations (CSOs) can be diverse, their role as intermediaries between the government and citizens is well-established and summarized in Table 8.3 (De Nigris & Giuliano, 2023). These entities advocate for communities, enhance government accountability, and safeguard citizens' rights. These organizations also play a crucial role in advocating for policy modifications and shaping policy development by interacting with government officials, policymakers, and other key stakeholders to align policies with social and environmental justice principles. They also lobby for policies prioritizing vulnerable groups, such as low-income and rural communities, as well as provide a platform for public engagement in the energy transition process, allowing communities to express concerns, offer feedback, and participate in decision-making. Furthermore, these organizations contribute technical expertise to inform policy development and implementation.

Table 8.3 Role of NGOs/CSOs in RE development

Collaborate with the private energy sector on specific projects to implement energy savings, efficiency measures, and renewable energy schemes
Directly promote, commercialize, and manage clean and renewable energy facilities
Provide education and training on clean energy transition, energy efficiency, and conservation practices
Conduct research and development for innovations in renewable energy and energy efficiency
Exert pressure, offer consultancy, and actively participate in energy and environmental law development processes
Raise public awareness and sensitize people to the energy transition through communication efforts
Collect data and monitor and assess energy efficiency, renewable energy usage, and related quality/efficiency metrics

Source De Nigris & Giuliano (2023, pp. 17–20)

Suppliers and Community Members

External parties such as suppliers and community members are essential in a sustainable REEP. The sustainability of an RE program depends mainly on establishing local businesses equipped to supply products, services, and spare parts. Hence, to successfully meet the needs of rural households, rural enterprises must be capable of delivering RE solutions. However, these entrepreneurs face significant challenges: high operational costs arise from the need to travel long distances and are often compounded by limited transportation infrastructure. Additionally, suppliers may need more access to financial resources, face inconsistent demand, and contend with barriers to accessing advanced training or technical support. Overcoming these challenges will require targeted support, strategic partnerships, and policies prioritizing local enterprise development and addressing rural logistical and financial constraints (Alliance for Rural Electrification, 2014).

Citizens play an essential role in RE electrification initiatives as the end users. Through increased community generation and self-consumption, citizens are becoming energy producers. People are installing solar panels in their houses and investing in energy-efficient appliances. At the community level, individuals are coming together to initiate community-owned RE projects across technologies such as solar and wind power, biogas digesters, and biomass power and heat (IRENA, 2018). Community-based energy communities can help address numerous challenges power systems face, including losses, grid congestion, and the need to accommodate growing peak demand.

Stakeholder Consultation for Five RE-Based Mini-Grids in Island Communities in Ghana

Context and Roles of the Key Stakeholders

Ghana is a West African country, bounded to the north by Burkina Faso, to the south by the Gulf of Guinea, to the east by Togo, and to the west by Ivory Coast. The population is approximately 30.8 million and had a US$ 2238.2 GDP per capita in 2022 (The World Bank, 2024b). As of the end of 2022, 72% of the rural population had access to electricity compared to 30.7% across the Sub-Saharan Africa region (The World Bank, 2024c). The Government of Ghana (GoG) determined that due to technical and economic reasons, the remaining unserved rural locations would be best served with RE mini-grids. GoG initiated three main interventions to tackle rural electrification, which are the National Electrification Scheme, the Self-Help Electrification Project, and, recently, the Ghana Energy Development and Access Project (GEDAP).

The implementation of the five-pilot RE-based mini-grid projects was a component under the multi-donor GEDAP led by the World Bank. The project provided electricity to approximately 3500 residents in the communities of Kudorkope, Aglakope, Atigagome, Wayokope, and Pediatorkope, and had an aggregated installed capacity of 1.7 MW. The key stakeholders involved in the project were the Ministry of Energy (MoEn), Energy Commission (EC), Public Utility and Regulatory Commission (PURC), Trama TechnoAmbiental (TTA), Volta River Authority (VRA), Kumasi Institute of Technology, Energy and Environment (KITE), and the World Bank. A steering committee oversaw the smooth implementation of the project. Table 8.4 gives details of the roles played by each stakeholder, including collaboration among the stakeholders.

Interview Session

As mentioned in Section 2, we organized a stakeholder consultation to identify key issues in mini-grid development in Ghana and discussed how collaboration among the key stakeholders can solve the various issues that arise in the implementation of REEPs. We use Fig. 8.4 to illustrate a word

Table 8.4 Roles played by key stakeholders during the implementation of five mini-grids in Pediatorkope in southern Ghana

Activity	*Details of Roles Played by Each Stakeholder*
Site selection	The site selection involved a collaborative effort among KITE, EPA, TTA, and the MoEn, in consultation with VRA. With the MoEn coordinating, KITE conducted a pre-feasibility study to identify the suitability of the earmarked sites for mini-grid development. TTA used the findings to conduct a more detailed user demand and willingness-to-pay analysis to accurately design the systems. The key criteria used to select a site was that the community should have a high population density with the population not less than 500 residents.
Design and construction	TTA designed and constructed the system in collaboration with VRA, while the World Bank provided technical assistance. The system in Pediatorkope is a hybrid PV/wind/generator and the other four are hybrid PV/generator systems.
Financing	The project was financed using a grant from the global environmental facility and a concessional loan facility. Other financiers included the African development Bank (AfDB), the Africa catalytic growth fund, the state secretariat for economic affairs (SECO, Switzerland), the GoG, and the global partnership on output-based aid (GPOBA).
Operation and management (O & M)	Representatives of the communities were integrated into all activities, including social studies, construction works, low voltage distribution grids installation, end-user connections, and system management, as an energy committee. Community dwellers have been trained and engaged in the O & M phase, creating jobs in customer relations, service fee collection, service card recording, troubleshooting, and energy efficiency assessment services. The systems were installed and commissioned by TTA on July 1, 2016. They assumed the O & M of the systems for an additional two years before handing it over to VRA in 2018.

(continued)

Table 8.4 (continued)

Activity	*Details of Roles Played by Each Stakeholder*
Tariff design and collection method	The energy daily allowance (EDA) concept is used to ensure load management. On a pre-paid monthly basis, the beneficiaries pay an amount based on their pre-contracted tariff class, which is determined by their monthly expected energy and power needs. Smart electric meters are used to disconnect the user when the EDA was exhausted.

Source Trama TecnoAmbiental (2018, p. 4–5)

frequency cluster of the most frequently mentioned words by the interviewees. The figure shows that "government," "renewable," "energy," "private," and "power" are commonly occurring words.

Fig. 8.4 Cluster of frequently occurring words during the interview sessions. *Source* Authors (2025)

Challenges Identified by the Experts

Further examination of the interview transcripts uncovered several challenges to sustainable mini-grid development. Policy and regulatory matters emerged as the top concern, mentioned by eight of the seventeen experts, indicating that the experts perceive inadequate or poorly implemented policies and laws as the primary obstacle to RE development in Ghana, particularly for mini-grid systems. These sentiments were summarized by an academic expert:

> As at now, there aren't any critical law[s] that affect negatively off-grid systems in Ghana. However, the implementation of the existing ones is the challenge. Some of the laws have been revised but [are] yet to [be] implemented.

The interviewees equally ranked lack of private sector engagement, tariff and subsidy issues, and the government's top-down approach to mini-grid development as the second most significant challenge. These frequent mentions suggest strong apprehensions about the private sector's role and difficulties in RE projects, including profitability and government interference. An expert who took part in the pre-feasibility study commented on the need to involve private sector, saying:

> During the rural electrification planning, we went around the whole country, and we took some data including GPS location et cetera. And some of them are good for mini grids. You ask yourself can the government really do all this, no.

The experts also emphasized that current pricing mechanisms are unsustainable for mini-grid projects, especially in rural areas. One expert said:

> These off-grid projects are inherently expensive and for an investment to be worthwhile, there is the need to let these private developers charge cost reflective tariffs.

While rural electrification rates have improved, mini-grid projects have not seen a significant uptake by public utilities and investors. Heavy reliance on donor funding has restricted private sector participation, as heavily subsidized tariffs often fail to reflect true operational costs. According to one engineering expert:

> Private people have difficulty in sustaining the business because capital investment and operational cost is also very, very high. In Ghana, the amount that they pay as revenue, that is, bills that they pay to the facility per month, when put together is not even able to even accommodate the maintenance people who go there to maintain the plant. So, it is expensive to put the facility together and it's also very expensive to run it. Meanwhile, the revenue that comes out of it is inadequate. So, it is extremely difficult for private people to go into these areas.

Therefore, there is a call for the government to adjust tariffs and broaden subsidies to cover more RE components, ensuring economic viability. The experts frequently criticized the government-led approach to RE development, with many participants advocating for hybrid models or increased private sector involvement to enhance efficiency and long-term sustainability.

Six experts discussed financial aid and coordination challenges, emphasizing difficulties in accessing funds and the need for improved institutional coordination. Developers face bureaucratic obstacles, and insufficient financial support hinders project initiation and sustainability. The expert stakeholders identified scarcity of commercial activities in rural communities as a critical threat to off-grid RE project sustainability, underscoring the importance of productive electricity use in rural areas. Without fostering electricity-dependent local businesses and industries, such projects may not be viable. The experts also noted the impact of project abandonment following government changes, resulting in numerous unfinished RE projects. They strongly advocated for continuity policies and binding contracts to address this issue.

The lack of strong partnerships has meant limited R&D in RE technologies, hampering innovation and adaptation to local conditions. One expert opined that:

> There's not so much collaboration between industries and the academia. First, there are not so many industries. The industries are a bit detached. What they are doing is sometimes quite different from what the universities are doing. You train people and they go out there and they don't know what is happening in industry.

The stakeholder interviews revealed several factors that impact the successful microgrid implementation, depicted in Fig. 8.5. This figure

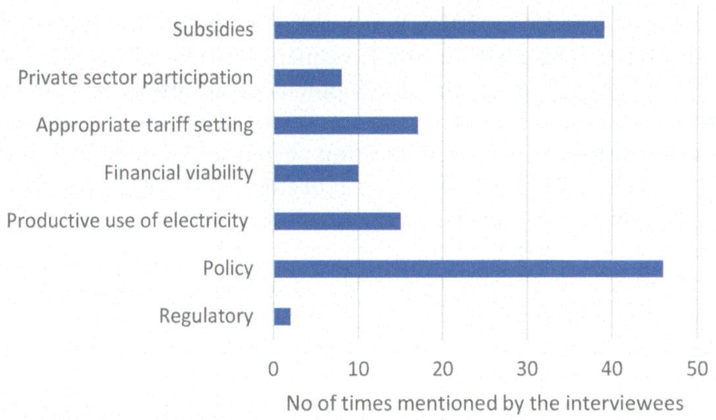

Fig. 8.5 Stakeholders' views on factors that impact successful microgrid implementation. *Source* Authors (2025)

shows that the expert vastly considers effective policies and subsidies as crucial factors for advancing RE development in Ghana.

Comparative Analysis of Stakeholder Roles in RE-Based Mini-Grid Development in Sierra Leone

Sierra Leone is one of the least electrified countries in the world. In 2023, more than 60% of the country's 8.79 million residents lived in rural communities with only 5% having access to electricity (The World Bank, 2024a). Unlike the GoG, the Government of Sierra Leone (GoSL) implemented legislative and regulatory reforms to encourage private sector involvement in the mini-grid energy sector, aiming to address the electricity access crisis. The National Energy Plan and the Electricity and Water Regulation Commission Act, enacted in 2009 and 2011, respectively, initiated these reforms (Holzigel & Bravard, 2020). The Public-Private Partnership (PPP) Act of 2014 established prerequisites for private sector participation in crucial areas, including asset ownership, PPP agreement content, and governing laws.

Mini-grids were first mentioned in Sierra Leone's regulatory framework in the 2015 National Renewable Energy Action Plan (NREAP), which outlined the rural electrification strategy and mini-grids' role in the energy landscape. This was later reinforced through the 2015

Action Agenda and the 2016 Renewable Energy Policy (Holzigel & Bravard, 2020). The 2030 Sector Roadmap, introduced in 2017, identified challenges hindering the development of an attractive market for the private sector and provided key recommendations to overcome them, such as establishing a rural electrification agency and a mini-grid funding facility, as well as implementing cost-reflective tariff structures (Irungu, 2024). In contrast to Ghana, Sierra Leone adopted a mini-grid regulation framework in 2019, aimed at simplifying RE project requirements, reducing associated development costs, and setting principles for tariff determination methodology.

A key stakeholder involved in mini-grid development in Sierra Leone is the Ministry of Energy (MoE), which is responsible for Comparative Analysis of Roles Played by Stakeholders in RE-Based Mini-Grid Development in Sierra Leone and the development of policies and programs such as the Chiefdom Mini-grids Electrification Project, Barefoot Solar Electrification Project, and the Seven Districts Headquarter Towns Electrification Project (Conteh et al., 2021). The MoE's Renewable Energy Unit oversees RE and rural electrification efforts. The MoE is supported by crucial public institutions like the Sierra Leone Electricity and Water Regulatory Commission (SLEWRC) and the PPP Unit. SLEWRC regulates the power sector by issuing generation licenses, approving and setting tariffs, and formulating and monitoring quality and compliance with regulatory frameworks (Holzigel & Bravard, 2020).

The mini-grid tariff framework in Sierra Leone, defined by SLEWRC, includes provisions for cost-reflective retail tariffs, market-entry procedures, contractual agreements, main grid arrival scenarios, technical and service standards, and guidelines and licenses for mini-grid projects based on capacity and setting. To ensure tariff affordability, reflect service costs accurately, and assess the market, Sierra Leone employs a split asset delivery model where household connections are financed and owned by the MoE, while the private sector finances and owns generation assets (Holzigel & Bravard, 2020; Irungu, 2024). The PPP division is tasked with boosting private sector involvement and offering transactional assistance to the MoE. It receives support from the National Public Procurement Authority (NPPA), which oversees public bidding processes, including those for mini-grids. Three private mini-grid operators—Off-Grid Power (OGP/PowerGen), Power Leone (Energicity), and Winch Energy—were chosen by NPPA through a competitive selection process to engage in a PPP with the GoSL.

While mini-grid development in Sierra Leone has been primarily grant-funded, private companies manage the operations. Key international organizations supporting mini-grid electrification include the UK's Foreign Commonwealth and Development Office and Department for International Development, along with the United Nations Office for Project Services. These entities backed the major Rural Renewable Energy Project (RREP), which aimed to create a conducive environment for a private sector-driven mini-grid market. RREP's goal is to provide approximately 5 MW of RE to rural areas through about 94 solar mini-grid installations under PPPs (Liu & Bah, 2021). Other developmental agencies that support the country's RE development include the Ecowas Centre for Renewable Energy and Energy Efficiency, Sustainable Energy for All, the U.S. Agency for International Development, the United Nations Development Project, and the German Agency for International Cooperation.

Table 8.5 highlights the main features of mini-grid development in Ghana and Sierra Leone. The comparison reveals that while the GoG solely drives mini-grid development in Ghana, the GoSL permits private sector participation. The analysis also shows that mini-grid regulations and cost-reflective tariffs facilitate smooth PPP implementation. Notably, both nations heavily rely on donor support for project financing.

Table 8.5 Key characteristics of mini-grid development in Ghana and Sierra Leone

Item	*Ghana*	*Sierra Leone*
Business model	Government-driven approach	Public-private partnership
Tariff	Uniform National Tariff	Split asset delivery model
Rural electrification agency	No	No
Renewable energy fund	Yes	No
Mini grid regulation	No	Yes
Donor funding	Yes	Yes

Source Authors (2025)

Key Insights and the Way Forward

Stakeholder collaboration is pivotal for the successful deployment of renewable energy-based mini-grids in rural communities. As evidenced in the cases of Ghana and Sierra Leone, a synergistic approach involving government entities, private sector actors, international donors, regulatory bodies, and community members can overcome financial, technical, and operational challenges.

Governments provide the regulatory framework and policies to create an enabling environment for investment and sustainability. Private sector participation, supported by transparent regulations and innovative financing mechanisms, ensures efficiency, scalability, and cost-effectiveness. International donors and development agencies address funding gaps and offer technical assistance, while regulatory bodies ensure project compliance and cost-reflective tariffs.

Community engagement is central to the acceptance and long-term success of mini-grids, fostering local ownership and building trust. Projects integrated with community-driven objectives are more sustainable, economically impactful, and socially beneficial. Additionally, productive use of electricity in income-generating activities enhances payment compliance and supports economic development.

Collaboration among stakeholders ensures that mini-grids are not only technically and financially viable, but also socially inclusive and environmentally sustainable. By leveraging the strengths of diverse actors, rural electrification projects can achieve long-lasting impacts, addressing energy access and contributing to broader development goals.

We have drawn the following recommendations on how stakeholders can contribute to successful remote area RE electrification from the study:

- Government and Policy Makers:
  - Create clear and consistent policies and regulations to promote renewable energy mini-grid development, ensuring cost recovery and long-term viability.
  - Establish renewable energy-specific regulatory bodies to streamline project approvals, enforce compliance, and manage risks associated with the arrival of centralized grids.

  - Provide fiscal incentives such as tax exemptions, subsidies, and grants to lower the financial burden on project developers and communities.

- Private Sector and Investors:
  - Engage in PPPs to enhance efficiency and sustainability in project implementation.
  - Offer innovative financing models such as partial subsidies and guarantees to mitigate investment risks and encourage broader participation.
  - Develop scalable and adaptable renewable energy systems that can expand with increasing community demand.

- International Donors and Development Agencies:
  - Provide concessional loans, grants, and technical support to reduce financial and technical barriers to project deployment.
  - Facilitate capacity-building initiatives, including training programs for local operators and technicians.
  - Support feasibility studies, impact assessments, and technical designs to ensure project success.

- Local Communities:
  - Actively participate in all stages of the project—from planning and design to implementation and maintenance—to foster ownership and long-term sustainability.
  - Use RE for productive purposes, such as agricultural processing, cold storage, and small-scale businesses, to enhance economic development and ensure tariff payment compliance.
  - Establish community energy committees to manage operations, maintain systems, and promote energy-efficient practices.

- Non-Governmental Organizations and Civil Society Organizations:
  - Advocate for inclusive policies that prioritize rural electrification and equitable energy access.
  - Provide training and education programs to enhance community awareness of renewable energy benefits and technical understanding.
  - Monitor and evaluate projects to ensure alignment with social, environmental, and economic goals.

- Utilities and Technical Experts:
  - Collaborate with independent power producers and community stakeholders to design and construct technically robust renewable energy systems.
  - Ensure system reliability and performance through advanced grid management techniques and regular maintenance protocols.
  - Incorporate hybrid renewable energy systems to address intermittency and enhance overall energy security.
- Academia and Research Institutions:
  - Conduct research to identify innovative solutions for challenges like cost efficiency, system durability, and grid integration.
  - Provide training for local stakeholders on renewable energy technologies, operational best practices, and system optimization.
  - Develop frameworks for assessing and scaling successful mini-grid models to other remote communities.

These recommendations, based on the case studies and stakeholder roles outlined throughout the chapter, emphasize the need for coordinated, multi-stakeholder collaboration to achieve effective, sustainable renewable energy electrification in remote areas.

References

Abdmouleh, Z., Alammari, R. A., & Gastli, A. (2015). Review of policies encouraging renewable energy integration & best practices. *Renewable and Sustainable Energy Reviews, 45*, 249–262.

African Development Bank. (2015). *Evaluation of the financial and economic combination of SHS and mini-grid systems*. African Development Bank.

African Development Bank. (2016). *Green mini-grids in sub-saharan Africa: Analysis of barriers to growth and the potential role of the African Development Bank in supporting the sector*. https://greenminigrid.afdb.org/sites/default/files/Media/GMG%20MDP%20%20Document%20Series%20N1.compressed.pdf

Ali, A., Li, W., Hussain, R., He, X., Williams, B., & Memon, A. (2017). Overview of Current Microgrid Policies, Incentives and Barriers in the European Union, United States and China.*Sustainability, 9*(7), 1146. https://doi.org/10.3390/su9071146

Alliance for Rural Electrification. (2014). *Hybrid Mini-Grids for Rural Electrification: Lessons learned*. http://www.ruralelec.org/sites/default/files/hybrid_mini-grids_for_rural_electrification_2014.pdf

Alvial-Palavicino, C., Garrido-Echeverría, N., Jiménez-Estévez, G., Reyes, L., & Palma-Behnke, R. (2011). A methodology for community engagement in the introduction of renewable based smart microgrid [article]. *Energy for Sustainable Development, 15*(3), 314–323. https://doi.org/10.1016/j.esd.2011.06.007

Azimoh, C. L., Klintenberg, P., Mbohwa, C., & Wallin, F. (2017). Replicability and scalability of mini-grid solution to rural electrification programs in sub-Saharan Africa. *Renewable Energy, 106*, 222–231. https://doi.org/10.1016/j.renene.2017.01.017

Conteh, F., Takahashi, H., Hemeida, A. M., Krishnan, N., Mikhaylov, A., & Senjyu, T. (2021). Analysis of hybrid grid-connected renewable power generation for sustainable electricity supply in Sierra Leone [article]. *Sustainability (Switzerland), 13*(20), Article 11435. https://doi.org/10.3390/su132011435

De Nigris, M., & Giuliano, F. (2023). The Role of Organised Civil Society in the Implementation of the Renewable Energy Transition and Renewable Energy Communities: A Qualitative Assessment. *Energies, 16*(10). https://doi.org/10.3390/en16104122

Edomah, N., Ndulue, G., & Lemaire, X. (2021). A review of stakeholders and interventions in Nigeria's electricity sector. *Heliyon, 7*(9), 2–6. https://doi.org/10.1016/j.heliyon.2021.e07956

Haldar, A., Sethi, N., Jena, P. K., & Padhan, P. C. (2023). Towards achieving Sustainable Development Goal 7 in sub-Saharan Africa: Role of governance and renewable energy [article]. *Sustainable Development*. https://doi.org/10.1002/sd.2521

Holzigel, J., & Bravard, M. (2020). *Sierra Leone: A Cost-Reflective Mini-Grid Tariff Framework* D. G. f. I. Z. G. https://www.get-transform.eu/wp-content/uploads/2021/01/Success-in-Rural-Electrification-Case-Study_Sierra-Leone.pdf

IRENA. (2015). *Africa 2030: Roadmap for a Renewable Energy Future*. https://www.irena.org/-/media/Files/IRENA/Agency/Publication/2015/IRENA_Africa_2030_REmap_2015_low-res.pdf

IRENA. (2018). *Policies and regulations for rewable energy mini-grids*. IRENA.

IRENA. (2022). *Renewable energy roadmap for Central America: Towards a regional energy transition*. International Renewable Energy Agency.

IRENA, & Africa Development Bank. (2022). *Renewable energy market analysis: Africa and its regions*. International Renewable Energy Agency.

IRENA, & African Development Bank. (2022). *Renewable energy markey analysis: Africa and its regions*. International Renewable Energy Agency.

IRENA, & CPI. (2023). *Global landscape of renewable energy finance*. International Energy Agency.

Irungu, S. (2024). Bridging the energy gap in Africa: Lessons from Sierra Leone's mini-grids sector. *Africa Energy*.. https://africasustainabilitymatters.com/bridging-the-energy-gap-in-africa-lessons-from-sierra-leones-mini-grids-sector/

Liu, Y., & Bah, Z. (2021). Enabling development impact of solar mini-grids through the community engagement: Evidence from rural Sierra Leone [article]. *Energy Policy, 154*, Article 112294. https://doi.org/10.1016/j.enpol.2021.112294

Ritchie, H., & Rosado, P. (2020). Electricity mix. *Our World in Data*. https://ourworldindata.org/electricity-mix

The World Bank. (2024a). Access to Electricity, Rural (% of Rural Population) - Sierra Leone. The World Bank. from https://data.worldbank.org/indicator/EG.ELC.ACCS.RU.ZS?locations=SL&start=1998&view=chart. Accessed November 22, 2024.

The World Bank. (2024b). GDP per capita of Ghana. The World Bank. from https://data.worldbank.org/indicator/NY.GDP.PCAP.CD?locations=GH. Accessed November 19, 2024.

The World Bank. (2024c). *Ghana's rural electrification rate*. from https://data.worldbank.org/indicator/EG.ELC.ACCS.RU.ZS?end=2022&locations=ZG-GH&start=2022&type=shaded&view=bar&year=2018. Accessed November 19, 2024.

CHAPTER 9

The Multi-Territorial Governance (MTG) Model: A Pathway for Self-Determined Indigenous Nations and Settler-States Collaboration in Energy Transitions

Fabienne Rioux-Gobeil and Annick Thomassin

Introduction

In Canada and Australia, Indigenous nations and organizations are actively participating in energy transition processes to ensure energy security and foster economic development. This engagement is bolstered by commitment from the governments of both countries to decentralized approaches, which have created opportunities for a wider array of participants. However, the involvement of Indigenous groups with traditional and/or treaty rights in these decentralized systems presents unique

F. Rioux-Gobeil (✉) · A. Thomassin
Australian National University, Canberra, Australia
e-mail: fabienne.rioux-gobeil@anu.edu.au

A. Thomassin
e-mail: annick.thomassin@anu.edu.au

T. Walker et al. (eds.), *Wiring the Future*, Palgrave Studies in Emerging Risk Management and Sustainable Finance,
https://doi.org/10.1007/978-3-032-14717-2_9

regulatory and coordination challenges, as well as opportunities. Given that research on the decentralization of energy systems has predominantly focused on community-led, place-based, or collaborative initiatives (Colton et al., 2016; Gunningham & Bowman, 2016; Bauwens, 2017; Nepal & Menezes, 2017; Avila, 2018; Godin, 2020; Winfield et al., 2021; Mcmaster et al., 2023); we questioned whether these governance models genuinely respect Indigenous claims to self-determination.

Recognizing that these frameworks often do not adequately include and promote the self-determination aspirations of Indigenous nations, we propose a new governance model for energy transition in this chapter. We have named it the Multi-Territorial Governance (MTG) model. We have designed the MTG model to facilitate the emergence and empowerment of renewable energy projects led by self-determined Indigenous nations. As this chapter is part of a broader research initiative that compares the Atikamekw Nation Council's bioenergy project in Canada with the Garawa's solar energy project in Australia, our focus is specifically directed toward these two cases.

This chapter has three key objectives: First, it serves as a theoretical foundation for our broader research aimed at proposing pathways to decolonize energy transition processes. Second, it traces the development of our MTG model, providing essential background on our thought process for the audience. Third, it outlines the potential impacts that such a model could have on energy transitions and Indigenous nations' self-determination aspirations, elaborating on how various stakeholders could effectively utilize the MTG model and why they would be compelled to do so.

In developing our MTG model, we start by examining the Atikamekw Nehirowisiw and Garawa case studies to identify which elements of the Canadian and Australian energy governance systems facilitate Indigenous participation in line with their claims to self-determination, and which elements do not. We then review the literature on hierarchical and polycentric decentralization theories to underscore the gaps our MTG model can address and the elements that can inform its development. Next, recognizing that governance models are rooted in relationships and networks, we explore which aspects of international relations, Indigenous relationalism, and entangled territorialities can serve as foundational support for our MTG model. In the final section, we outline the characteristics of the MTG model. We assert that our MTG model can provide a framework to assist Indigenous groups and nations, policymakers, and energy developers in decolonizing the processes associated with energy transition.

The Canadian and Australian Contexts: A Comparative Case Study

This section presents our two case studies, along with the energy governance contexts that surround them. As part of a broader comparative analysis, we draw upon both theoretical frameworks and empirical insights. The cases of the Atikamekw Nehirowisiw and the Garawa peoples are already enriched by a literature review centred on Indigenous rights and energy transition justice, published in a paper titled "A Just Energy Transition for Indigenous Peoples: From Ideal Deliberation to Fairness in Canada and Australia" (Rioux-Gobeil & Thomassin, 2024). We encourage readers to consult this work for additional information. Moreover, we collected ethnographic data through participant observation, document and archival research, and semi-structured interviews during our fieldwork in Borroloola, home of the Garawa people, and on Nitaskinan, the claimed territory of the Atikamekw Nehirowisiwok. Given that the chapter's objective is to develop a governance model that honors Indigenous self-determination, we must first clarify the claims made by the Atikamekw Nehirowisiw and Garawa nations. To achieve this, we base our analysis on two governance documents: The Atikamekw Declaration of Sovereignty (2014) and the Garawa Land and Sea Country Plan (2014).

In Canada, through the Indian Act, the federal government relocated Indigenous peoples to designated areas known as reserves to assert control over their territories and resources (Coulthard, 2022; Whyte, 2017, 2018). Starting in 1895, the Atikamekw Nehirowisiwok gradually moved to three communities—Manawan, Wemotaci, and Opitciwan—which had been "reserved" for them by the Canadian government (Poirier et al., 2014). In 1979, the Atikamekw Nehirowisiwok initiated their native title negotiations with both the federal and provincial governments (Poirier, 2017). After over four decades of unsuccessful negotiations (Dussart & Poirier, 2017), and following the landmark recognition of native title by the Supreme Court for the Tsilhqot'in nation in June 2014 (Caron, 2016), the Atikamekw Nehirowisiw nation self-declared their sovereignty over an 80,000 km^2 territory (see Fig. 9.1

below) on September 8, 2014 (Conseil de la Nation Atikamekw, 2014; Cousineau, 2014).

In the Atikamekw Declaration of Sovereignty (2014), the Atikamekw Nehirowisiwok request to be recognized as a nation: "We are not Canadian, we are not Québécois, we are Atikamekw Nehirowisiw, and we are from Nitaskinan." Nitaskinan is where the Atikamekw Nehirowisiw language and culture are rooted and where the people and their ancestors have lived since time immemorial. They hold responsibilities toward the Nitaskinan, its resources, and its peoples, and must be able to fulfil them. Furthermore, they have traditional political institutions, such as the ka nikaniwitc (territory chief), and want them recognized. The Declaration demands that communication with the Canadian and Québec governments be based on principles of international relations. "International" refers here to the relationships between the Atikamekw Nehirowisiw, Québecoise and Canadian nations. As such, developers must obtain consent from the Atikamekw Nehirowisiw Nation for any projects developed on the Nitaskinan. Finally, the Declaration points to the Atikamekw Nehirowisiw's contribution to Québec's and Canada's economic development and rightfully asserts that the federal and provincial governments should acknowledge and respect these contributions.

In Australia, the Garawa's ancestral Country[1] stretches from the southwest Gulf of Carpentaria in the Northern Territory to Massacre Inlet in Queensland (see Fig. 9.2 below) (Gangalidda Garawa Native Title Aboriginal Corporation, 2014). Following the period of Wabulinji (Wild Time) and the significant expansion of pastoralism in the region since the mid-1800s, numerous Garawa individuals relocated away from their ancestral land. In the 1970s, a movement began among the Garawa people who sought to return to their Country, known as the Homeland Movement.[2] However, they encountered significant challenges in achieving this goal due to a lack of government support (O'Bryan et al., 2024). Under the Commonwealth Aboriginal Land Rights Act 1976, the

[1] Country is an all-encompassing Aboriginal-English concept referring to the sentience and intentionality of the land, the sea and wider world to which Aboriginal peoples are genealogically and spiritually connected (Morphy & Mahood, 2019).

[2] The Homeland Movement began in the 1970s, fueled by Aboriginal peoples' commitment to protect and preserve their people, culture, language, and Country through the return to their clan homelands. The Yolŋu people initiated the movement after a lengthy struggle for land rights in the late 1950s, deciding to return to their homeland to protect it from bauxite mining development (O'Bryan et al., 2024).

federal government established the Waanyi-Garawa Aboriginal Land Trust (ALT) in 1985, followed by the Garawa ALT in 1992. These declarations granted the Garawa Traditional Owners (traditional landowners) exclusive land rights over a part of their ancestral territory, as shown in Fig. 9.2 (Gangalidda Garawa Native Title Aboriginal Corporation, 2014; North Australian Indigenous Land and Sea Management Alliance, 2020).

In 2014, the Gangalidda and Garawa Native Title Aboriginal Corporation and the Carpentaria Land Council Aboriginal Corporation published the Garawa Land and Sea Country Plan (2014) to promote economic development and partnerships on Garawa Country while respecting Garawa people's rights, values, and interests. The plan emphasizes the importance of intergenerational knowledge transmission, drawing on Garawa ancestral ways of knowing to manage and protect the land and sea. It explains how their governance structure is embedded in their responsibilities and relationships with both people and the natural world. The authors anticipate that this plan will enhance Garawa claims to governance on the land, making consultation with Traditional Owner groups compulsory for any project development on their land. It is also important to recognize the traditional and inherited roles of the Junggayi (land manager) and Minggirringi (Traditional Owners) in ensuring that the land is properly cared for in all decision-making processes. Furthermore, the Garawa Indigenous Protected Area (IPA)[3] must be respected. The Land and Sea Country Plan marks the first step in collaborative work on Country by encouraging mutual understanding and clearly establishing the Garawa people's interests and aspirations.

With this understanding, we can now assess the power of these two documents for the participation of the Atikamekw Nehirowisiw and Garawa nations in Canadian and Australian energy transition governance systems.

Exploring Energy Governance in Canada: A Closer Look at the Atikamekw Nehirowisiw Participation

In Canada, energy production and distribution are provincial and territorial jurisdictions, and the governance structure for energy is one of the

[3] The Indigenous Protected Areas program was created in 1997. It supports First Nations communitiesto manage areas of Land and Sea Country in accordance with Traditional Owners' objectives (National Indigenous Australians Agency, n.d.).

Fig. 9.1 The Nitaskinan and the Atikamekw Nehirowisiw Reserves. *Source* Magalie Petiquay, Conseil de la Nation Atikamekw, 2024 (partner in this research)

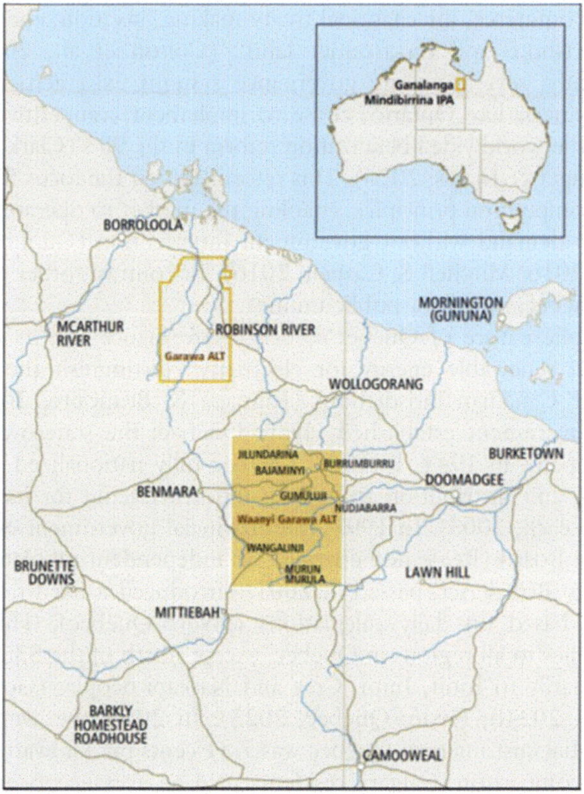

Fig. 9.2 The Garawa Ancestral Country and Aboriginal Land Trusts. NAILSMA (2020) *Waanyi and Garawa Land Use and Economic Diversification Plan*. Available at: https://crcna.com.au/wp-content/uploads/2024/05/waanyigarawalanduseplan_nov2020.pdf

most decentralized in Western countries (Colton et al., 2016). In the provinces and territories, energy responsibilities are related to the production, transmission, and sales of electricity, trade and commerce, as well as the exploration, development and management of energy resources, all of which must occur inside their respective borders (Colton et al., 2016). However, the federal government does have some responsibilities in the energy sector, in such areas as "interprovincial and international trade and commerce, international treaty-making, taxation, energy development offshore and on frontier land" (Colton et al., 2016, p.3). Consequently, several energy governance systems exist across Canada. Some provinces, like Ontario, chose to implement competitive markets following the worldwide liberalization reform in the 90 s (Clark & Leach, 2005; Stringer & Joanis, 2023). This reform shifted the focus from regulation to competition principles, enabling the market to operate based on supply and demand without government intervention (Gunningham & Bowman, 2016; Mitchell & Casben, 2016). In contrast, other provinces, such as Québec, opted for public utilities.

The public nature of Québec's energy governance and its near-total reliance on renewable energy for electricity[4] distinguish the province from other Canadian jurisdictions (Fagoaga & Beaucaire, 2024). The Québec government established Hydro-Québec, the state-owned electricity company, in 1944. In 1963, Québec fully nationalized electricity production and distribution to ensure uniform pricing for its residents (Clark & Leach, 2005). In 1997, the provincial government established the Energy Board (Régie de l'énergie), an independent regulatory body. The Energy Board Act, passed in 2001, introduced a fixed-rate pricing obligation based on daily calculations (Hydro-Québec). This pricing system applies to all regions in Québec, except north of the 53rd parallel, home primarily to Inuit, Innu, Cree and Naskapi peoples (Government of Canada, 2024b; Hydro-Québec, 2023). In 2023, the average price for residential customers in Québec was 7,81 cents per kilowatt-hour (c/kWh). In comparison, Calgary residents paid an average price of 29,80 c/kWh and Toronto residents an average of 13,88 c/kWh in 2023 (Hydro-Québec, 2023).

Québec has not applied a competitive approach to its energy markets (Clark & Leach, 2005). By the end of the 1990s, however, in response to the liberal reforms occurring in various Western countries and to facilitate electricity sales outside the province, it was necessary to modify its governance structure (Godin, 2019). The government and Hydro-Québec needed to find a way to sell electricity while maintaining the social benefits of public utilities, such as low rates for consumers. In 1997, Hydro-Québec separated its three main sectors: distribution (HQD), production (HQP), and transportation (HQT). Consequently,

[4] Ninety-four per cent of Québec's electricity is from hydroelectricity (Canada Energy Regulator, 2024).

private companies are now allowed to produce electricity and sell it to HQD, while Hydro-Québec maintains a monopoly on distribution (Fagoaga & Beaucaire, 2024). In 2000, the Québec government and Hydro-Québec created a patrimonial energy bloc of 165 TWh entirely produced by HQP and sold to HQD at a fixed rate of 2,79c/kWh (Hydro-Québec XE "Hydro-Québec", 2022). The electricity generated outside this bloc (post-patrimonial) by private, community-led, or cooperative entities must undergo the government's call for projects and comply with specific criteria, including local and Indigenous participation or investment percentage.

Autonomous Networks (Réseaux Autonomes), another division of Hydro-Québec, is responsible for providing electricity to areas not connected to the national grid (Hydro-Québec, 2017), including the Atikamekw Nehirowisiw community of Opitciwan, which is located 250 km from the nearest town (Blais, 2024). In Canada, large-scale electricity production is generated from 70% of renewable sources (Natural Resources Canada, 2024; Stringer & Joanis, 2023). However, across all provinces and territories, small-scale and off-grid electricity is mainly generated from diesel generators (Stringer & Joanis, 2023). Governments implemented off-grid systems in remote and rural areas due to the limited reach and diversity of the provincial networks. Studies have since shown that these generators contribute significantly to Canada's greenhouse gas emissions and are costly for communities and governments (Stringer & Joanis, 2023). In Québec, the government is required to subsidize the price difference for consumers, as per the Energy Board Act (Clark & Leach, 2005). As a result, provincial and federal governments are recognizing a significant incentive to shift from diesel generators to renewable energy in remote and rural communities, mainly inhabited by Indigenous peoples.

Most, if not all, of the government's funding in Indigenous energy transitions aims to replace diesel with renewables in off-grid systems (Hydro-Québec, 2017; Laboucan-Massimo, 2022b; Government of Canada, 2023, 2024a). For communities replacing their off-grid diesel system with renewable energy, it is becoming increasingly common to own up to 100% of the project share (Laboucan-Massimo, 2022). Such a transition provides them with essential energy and financial security benefits (Stringer & Joanis, 2023). For those connected to the national grid and interested in developing renewable projects, engaging in existing energy transition initiatives proposed by developers tends to offer a smoother path. This is largely due to the fact that self-determined projects tend to be significantly more expensive than forming partnerships with Hydro-Québec, which also offers valuable technical support during implementation. Furthermore, project promoters and governments are recognizing the benefits of collaborating with Indigenous communities for social acceptability, leading to increased engagement in such partnerships (Hoicka et al., 2021). In fact, energy stakeholders across Canada are now establishing clean energy targets that incorporate collaboration with Indigenous peoples (Laboucan-Massimo, 2022).

The Atikamekw Nation Council's (ANC) bioenergy project (2021), situated on Nitaskinan, is an example of a self-determined Indigenous initiative. Although the ANC is headquartered in Capetciwotakanik (La Tuque), it represents the entire Atikamekw Nehirowisiw nation. The bioenergy project is a good example of how the ANC is seriously applying its Sovereignty Declaration, despite it not being formally recognized by any government. The bioenergy project was created in 2017 by Patrice Bergeron and Patrice Mangin from the La Tuque municipality and the Université de Québec à Trois-Rivières. During the same year, Constant Awashish, the Chief of the ANC, requested a meeting with Bergeron and Mangin to discuss the project's development on Nitaskinan. In our interview, Chief Awashish stated: "There will be no more projects on our territory without consulting us and without our participation" (2023). For years, the bioenergy project encountered significant resistance from the forestry industry and certain political figures (Tremblay, 2019). Bergeron and Mangin believed that the ANC would be more effective in navigating this resistance than the municipality of La Tuque, which is why they agreed to pass leadership of the project to the ANC (Tremblay, 2022). The primary challenge now lies in securing funding and establishing a collaborative partnership with a promoter.

The province of Québec exemplifies centralized energy governance, particularly within the context of Canada, one of the most decentralized countries in the Western world. Québec plays a crucial role in developing and implementing major energy projects and policies, mainly through the province-owned entity Hydro-Québec, which is characteristic of a centralized system (Aykut and Evrard, 2017). The government's pivotal role in establishing regulations that promote and safeguard social benefits is particularly inspiring for our MTG model. For instance, the criteria for Indigenous participation in the government's call for projects are essential as they serve as a crucial incentive for developers. However, the Québec example has also shown that diversifying both production and producers can be complex in a centralized structure. This complexity limits opportunities for Indigenous peoples to pursue self-determined renewable energy projects and connect them to the national grid.

Exploring Energy Governance in Australia: The Garawa People and Original Power Are Taking Action

In Australia, control over the energy sector is shared between the states and territories and the federal government. The state and territory governments are responsible for mineral leases and operations, environmental assessments and regulations, regional planning, and infrastructure (Mitchell & Casben, 2016). The federal government is in charge of taxation, export controls, examination and approval of foreign investment proposals, native titles, and environmental matters of national significance (Mitchell & Casben, 2016). The effect of these overlapping jurisdictions, described by Mitchell and Casben (2016), can lead to complex policy applications, as projects may require multiple approvals, licenses, permits, and authorizations from various levels of government, resulting in increased time, costs, and resource consumption. However, Gunningham and Bowman (2016) argue that this governance model is practical, as it enables states and territories to implement more efficient energy transition strategies when the federal government falls short, and vice versa.[5] This dynamic can also help mitigate the short-term focus often exhibited by elected parties, which can conflict with the long-term vision needed to effectively address climate change (Gunningham & Bowman, 2016). To address these challenges, the Australian Energy White Paper suggested further deregulating the market by eliminating regulations deemed unnecessary by the government (Gunningham & Bowman, 2016; Mitchell & Casben, 2016; Parliament of Australia, n.d.; Whitmore & Hopkin, 2015).

[5] The Voice to Parliament in South Australia serves as an example. While the national referendum for the Voice to Parliament was rejected by a majority vote in 2023, South Australia took a different direction. The South Australia's First Nations Voice to Parliament was established in 2023. This body, created through legislation, is an elected representative assembly for Aboriginal and Torres Strait Islander peoples within the state (South Australian First Nations Voice to Parliament, 2024).

In the 1990s, Australia implemented liberal reforms in its energy governance system, aligning it more closely with practices observed in most Canadian provinces, while differing from the governance structure in Québec. To facilitate this reform, in 1998, the Commonwealth and state governments established the National Electricity Market (NEM) as a distribution system that encompasses Queensland, New South Wales, the Australian Capital Territory, Victoria, and South Australia (Nepal & Menezes, 2017). Accordingly, the Northern Territory developed the Interim NT Electricity Market (INTEM) in 1992, and Western Australia developed the Wholesale Electricity Market (WEM) in 2006 (Nepal & Menezes, 2017).

Within the NEM, private utilities are allowed to connect their production to the grid freely, provided they adhere to safety and stability requirements (Blakers, 2024). Power generators submit bids, and prices are determined every half hour by the Australian Energy Market Operator (AEMO) to align with demand (Nepal & Menezes, 2017). An annual market price cap,[6] established by regulatory bodies, governs this pricing (Australian Energy Market Commission, 2024). In addition to this cap, rates fluctuate according to supply and demand, resulting in significant price variability for consumers, especially between urban centers and rural areas. Within this competitive framework, the government plays a crucial role in ensuring that renewable energy sources can compete with traditional resources through appropriate regulations. In that vein, the Commonwealth government created the Australian Renewable Energy Agency (ARENA) to provide funding for the development and implementation of renewable energy projects (Gunningham & Bowman, 2016).

As in Canada, many renewable energy funds and policies for Indigenous peoples aim to transition energy distribution in remote communities from diesel generators to more sustainable and affordable energy sources (Lloyd et al., 2000; Riley, 2021). In the Northern Territory, where Aboriginal people make up 31% of the population (Australian Bureau of Statistics, 2021), the Indigenous Essential Services, managed by the Power and Water Corporation, operates the largest fleet of diesel generators in Australia (Nepal & Menezes, 2017). Government-led programs

[6] The price cap is the maximum price that can be reached during trading. The price was established at $16,600/MWh from 1 July 2023 to 30 June 2024 (Australian Energy Market Commission (AEMC), 2024).

were established in the late 1980s to facilitate this transition from diesel to renewable sources; however, these initiatives were short-lived due to the lack of local involvement and a "one size fits all" approach (Lloyd et al., 2000; Riley, 2021). As highlighted by Nursey-Bray et al. (2019), top-down policies and initiatives often neglect cultural contexts, which can impede Indigenous communities'participation. In recent years, there has been a notable shift as bottom-up initiatives gain traction and are increasingly implemented within Indigenous communities across Australia (First Nations Clean Energy Network, 2024). While social acceptability and local participation are more pronounced in this framework, securing adequate funding remains a significant challenge, similar to the experience in Canada.

In this context, non-governmental organizations (NGOs) play a key role in facilitating Indigenous energy transitions (Bellchambers, 2023). These organizations help bridge regulatory gaps and offer financial, technological, and legal support to communities shifting toward renewable energy sources. Original Power, an Aboriginal-run NGO, assists communities in implementing self-determined renewable energy projects throughout Australia. Currently, Original Power is collaborating with the Garawa, Marra, Yanyuwa, and Gudanji peoples on the *Ngardara*[7] community-led microgrid solar project, aimed at reducing the community's reliance on diesel generators (Ngardara Cooperative, 2023). This initiative is located in Borroloola, partially situated on Garawa's ancestral Country. The Garawa Land and Sea Country Plan (2014) could serve as a guiding framework for the Garawa people, identifying the type of partnerships they seek in initiatives such as Borroloola's Ngardara solar microgrid project. Originating from the Borroloola community, the microgrid project secured financial and technical support from Original Power, resulting in the establishment of the Ngardara Cooperative—Australia's first Indigenous clean energy cooperative (Original Power, n.d.). The Ngardara Cooperative, Original Power, and key energy stakeholders, including ARENA and Power and Water Corporation) are establishing this collaboration through Partnership Agreements. Such agreements are setting a precedent in Australia, paving the way for other Indigenous communities.

[7] "*Ngardara*" means "*sun*" in Garawa language.

In Australia, despite energy management being centralized within a few electricity markets—namely the NEM, the WEM, and the I-NTEM—the decentralization of energy sources and producers appears to be more straightforward than in Québec. Decentralized energy systems, characterized by their openness to a diverse range of participants, allow local authorities to emerge as significant actors, which is particularly promising for the Borroloola community. While this decentralized framework supports tailored responses to the unique needs of local communities, it also presents challenges due to limited government intervention (Chabrol & Grasland, 2014). Such a lack of support can lead to price variability across different regions and complicate the implementation of sustainable renewable energy projects, often resulting from inadequate financial and technological services. It is encouraging to observe NGOs stepping in to provide solutions that help non-traditional entities pursue renewable energy initiatives. However, their influence remains considerably less significant than that of government bodies. Ultimately, given that cohesion and coordination can be challenging in decentralized systems plagued by overlapping policies, the proposed Partnership Agreements, as introduced by Original Power and the Ngardara Cooperative presents a promising path forward.

Literature Review—Decentralized Governance Systems

This section reviews the literature to identify elements of existing decentralized governance models that can support or hinder the integration of projects developed by Indigenous nations, thereby informing the development of our MTG model. A comparative analysis of Canada and Australia reveals that decentralized governance models foster a greater diversification of energy sources, energy management structures—encompassing public, private, and cooperative frameworks—and key energy stakeholders. This suggests an increased potential for Indigenous nations to engage actively in such systems. Given that decentralization has led to the emergence of various governance models, it is useful to draw upon Winfield et al.'s (2021) hierarchical and polycentric perspectives to effectively navigate this diversity.

Hierarchical Decentralization & Community Energy Planning

The hierarchical view concerns the relationships between different levels of government. The Community Energy Planning (CEP) governance model exemplifies strategic energy planning in a local context (Winfield et al., 2021). CEPs play a crucial role in developing projects tailored to local needs and requirements. In this model, higher levels play a crucial role in providing technological and financial support, as well as in planning and coordination. Therefore, they significantly impact CEPs that depend on these upper levels (St. Denis & Parker, 2009). CEP's approach to involving various policy frameworks in the energy sector is essential, as it aligns more closely with local realities. However, this hierarchical perspective is not suitable for Indigenous peoples who aim to develop projects as self-determined nations. Rather than honoring Indigenous national self-determination claims, the hierarchical approach to decentralization limits their aspirations to the community level, making them dependent on higher levels of decision-making (Laboucan-Massimo, 2022). Further, Indigenous community governance often relies on political administrations created by Canadian and Australian governments. The hierarchical approach, therefore, overlooks crucial aspects in addressing interactions between the governments of different nations.

Polycentric Decentralization & Place-Based Approach

The polycentric view focuses on multiple, overlapping, and horizontal spheres of authority (Ostrom et al., 1961; Winfield et al., 2021). It supports the execution of territorialized projects, meaning initiatives specifically designed for and implemented in particular territories. This model is well-suited for projects tailored to the cultural contexts of Indigenous nations. Due to the territorial scale at which they are implemented, the territorial and place-based approaches contribute to the cultural legitimacy of the projects (Bellchambers, 2023). As Beuret and Cadoret (2011) assert, the territorial scale is not defined by administrative borders, such as regions or municipalities, but by networks and connections between biological, cultural, political, social, economic, and historical elements. As such, the strength of the polycentric view lies in the network it creates among multiple decision-making spheres.

This, nevertheless, poses significant challenges in terms of project coherence and group cohesion (Winfield et al., 2021). According to

Winfield et al. (2021), the state should act as a coordinator to minimize inconsistency and miscommunication among all spheres of authority. For Bauwens (2017), the proper functioning of polycentric systems relies on higher-level of governance units and top-down policies. O'Neill (2015) likewise argues that there must be a coordinator to mediate and balance the unequal power relations, although, she also highlights an obvious concern about a paternalist settler-state setting the rules for Indigenous peoples. In a governance structure that respects Indigenous self-determination aspirations, we must consider coordinating all spheres of authority without a higher decision-making scale.

Our literature review has called attention to the levers and barriers influencing the participation of Indigenous nations in energy transition processes within decentralized governance frameworks. This analysis has allowed us to pinpoint the gap that our MTG model could effectively fill. In a hierarchical framework, initiatives led by Indigenous communities are often designed to be community-based. However, at the community level, accountability often rests with higher government tiers, a system rooted in colonial policy. Thus, this approach does not necessarily align with the aspirations of Indigenous peoples for recognition as distinct nations. In this context, a polycentric approach and territorialized projects present more compelling alternatives. These initiatives are implemented on a territorial scale, honoring Indigenous cultures, governance systems, and traditional territories. Nevertheless, the need for a higher authority to facilitate coordination is at odds with Indigenous aspirations to move beyond historical power dynamics with settler states. This highlights the need to develop a new governance framework that effectively addresses the complexities of Indigenous energy transitions. Additionally, it suggests that our MTG model should align with the principles of international relations, embrace multiple spheres of authority, recognize diverse ontologies, and support territorialized initiatives.

Theories of Relational Network Organization

Recognizing the crucial role that relational networks play in shaping governance systems, this section aims to establish the theoretical foundations of our MTG model based on relational theories. Our focus on the theories of international relations, Indigenous relationalism, and entangled territorialities enables us to anchor the MTG model within nation-to-nation networks—specifically, the relationships between Indigenous nations and settler states—as well as the interconnections among humans, other-than-human entities, various generations, and territories. We assert that these relational theories are significant due to the nationhood claims made by the Garawa and Atikamekw Nehirowisiwok peoples, the essential insights of Indigenous relational ontologies in their pursuit of self-determination, and the territorialization of projects.

International Relations: Indigenous Nations and Settler-States' Relational Networks

According to the realist school of international relations (IR), global relations are primarily rooted in power and security objectives. According to Benjamin Frankel (1996), international systems are characterized by anarchic dynamics due to the absence of a decision-making structure above national states. Because of this, states' primary goal is to protect or enhance their national security and sovereignty, leading to conflicting relationships between them (Frankel, 1996; Lawson, 2017). In the context of Indigenous nations living within borders set by settler states, federal, provincial, and/or state governments have imposed themselves as the highest levels of authority. Additionally, these authorities do not recognize all Indigenous sovereignty claims, which is a central tenet of international relations. This refusal to recognize, leading to unclear definitions of relationships and Indigenous sovereignties, is creating conflicts despite the existence of a higher decision-making structure.

According to the liberal framework, these international relationships are driven by more than just the pursuit of gain and the fear of loss (Keohane & Nye, 1973). The concept of international society suggests that nations can agree on common rules and shared interests based on law and morals (Bull & Watson, 1984). Nonetheless, perpetuated inequities between Indigenous peoples and settler states demonstrate that law and morals alone are insufficient for nations to adhere to non-restrictive international society norms. To move toward a highly connected international society that includes Indigenous nations, it is essential to reconsider the nature of relationships between them and the settler-state.

Similarly, the solidarist view asserts that a collective will can be motivated by the need to take purposeful action, such as combating climate change (Dunne, 1998; Lawson, 2017). This is one of the most important aspects of the energy transition: working together toward a common objective, driven by our respect for past, present, and future generations, as well as for nature and ourselves.

From a constructivist view, states' identities are shaped by their inhabitants' identities and vice versa. State identities are changing through the creation or improvement of policies and relationships (Lawson, 2017). For instance, a state could shift from being a major producer of fossil fuels to leading the way in the energy transition, or it could progress from a controlling settler state to supporting Indigenous self-determination. On the other hand, state identities influence the nature and the implementation of policies.

The postcolonial approach critiques the role of colonial states and institutions in IR scholarship (Lawson, 2017). It also provides a critical analysis of how this field has influenced colonialism in countries such as Australia and Canada (Davis & Blackwell, 2023). In order to decolonize practices, postcolonial argues that IR needs to actively incorporate a wider range of perspectives in its decision-making processes (Davis & Blackwell, 2023). Furthermore, the field must create opportunities to fundamentally change the colonial methods of knowledge production (Capan, 2017). Decolonization in IR is also a call to challenge Western/non-Western and modern/traditional dualities (Capan, 2017; Davis & Blackwell, 2023).

Indigenous Relationalism: Embracing Reciprocal Responsibilities to all Beings

Many Indigenous ontologies are relational (Brigg et al., 2022; Million, 2015; Moreton-Robinson, 2016). Brigg et al. (2022) explain that relationalism places a high importance on relationships grounded in respect for others. It emphasizes the importance of connection to both living and non-living entities, as well as to place and time. It ensures that the projects are not only aligned with each other, but also with cultural, political, economic, and territorial characteristics, as well as the urgent requirements arising from the climate crisis. As explained by Quandamooka academic Aileen Moreton-Robinson (2016), history and culture are what connect Indigenous nations to their traditional territory. Many Indigenous nations remained connected to their traditional territories despite governments forcing them into boarding schools and reserves. For instance, the Garawa and the Atikamekw Nehirowisiwok peoples both have a profound connection to their ancestral territory, which is intricately woven into their culture and history.

These connections come with certain responsibilities, as Moreton-Robinson (2016) suggests. These responsibilities include the need for knowledge, action, respect, and reciprocity. According to Potawatomi professor Kyle Whyte (2018), it is imperative that all beings possess both the right and the capability to fulfill their reciprocal responsibilities toward living and non-living entities. Just as water is responsible for hydrating humans, reciprocally, humans are responsible for protecting it. Indigenous peoples' responsibilities are integral to their identity (Whyte, 2018). However, ongoing colonial processes have affected Indigenous peoples' ways of living and their capacity to fulfill these reciprocal responsibilities.

As Kombumerri and Wakka Wakka, Mary Graham et al. (2022) explain, Australian Aboriginal relationalists' dispositions are grounded in Country as the source of Law. Indeed, they are drawing upon their ancestral land, which holds a wealth of cultural, social, and historical significance, to shape and define their community's political structure and governance. The presence of higher levels of authority beyond the Country is, in some ways, less essential. Each individual is expected to regulate themselves through their responsibilities toward their community and Country. While being autonomous, they remain interconnected through their responsibilities to one another and the shared territory (Brigg et al., 2022). This autonomy can be beneficial in addressing the challenges associated with coordinating the multiplication and decentralization of energy systems.

Territorialities Entanglement: Coexisting Territorialized Projects

As Anishinaabe Professor of Law John Borrows poignantly explains: "To be alive is to be entangled in relationships not entirely of our own making" (Borrows, 2017, p.vii). The energy transition encompasses the involvement of diverse nations and development perspectives that coexist within a shared physical space. The concept of territoriality holds significant importance as it shapes the interactions between various actors and influences how they understand and manage their territories. The notion of relationships also plays a central role in defining the territorial construct of any group. In other words, territories and peoples are mutually constituted (Éthier, 2020; Poirier, 2017).

According to Poirier (2017), the territorial construct, or territoriality, relies on how people act, experience, and conceive the territory. Consequently, several population groups with different territorialities can coexist within overlapping geographies. Poirier (2017) highlights a disparity between the Atikamekw Nehirowisiw territoriality and that of the Québec government. While the Québec government considers all groups of actors as "users" of the territory, the Atikamekw Nehirowisiwok people prefer to see themselves as its "guardians." Many Indigenous peoples view the territory not as something to own, but as a shared living space that includes both humans and non-humans, thereby valuing coexistence (Poirier, 2017). Making visible the various territorialities at play in energy transition processes can positively contribute to genuinely considering the perspectives and interests of Indigenous peoples. This, in turn, enables developers to implement and fulfil various projects, ideas, and responsibilities.

The geographic coexistence of these various perspectives can be described as "entangled territorialities"(Poirier, 2017). The ongoing processes of colonialism have engendered and sustained complex relationships between settler states and Indigenous nations characterized by unequal power dynamics. The territorialities of Canada and Australia are rooted in the dominance of the natural environment and Indigenous populations (Whyte, 2018). Consequently, the entanglement of settler states and Indigenous ontologies has been characterized by the domination of one over the other. These uneven relationships and entangled territorialities underpin the current energy transition and frame what is possible or not in this space. Non-hierarchical relationships within the energy system would empower entities to safeguard their territory while upholding their cultural and political values.

From our analysis of the theories of international relations, Indigenous relationalism, and entangled territorialities—especially in relation to the Garawa Land and Sea Country Plan and the Atikamekw Declaration of Sovereignty—four key pillars have emerged: non-hierarchical relationships, territorial scale, a clear sense of national identity, and strong reciprocal connections. These pillars provide a framework that allows us to systematically compile and organize the key insights we have gathered from previous sections.

Discussion: The Multi-Territorial Governance (MTG) Model

Our theoretical analysis has yielded insights that are instrumental in crafting a model that honors Indigenous self-determination aspirations while aligning with energy transitions' objectives. We have organized these findings into four key pillars: non-hierarchical relationships, territorial scale, clear national identity, and strong reciprocal connections. In Table 9.1, which illustrates our MTG model, we present these findings in accordance with the four pillars, along with the theoretical frameworks from which they originated: international relations, Indigenous relationalism, and entangled territorialities.

The findings presented in Table 9.1 outline actionable steps to foster sustainable energy transitions that respect and promote Indigenous self-determination. This aligns with our objective to provide tools and methods that assist decision-makers, promoters, and investors in navigating the shift toward a more equitable and effective energy future. In the final two sections, we examine the potential contributions of the MTG model in terms of equity and efficiency, providing concluding arguments for its relevance and utility.

Aiming for Indigenous Nations' Self-Determination

Non-Hierarchical Relationships

In their Declaration of Sovereignty, the Atikamekw Nehirowisiwok people clearly demand recognition as a nation. In their Land and Sea Country Plan, the Garawa people request that developers acknowledge Garawa governance and consult with Traditional Owners for any projects they wish to undertake on Garawa Country. Therefore, in a decentralized energy system involving multiple Indigenous nations and a settler colonial state, the requirement for a higher scale of authority is a topic that necessitates careful consideration. While many scholars view such an entity as essential for coordination, cohesion, and coherence (Bauwens, 2017; Winfield et al., 2021), we argue that such a hierarchical system can perpetuate colonial asymmetrical power relations.

Nevertheless, the IR's idea that the external environment impacts the nature of relationships is relevant to our proposed MTG model. In the

Table 9.1 Conceptual foundations of the multi-territorial governance (MTG) model

Basis of the MTG model	*International relations*	*Relationalism*	*Entangled territorialities*
Non-hierarchical relationships	– Act on the nature of relationships by modifying the external environment – Decolonize relationships by adding a greater diversity of voices in decision-making – Decolonize relationships by acting on the ontological dualities	– Develop and maintain relationships by respecting reciprocal responsibilities – Develop relationships grounded in regards for others (livings and non-livings)	– Modify the nature of relationships by acting on the territorialities – Modify the nature of relationships by acting on the intensity of entanglement
Territorial scale	– Acknowledge non-Western conceptions of sovereignty	– Connect to the territory through history and culture – Respect the territory as the source of law and political order	– Act on the territorialities by modifying relationships with livings and non-livings – Act on the territorialities by modifying the intensity and the nature of entanglement
Clear national identity	– Act on nations' identities to influence international relations – Modify nations' identities through rules and agreements-making	– Build and modify identities by fulfilling reciprocal responsibilities – Be a citizen autonomous and self-regulated by respecting the fulfillment of reciprocal responsibilities	– Understand nations' territorialities to understand nations' identities – Respect the foundations of other nations' identity

(continued)

Table 9.1 (continued)

Basis of the MTG model	*International relations*	*Relationalism*	*Entangled territorialities*
Strong reciprocal connections	– Motivate collective will through purposive actions – Create an inter-national society	– Understand and respect our connections with livings, non-livings, place, and time	– Modify the intensity of territorialities entanglement by modifying the nature of relationships (asymmetrical, etc.)

realist school of IR, the need for a higher level of authority is justified to prevent anarchy, conflicts, and chaotic interactions between nations. In the MTG model, Canada and Australia are tasked with fostering symmetrical, respectful, and decolonized international relationships by recognizing Indigenous rights to sovereignty and self-determination.

Postcolonial scholars in IR argue for a greater diversity of voices in decision-making processes. Canada, Australia, and all Indigenous nations' perspectives need representation and inclusion in energy transitions. Notably, the ontological perspectives of Indigenous peoples, including their territorialities, provide valuable insights for fostering non-hierarchical relationships. The relationalist concept of reciprocity, in particular, places living and non-living entities as equally important. Further, as postcolonial scholars have pointed out, addressing ontological dualities is essential for decolonizing international relations (Capan, 2017; Davis & Blackwell, 2023). A critical examination and upfront discussion of the entanglement between settler states and Indigenous peoples in the energy transition may help shift the current asymmetrical dynamics and contribute to more balanced and equitable relationships.

Territorial Scale

Territorial scale is not bound by political-administrative borders but by its connection to a specific nation, governance model, geographic characteristics, and more. The Atikamekw Declaration of Sovereignty is a good example of a different perspective on sovereignty based on recognition and its everyday performance through tradition and culture

(Thomassin et al., n.d.). Both the Atikamekw Nehirowisiwok and the Garawa peoples assert their rights to protect and occupy their traditional lands. They define Nitaskinan and Garawa Country by their historical presence, traditional knowledge, and reciprocal responsibilities toward the land. Territorial governance entities, such as the Atikamekw Nehirowisiw ka nikaniwitc or the Garawa Junggayi and Minggirringi, possess the authority, knowledge, and respect of their peers required to agree on or develop projects on their territory based on the specific characteristics of their environment.

The polycentric approach draws attention to the interconnections that form a network of diverse decentralized projects. For example, place-based approaches highlight the importance of establishing strong connections to the local territory. In the MTG, we emphasize the territory's political, historical and cultural significance, drawing inspiration from Indigenous territorialities and relationalism. For Aboriginal relationalists in Australia, the political order is intrinsically tied to their relationship with Country. In this context, sovereignty and self-determination are deeply rooted in the connection to the territory and the respect of territorial governance entities. Thus, the MTG recognizes that the development of territorialized projects can transform and shape the entanglement of all actors involved, which can subsequently influence the nature of their territorialities. For instance, symmetrical power relations encourage mutual influence among the territorialities of all participants. Therefore, as Canada and Australia increasingly adopt Indigenous relationalist perspectives, their interpretations of territory will evolve.

Clear National Identity

For the Garawa people, cultivating a mutual understanding of the identities of all actors is vital for progressing toward collaborative efforts that honor their rights and responsibilities in caring for Country. Similarly, for the Atikamekw Nehirowisiwok people, attaining sovereignty in Nitaskinan requires recognition as a significant economic and political entity by other actors. Establishing a distinct national identity is crucial to reshaping external perceptions of one's nation. By fostering a positive and well-informed view of one another, relationships between actors can be greatly enhanced. Moreover, clarifying each entity's roles, connections, responsibilities, and aspirations can diminish the need for higher authorities and promote cohesion and coordination among actors and their respective projects.

In Aboriginal relationalism, identities are shaped and transformed through the fulfillment of reciprocal responsibilities toward both living and non-living entities. As a result, everyone plays a role in upholding the political order. In contrast, for IR scholars, national identities can be altered through rules and agreements. Therefore, we argue that within the MTG model, a nation's ability to meet its reciprocal responsibilities can also be defined and transformed in this manner. However, while national identities may evolve due to agreements or territorial conflicts, external actors should regard the essence of identity, along with its associated perspectives and interests, as fundamental and enduring values.

Strong Reciprocal Connections

Establishing a strong network is essential to ensure coherence among multiple projects and actors in polycentric decentralization. Inspired by the Atikamekw Nehirowisiwok and Garawa peoples' perspectives and relationalist ideas, our MTG model emphasizes the importance of strong reciprocal connections to both living and non-living beings, as well as place and time. IR scholars posit that purposive actions are the driving force behind collective will (Dunne, 1998; Lawson, 2017). Understanding our connections with the living and non-living world, and the reciprocal responsibilities we must fulfill, gives purpose to our actions. Self-regulated and autonomous actions, grounded in consideration for others, foster a stronger connection among individuals.

The connections between nations are multifaceted and extend beyond aspirations for power and security. This suggests that in the MTG, a non-hierarchical relationship structure is possible, but Indigenous nations and other actors must be connected through a collective will. Failure to address the power imbalance will perpetuate the unequal distribution of decision-making influence, depriving others of their rights to fulfill their reciprocal responsibilities and to establish meaningful connections with living and non-living entities, as well as with place and time.

Navigating Sustainable Energy Transitions

In this section, we examine the potential impact of the MTG model on the energy transition. We explore its relevance for expanding decentralized renewable energy projects while maintaining equity, coherence and transparency. Firstly, as Gunningham and Bowman (2016) observe, long-term policy planning and policy stability are crucial in the energy transition. It is essential to incorporate longstanding Indigenous territorial and political claims into the implementation of energy transition initiatives. Many of these claims have been under negotiation for decades and significantly influence policymaking in Canada and Australia. By acknowledging these claims and integrating them into current energy policies, we can enhance long-term policy stability and relevance while supporting a more equitable energy transition process.

As mentioned above, the combined use of federal, provincial, and local policy tools provides more opportunities to strengthen energy transition strategies and results (Gunningham & Bowman, 2016). For example, if the elected federal party fails to implement sufficiently strong environmental policies, the state, provincial, or local governments can step in. This is why we argue that incorporating the territorial scale into energy transitions in Canada and Australia can have a positive impact. Doing so would introduce additional and potentially more stringent targets, strategies, policies, and regulations.

Indigenous peoples' rights and claims to be consulted and to give or withhold consent for projects on their territories give them an important role in assessing and approving projects. This can create an additional comprehensive environmental assessment, allowing for a more thorough examination of the potential socio-ecological impact of a proposed project.

Furthermore, establishing a clearer understanding of Indigenous national identity and their relationships with Canada and Australia would bring about increased certainty for projects developed on their territories, possibly leading to fewer disagreements and protests (Nolan, 2024). Colton et al. (2016) suggest that stakeholders are beginning to see this as an opportunity to engage in collaborative efforts with Indigenous nations on projects. However, within the MTG, rather than pursuing partnerships with Indigenous communities solely for social acceptability, project developers would be required to obtain consent from Indigenous authorities for the implementation of their projects and align their proposals with Indigenous peoples' own projects, ontologies, and self-determination goals.

The outcomes of renewable energy projects led by the Australian government in Aboriginal and Torres Strait Islander communities suggest that top-down initiatives are often less well-received by residents, resulting in a lack of long-term sustainability. In contrast, territorialized projects undertaken by culturally recognized organizations that are rooted in the local context tend to enhance social acceptability (Colton et al., 2016). This place-based cultural legitimacy empowers the developers and their projects, increasing their likelihood of longevity (Bellchambers, 2023). Now that Indigenous nations and communities are spearheading more renewable energy projects, the successful outcomes demonstrate the positive impact of grassroots and territorialized efforts.

The main distinction between private and public power production is that the former is primarily driven by economic incentives, where higher electricity consumption leads to greater profits. In contrast, one of the strengths of public utilities is their focus on societal objectives. The decentralization of energy systems and the multiplication of private renewable energy projects in Canada and Australia raise concerns about the potential loss of some of the social benefits of public utilities. The MTG emphasizes the importance of reciprocal responsibilities, which should serve as a motivating factor for taking social and environmental considerations into account when implementing energy transition projects.

Conclusion

In Canada and Australia, governments primarily initiate and fund energy transition projects for Indigenous nations to help them move from off-grid diesel generators to renewable energy sources. However, these top-down initiatives often lack the necessary cultural considerations, which can undermine their sustainability. While an increasing number of grassroots initiatives are emerging within Indigenous communities, securing funding for these projects remains a significant challenge. To support the implementation of self-determined projects on their traditional territories and to foster respectful partnerships that align with their nations' perspectives and interests, the Garawa (Australia) and the Atikamekw Nehirowisiwok (Canada) peoples have published two important documents. The *Garawa Land and Sea Country Plan* (2014) and the *Atikamekw Declaration of Sovereignty* (2014) explicitly articulate each nation's national and cultural identity and outline their expectations for collaborative efforts on their ancestral territories. These key documents informed our elaboration of a new governance model framed around and supporting Indigenous peoples' aspirations for self-determination: the

Multi-Territorial Governance (MTG) model. The objective of this chapter was twofold. First, we aimed to guide the audience through our thought process in developing the MTG model. Second, we presented arguments illustrating how the MTG model can serve as a guide for energy transition developers, decision-makers, and investors.

Our review of the literature about decentralized energy transition governance systems has brought attention to the absence of models that could fundamentally change the collaborative nature between self-determining Indigenous nations and settler states. In decentralized frameworks, which encompass a diverse array of stakeholders, the necessity for a higher decision-making authority tasked with ensuring coordination and cohesion can pose challenges to the promotion and protection of Indigenous peoples' self-determination claims. The literature review also underscored the importance of recognizing the relationships between Indigenous nations and settler states as inter-national in nature, integrating Indigenous ontologies, and considering the intricate entanglements of territorialities on overlapping lands. Consequently, we investigated theories of international relations, Indigenous relationalism, and entangled territorialities to identify potential foundations for our MTG model. Our analysis revealed four critical areas requiring attention, which we refer to as the four pillars of the MTG model: non-hierarchical relationships, territorial scale, a clear national identity, and strong reciprocal connections. We organized our findings around these pillars and presented them as purposeful actions. Thus, the MTG model offers actionable tools for implementing a decentralized energy transition that prioritizes Indigenous rights to self-determination at its core.

In addition to supporting the self-determination aspirations of Indigenous peoples, we believe that the MTG could positively impact the energy transition. By granting decision-making power at the territorial level, new policies, goals, regulations, and assessments can be introduced. This level of governance is intricately linked to culture, history, and nature, which can lend cultural legitimacy and sustainability to policies and projects. Clarifying national identities may enhance the interactions between Indigenous nations and settler-states, fostering a climate of certainty for investments and initiatives in shared spaces. We recognize that implementing a new governance model that supports Indigenous self-determination aspirations, such as the MTG, presents significant challenges, particularly in settler-states like Canada and Australia, where some assimilationist policies are still in effect. Nevertheless, we contend that the urgent need to transition to a different energy system represents an opportunity that many Indigenous nations may wish to embrace. Future research inspired by these findings should delve deeper into the literature on co-management, as it may provide valuable insights for a more project management-oriented perspective.

References

Australian Bureau of Statistics. (2021). Estimates of Aboriginal and Torres Strait Islander Australians. from Australian Government website: https://www.abs.gov.au/statistics/people/aboriginal-and-torres-strait-islander-peoples/estimates-aboriginal-and-torres-strait-islander-australians/latest-release. Accessed March 16, 2023.

Australian Energy Market Commission (AEMC). (2024). 2024–25 market price cap now available. Retrieved from https://www.aemc.gov.au/news-centre/media-releases/2024-25-market-price-cap-now-available

Avila, S. (2018). Environmental justice and the expanding geography of wind power conflicts. *Sustainability Science, 13*(3), 599–616. https://doi.org/10.1007/s11625-018-0547-4

Bauwens, T. (2017). Toward a polycentric low-carbon transition: The roles of community-based energy initiatives in enhancing the resilience of energy systems. In N. Labanca (Ed.), *Complex systems and social practices in energy transitions* (pp. 119–145). Springer.

Bellchambers, K. (2023). Governing country: A literature review of indigenous governance principles in indigenous ranger groups & indigenous protected areas. In CAEPR discussion paper (No. 303/2023). Retrieved from https://caepr.cass.anu.edu.au/research

Beuret, J. E., & Cadoret, A. (2011). Une gouvernance territoriale endogène de l'environnement: Contours et enjeux: Près de 300 initiatives soumises à une analyse comparative. *Geographie Economie Societe, 13*(4), 363–386. https://doi.org/10.3166/ges.13.363-386

Blais, S. (2024, August 28). *Opitciwan veut réduire ses GES de 75%.* La Presse. Retrieved from https://www.lapresse.ca/actualites/environnement/2024-08-28/remplacement-du-diesel-par-la-biomasse/opitciwan-veut-reduire-ses-ges-de-75.php

Blakers, A. (2024, January 15). PV and prices—The fast uptake of solar in Australia.Pdf. *PV Magazine.*. Retrieved from https://www.focustechnica.com/pv-and-prices-the-fast-uptake-of-solar-in-australia/?utm_source=Main+ICEDS+Database&utm_campaign=93f2e57862-EMAIL_CAMPAIGN_2024_02_08_05_52&utm_medium=email&utm_term=0_-93f2e57862-%5BLIST_EMAIL_ID%5D

Borrows, J. (2017). Challenging historical frameworks: Aboriginal rights, the trickster, and originalism. *Canadian Historical Review, 98*(1), 114–135. https://doi.org/10.3138/chr.98.1.Borrows

Brigg, M., Graham, M., & Weber, M. (2022). Relational indigenous systems: Aboriginal Australian political ordering and reconfiguring IR. *Review of International Studies, 48*(5), 891–909. https://doi.org/10.1017/S0260210521000425

Bull, H., & Watson, A. (1984). In H. Bull & A. Watson (Eds.), *The expansion of international society*. Clarendon Press.

Canada Energy Regulator. (2024). Profil énergétique des provinces et territoires. Retrieved from https://www.cer-rec.gc.ca/fr/donnees-analyse/marches-energetiques/profils-energetiques-provinces-territoires/profils-energetiques-provinces-territoires-quebec.html

Capan, Z. G. (2017). Decolonising international relations? *Third World Quaterly, 38*(1), 1–15. https://doi.org/10.1080/01436597.2016.1245100

Caron, J.-F. (2016). La déclaration de souveraineté de la nation atikamekw: les méandres de l'octroi d'un titre aborigène. *Canadian Ethnic Studies, 48*(1), 141–160.

Chabrol, M., & Grasland, L. (2014). Contraintes spatiales et enjeux territoriaux d'une déclinaison régionale de la transition énergétique: l'exemple de la région Provence-Alpes-Côte d'Azur. *VertigO, 14*(3), 1–21. https://doi.org/10.4000/vertigo.15657

Clark, R., & Leach, A. (2005). La réglementation de l'énergie au Québec. Retrieved from https://depot.erudit.org//id/002928dd

Colton, J., Fast, S., Gattinger, M., Gehman, J., & Winter, J. (2016). Energy projects, social Licence, public acceptance and regulatory Systems in Canada: A white paper. *SSRN Electronic Journal, 9*(20). https://doi.org/10.2139/ssrn.2788022

Conseil de la Nation Atikamekw. (2014). Déclaration de souveraineté d'Atikamekw Nehirowisiw. Retrieved from https://www.atikamekwsipi.com/fr/publications

Coulthard, G. S. (2022). In L. Éditeur (Ed.), *Peaux rouges, masques blancs - contre la politique coloniale de la reconnaissance*. University of Minnesota Press.

Cousineau, M.-È. (2014, September 9). *Les Atikamekw déclarent leur souveraineté*. Radio-Canada. Retrieved from https://ici.radio-canada.ca/nouvelle/683598/nation-atikamekw-declaration-souverainete-territoire

Davis, A. E., & Blackwell, J. (2023). Decolonising Australia's international relations? A critical introduction. *Australian Journal of Politics and History, 69*(3), 405–421. https://doi.org/10.1111/ajph.12947

Dunne, T. (1998). *Inventing international society: A history of the English school* (1st ed.). Macmillan Press Ltd.. https://doi.org/10.1057/9780230376137

Dussart, F., & Poirier, S. (2017). In F. Dussart & S. Poirier (Eds.), *Entangled territorialities - negotiating indigenous lands in Australia and Canada*. University of Toronto Press.

Éthier, B. (2020). Analyzing entangled territorialities and indigenous use of maps: Atikamekw Nehirowisiwok (Quebec, Canada) dynamics of territorial negotiations, frictions, and creativity. *Canadian Geographer, 64*(1), 32–48. https://doi.org/10.1111/cag.12603

Fagoaga, N., & Beaucaire, K. (2024). La privatisation de l'énergie éolienne et l'impact sur la mission d'Hydro-Québec. Retrieved from https://irec.quebec/publications/rapports-de-recherche/la-privatisation-de-lenergie-eolienne-et-limpact-sur-la-mission-dhydro-quebec

First Nations Clean Energy Network. (2024). First Nations Energy Projects. from https://www.firstnationscleanenergy.org.au/energy-projects. Accessed May 17, 2024.

Frankel, B. (1996). In B. Frankel (Ed.), *Roots of realism*. Routledge Taylor & Francis Group.

Gangalidda Garawa Native Title Aboriginal Corporation. (2014). Garawa land and sea country plan.

Godin, P. (2019). *Hydro-Québec 2020 Les défis du siècle de la transition énergétique*. Institut de recherche en économie contemporaine.

Government of Canada. (2023). Northern REACH program. Retrieved from https://www.rcaanc-cirnac.gc.ca/eng/1481305379258/1594737453888

Government of Canada. (2024a). Clean energy for rural and remote communities program. Retrieved from https://natural-resources.canada.ca/reducingdiesel

Government of Canada. (2024b). Indigenous communities in Quebec. Retrieved from https://www.sac-isc.gc.ca/eng/1634312499368/1634312554965

Gunningham, N., & Bowman, M. (2016). Energy regulation for a low carbon economy: Obstacles and opportunities. *Environmental and Planning Law Journal, 33*(2), 118–141.

Hoicka, C., Savic, K., & Campney, A. (2021). Reconciliation through renewable energy. *Energy Research & Social. Science, 74*. https://doi.org/10.1016/j.erss.2020.101897

Hydro-Québec. (n.d.). Helping you understand your electricity rate. Retrieved from https://www.hydroquebec.com/residential/customer-space/account-and-billing/understanding-bill/helping-understand-your-electrcity-rate.html

Hydro-Québec. (2017). *Programme Systèmes industriels—Réseaux autonomes et modalités propres à ceux-ci*. Retrieved from https://www.hydroquebec.com/data/affaires/pdf/aff-reseaux-autonomes-fevrier2017-fr.pdf

Hydro-Québec. (2022). *Plan d'approvisionnement 2023–2032* (No. HQD-1).

Hydro-Québec. (2023). *Comparison of electricity prices in major North American cities*. Retrieved from http://www.hydroquebec.com/publications/en/comparison_prices/index.html%5Cn, http://www.hydroquebec.com/publications/en/corporate-documents/comparaison-electricity-prices.html

Keohane, R., & Nye, J. (1973). Power and Interdependance. *Survival Global Politics and Strategy, 15*(4), 158–165. https://doi.org/10.1080/00396337308441409

Laboucan-Massimo, M. (2022). *Waves of Change: Indigenous clean energy leadership for Canada's clean, electric future*. Retrieved from https://climateinstitute.ca/wp-content/uploads/2022/02/ICE-report-ENGLISH-FINAL.pdf

Lawson, S. (2017). Theorizing international relations. In *International relations* (3rd ed., pp. 32–66). Polity Press. Retrieved from https://ebookcentral.proquest.com/lib/anu/reader.action?docID=4790367&ppg=137

Lloyd, B., Lowe, D., & Wilson, L. (2000). *Renewable energy in remote Australian communities (a market survey) (final repo)*. Australian CRC for Renewable Energy Ltd.

Mcmaster, R., Noble, B., Poelzer, G., & Menghwani, V. (2023). Local capacity for energy transition in northern and indigenous communities: Analysis of gwich'in communities in Northwest Territories. *Canada. Arctic, 76*(2), 143–159. https://doi.org/10.14430/arctic77183

Million, D. (2015). Epistemology. In S. Nohelani Teves, A. Smith, & M. Raheja (Eds.), *Native studies keywords* (pp. 339–346). University of Arizona Press. https://doi.org/10.2307/j.ctt183gxzb.28

Mitchell, A. D., & Casben, J. (2016). Natural resources and energy regulation in Australia: The energy white paper in context. In M. Matsushita & T. J. Shoenbaum (Eds.), *Emerging issues in sustainable development: International trade law and policy relating to natural resources, energy, and the environment* (pp. 3–25). Springer Japan. https://doi.org/10.1007/978-4-431-56426-3_1

Moreton-Robinson, A. (2016). Relationality: A key presupposition of an indigenous social research paradigm. In C. Andersen & J. M. O'Brien (Eds.), *Sources and methods in indigenous studies* (pp. 69–77). Taylor & Francis Group. https://doi.org/10.4324/9781315528854

Morphy, F., & Mahood, K. (2019). Representations of space and place. In B. Arthur & F. Morphy (Eds.), *The Macquarie atlas of indigenous Australia* (2nd ed.). Macquarie Dictionary Publishers.

National Indigenous Australians Agency. (n.d.). Indigenous protected areas (IPA). Retrieved from Australian Government website: https://www.niaa.gov.au/our-work/environment-and-land/indigenous-protected-areas-ipa

Natural Resources Canada. (2024). *Energy fact book 2024–2025*. Retrieved from https://energy-information.canada.ca/en/energy-facts/clean-power-low-carbon-fuels

Nepal, R., & Menezes, F. (2017). Regulatory reforms in small energy systems: Experience from Australia's Northern Territory electricity market. *Economic Papers, 36*(3), 300–316. https://doi.org/10.1111/1759-3441.12181

Ngardara Cooperative. (2023). Ngardara project. from https://www.ngardaraproject.org.au/. Accessed May 17, 2024.

Nolan, K. (2024, May 16). First Nations energy exclusion hurts Australia's global emission target. *National Indigenous Time*. Retrieved from https://nit.com.au/16-05-2024/11433/first-nations-energy-exclusion-hurts-australias-global-emission-target

North Australian Indigenous Land and Sea Management Alliance. (2020). *Waanyi and Garawa land use and economic diversification plan*. Developing Northern Australia. Retrieved from https://crcna.com.au/wp-content/uploads/2024/05/waanyigarawalanduseplan_nov2020.pdf

Nursey-Bray, M., Palmer, R., Smith, T. F., & Rist, P. (2019). Old ways for new days: Australian indigenous peoples and climate change. *Local Environment, 24*(5), 473–486. https://doi.org/10.1080/13549839.2019.1590325

O'Bryan, M., Fogarty, W., Mununggurr, B., & Brachtendorf, L. (2024). *Laynhapuy homelands housing report*. Australian National University.

O'Neill, L. (2015). The role of state governments in native title negotiations: A tale of two agreements. *Australian Indigenous Law Review, 18*(2), 29–42.

Original Power. (n.d.). Borroloola's Ngardara Solar Microgrid Project. Retrieved from https://www.originalpower.org.au/borroloola

Ostrom, V., Tiebout, C., & Warren, R. (1961). The Organization of Government in metropolitan areas: A theoretical inquiry. *American Political Science, 55*(4), 831–842. https://doi.org/10.2307/1952530

Parliament of Australia. (n.d.). Chapter 2-The Energy White Paper. Retrieved from https://www.aph.gov.au/Parliamentary_Business/Committees/Senate/Environment_and_Communications/Completed_inquiries/2004-07/energywhitepaper/report/c02

Poirier, S. (2017). Nehirowisiw territoriality: Negotiating and managing entanglement and coexistence. In F. Dussart & S. Poirier (Eds.), *Entangled territorialities: Negotiating indigenous lands in Australia and Canada* (pp. 212–234). University of Toronto Press. https://doi.org/10.3138/9781487513764-011

Poirier, S., Jérôme, L., & Société d'histoire atikamekw (Nehirowisiw Kitci Atisokan). (2014). Présentation: Les Atikamekw Nehirowisiwok: territorialités et savoirs. *Recherches Amérindiennes Au Québec, 44*(1), 3–10. https://doi.org/10.7202/1027875ar

Riley, B. (2021). Scaling up: Renewable energy on aboriginal lands in north West Australia. *Nulungu Research Papers, 6*(3), 1–23. https://doi.org/10.32613/nrp/2021.6

Rioux-Gobeil, F., & Thomassin, A. (2024). A just energy transition for indigenous peoples: From ideal deliberation to fairness in Canada and Australia. *Energy Research and Social Science, 114*, 1–8. https://doi.org/10.1016/j.erss.2024.103593

South Australian First Nations Voice to Parliament. (2024). About the Voice. Retrieved from https://www.firstnationsvoice.sa.gov.au/about

Denis, G., & Parker, P. (2009). Community energy planning in Canada: The role of renewable energy. *Renewable and Sustainable Energy Reviews, 13*(8), 2088–2095. https://doi.org/10.1016/j.rser.2008.09.030

Stringer, T., & Joanis, M. (2023). Decarbonizing Canada's remote microgrids. *Energy, 264*, 1–11. https://doi.org/10.1016/j.energy.2022.126287

Thomassin, A., Murray, M., & Rioux-Gobeil, F. (n.d.). *The everyday performance of indigenous sovereignties*. Anthropologica.

Tremblay, A. (2019, November 27). L'usine WestRock menacée par le projet BELT à La Tuque? In *Le Nouvelliste*. Retrieved from https://www.lenouvelliste.ca/2019/11/28/lusine-westrock-menacee-par-le-projet-belt-a-la-tuque-84b668d4c38119cf658d5c0fb1655a7d/

Tremblay, A. (2022, December 22). Bioraffinerie: «Le projet va avoir lieu». *Le Nouvelliste.*. Retrieved from https://www.lenouvelliste.ca/2022/12/23/bioraffinerie-le-projet-va-avoir-lieu-5fa026382f3ed23f740740ffc7f147c7/

Whitmore, J., & Hopkin, M. (2015, April 8). Energy white paper promises privatisation and lower prices: Experts respond. *The Conversation.*. Retrieved from https://theconversation.com/energy-white-paper-promises-privatisation-and-lower-prices-experts-respond-39853

Whyte, K. (2017). Indigenous climate change studies: Indigenizing futures, decolonizing the anthropocene. English Language Notes, 55(1–2), 153–162. doi:https://doi.org/10.1215/00138282-55.1-2.153.

Whyte, K. (2018). Settler colonialism, ecology, and environmental injustice. *Environment and Society: Advances in Research, 9*(1), 125–144. https://doi.org/10.3167/ares.2018.090109

Winfield, M., Morrissey Wyse, S., & Kaiser, C. (2021). Enabling community energy planning? Polycentricity, governance frameworks, and community energy planning in Canada. In K. Calvert & T. Gunton (Eds.), Canadian planning and policy (pp. 35–54). doi:10.24908/cpp.

CHAPTER 10

Indigenous Ownership in Canada's Energy Transition: A Pathway to Sustainable Infrastructure Development

Andre Poyser

Introduction

Contextualizing Indigenous Involvement in Energy Projects

Energy infrastructure has become a critical area of focus in the drive to transition the global economy from fossil fuels to renewable energy and electrify the future. Due to its rich natural resource base, Canada has long been a major player in global energy supply and is set to play an even greater role in the energy transition given its deposits of critical transition minerals. Canada already supplies 13 of the 35 rare earth elements (REEs) considered critical by the United States Department of Energy and, in addition to being the leading producer of potash, is among the top five producers of uranium, niobium, titanium, palladium,

A. Poyser (✉)
Auckland University of Technology, Auckland, New Zealand
e-mail: andre.poyser@aut.ac.nz

T. Walker et al. (eds.), *Wiring the Future*, Palgrave Studies in Emerging Risk Management and Sustainable Finance,
https://doi.org/10.1007/978-3-032-14717-2_10

aluminium, platinum, nickel, and graphite (Munzar, 2021a). While the People's Republic of China (PRC) dominates processing and refining of REEs, Canada has begun the process of building out its capabilities in this area through the implementation of the 2022 Canadian Critical Minerals Strategy. A key component of this strategy is the role of indigenous groups in the ownership of critical energy infrastructure and partnerships for the extraction of REEs on indigenous lands. This represents a major shift in resource and energy infrastructure development in Canada. Historically, the provincial and federal governments have excluded indigenous communities in Canada from participation in natural resource projects carried out within their territories and have often had to bear the negative social, environmental, and economic consequences of these major development projects.

A report commissioned by the Royal Commission on Aboriginal People in 1994 documents a complex history of the relationship between indigenous communities, energy infrastructure projects, and resource development in Canada. At the heart of this historical account is the tension between indigenous values relating to the land and Canada's need for industrial development, employment, and economic growth (DesBrisay, 1994). Such a clash of these objectives is typified in the 1974 Mackenzie Valley Pipeline Inquiry, which the government commissioned to investigate the social, environmental, and economic impact of a proposed gas pipeline that would run through the Yukon and the Mackenzie River Valley of the Northwest Territories.

The Mackenzie Valley Pipeline Inquiry, led by Justice Thomas Berger, was a landmark event in Canadian history, and profoundly influenced the ways in which energy companies and regulators consider Indigenous rights and environmental concerns in energy and infrastructure projects. The inquiry was established by the federal government to examine the potential impacts of a proposed natural gas pipeline that would stretch from the Beaufort Sea through the Mackenzie Valley to southern Canada. This project, which promised significant economic benefits, also posed serious risks to the environment and the Indigenous communities living in the affected regions.

The inquiry was unique in its approach, particularly in its engagement with the Indigenous peoples of the Northwest Territories and the Yukon. Over two years, Justice Berger conducted extensive hearings across 35 Indigenous communities, allowing thousands of Indigenous individuals to voice their opinions and concerns. These hearings revealed a deep and widespread opposition to the pipeline among Indigenous communities, not merely due to environmental concerns but also because of the potential cultural and social upheavals it would cause. Historically, the government has systemically excluded Indigenous communities in Canada from decision-making processes concerning resource development on their lands. The Mackenzie Valley Pipeline Inquiry highlighted this exclusion, bringing to light the broader context of Indigenous disenfranchisement in Canada. Indigenous peoples have long faced the encroachment of industrial projects on their lands without their consent, leading to environmental degradation, disruption of traditional ways of life, and socio-economic displacement. The inquiry's findings underscored that these communities were not just concerned about the immediate impacts of the pipeline but were also wary of losing control over their lands and resources, which are integral to their identity and survival.

Justice Berger's report, published in 1977, recommended a ten-year moratorium on pipeline construction in the Mackenzie Valley, arguing that Indigenous land claims needed to be settled before any further development could proceed. This recommendation was unprecedented at the time, marking a significant departure from the government's usual prioritization of industrial development over Indigenous rights. The report emphasized that the social costs of the pipeline—particularly the potential for cultural disintegration and increased dependency on wage labour—outweighed the economic benefits. The Mackenzie Valley Pipeline Inquiry serves as a pivotal moment in the history of Indigenous involvement in energy projects in Canada as it set a precedent for the inclusion of Indigenous voices in the planning and execution of large-scale infrastructure projects. This inquiry also catalysed the movement towards recognizing Indigenous land rights and their role as stewards of the environment. In the years following the inquiry, there has been a gradual but significant shift towards more equitable partnerships between the Canadian government, industry, and Indigenous communities in energy and resource projects.

Contextualizing Indigenous involvement in energy projects today requires an understanding of this historical exclusion and the profound impact it has had on Indigenous communities. The Mackenzie Valley Pipeline Inquiry illustrates the broader struggles Indigenous peoples have faced in asserting their rights and protecting their lands from unwanted industrial encroachment. It also highlights the importance of their involvement in the decision-making processes that affect their lands and futures. Today, Indigenous ownership and participation in energy projects have become essential to achieving both sustainable development and reconciliation, as exemplified in the inclusion of indigenous partnerships as a key component of the Canadian Critical Minerals Strategy. The lessons from the Mackenzie Valley Pipeline Inquiry continue to inform contemporary discussions about the rights of Indigenous peoples in Canada and their critical role in the country's energy transition. This historical context underscores the need for continued vigilance to ensure that Indigenous communities are not only consulted but are also active partners in the development of projects that affect their lands and livelihoods.

Purpose and Scope of Chapter

Building on the legacy of the Mackenzie Valley Pipeline Inquiry, this chapter will examine the evolving landscape of Indigenous ownership in energy projects across Canada. It will explore the legal, financial, and regulatory frameworks that have enabled Indigenous groups to secure equity stakes in major energy projects. Furthermore, this chapter will analyze case studies of successful Indigenous ownership deals, drawing on data from the Canada Energy Regulator and the Indigenous Resource Network.

Through this analysis, this chapter aims to shed light on the broader implications of indigenous ownership for Canada's energy transition. It will argue that Indigenous ownership not only contributes to more equitable and sustainable infrastructure development but also plays a crucial role in advancing indigenous self-determination and economic empowerment. By engaging indigenous communities as partners rather than passive stakeholders, Canada can foster a more inclusive energy transition that respects Indigenous rights and contributes to the long-term sustainability of the nation's energy infrastructure.

Literature Review

Much of the literature on indigenous equity in energy projects and infrastructure development has been in the field of law, where legal scholars have sought to document the legal instruments and arrangements that energy companies use to negotiate deals with indigenous communities. Isaac and Hoekstra (2020) focus on Impact Benefit Agreements

(IBAs) and argue that project proponents primarily employ IBAs as risk management tools. IBAs are contractual agreements between Indigenous communities and companies developing energy or infrastructure projects that may affect these communities. These agreements often include monetary provisions—such as revenue-sharing mechanisms, lump-sum payments, or funding for community programs—to secure financial benefits for Indigenous groups in exchange for project approvals. IBAs have grown out of a string of legal successes by various First Nations groups in the Canadian courts following the Haida and Taku Supreme Court decisions, handed down in 2004, which imposed a duty of consultation and accommodation of indigenous peoples prior to the initiation of development projects. The duty to consult doctrine, along with the free, prior, and informed consent doctrine, have become embedded into the legal ecosystem within Canada and have forced energy and infrastructure developers to proactively engage IBAs with indigenous communities in order to avoid the legal liability that may hinder their projects, particularly if they have progressed far into the development cycle (Coates, 2016). The establishment of the Canadian Truth and Reconciliation Commission in 2015 and the ratification of United Nations Declaration on the Rights of Indigenous Peoples (UNDRIP) by the Canadian Parliament in 2021 have further created a legal and political environment which prioritizes the needs and concerns of Canada's indigenous communities in energy and infrastructure development.

As environmental, social, and governance (ESG) concerns have become more prominent, energy and infrastructure developers have shifted from the risk mitigation motives which drove the use of IBAs towards a more partnership-oriented model which involves greater involvement of indigenous communities through ownership stakes (Warrier et al., 2021). The difference between IBAs and the emerging indigenous equity structures is the provision of long-term stable revenue as opposed to one-off payments and the involvement of indigenous communities in project governance and management. Legal scholars have described the shift towards equity ownership as an essential component of the reconciliation process, as it allows First Nations communities in Canada to fulfil their stewardship duty towards the land in keeping with their cultural traditions. Through engaged partnership in natural resource project development, First Nation leaders and elders ensure that sustainability, environmental protection, and indigenous rights are embedded and upheld through every stage of the project development life cycle (Nelson, 2019). Additionally, equity ownership provides significant resources for indigenous communities to undertake endeavours like capacity building—such as developing skills for project management, strengthening governance structures, and fostering technical expertise, training and education opportunities, and the revitalization of indigenous languages and cultural practices, all of which collectively contribute to environmental protection and combating climate change.

The literature has pointed to a variety of legal structures through which energy companies facilitate indigenous ownership in energy projects and infrastructure development. According to Warrier et al. (2021), the most common legal structure is a limited partnership arrangement. The limited partnership is generally formed by a single First Nation group or a consortium of these groups who invest through independently registered and managed investment vehicles, called special purpose investment vehicles (SPVs). The primary benefit of the limited partnership as a legal entity is the tax advantages it offers for indigenous communities. Canadian tax law grants tax exemption to any person or entity if such person or entity is a municipality in Canada performing a government function and 2016 ruling issued by the Canadian Revenue Agency (CRA) accorded municipal status to First Nation governments in accordance with an interpretation of the Indian Act which designates the indigenous people of Canada as separate and sovereign nations in their own right (Canada Revenue Agency, 2016).

Despite the tax advantages of limited partnership in a SPV, First Nation groups still face the challenge of financing the SPV. Legal scholars have documented that the main challenges include a lack of access to private or federal loans due to credit risk issues and requirements to obtain a guarantee or backstop support (Warrier et al., 2021). First Nation SPVs that seek to obtain financing without sufficient equity or underlying assets to serve as collateral for the loan are designated as having a high credit risk and, as such, are required to have a guarantor or other supporting backstop measures. These requirements have hindered many First Nation groups in Canada in advancing co-investment in energy and natural resources projects. Recent government support programmes have, however, alleviated some of these issues through the provision of loans and loan guarantees to First Nation entities seeking to engage in energy projects. These loan and financing support programmes are mainly provided through the First Nations Major Projects Coalition (FNMPC),

Alberta Indigenous Opportunities Corporation (AIOC), First Nations Finance Authority (FNFA), and the Canada Infrastructure Bank.

Other legal structures First Nations groups utilize to facilitate indigenous ownership of energy projects and infrastructure development include corporations, joint ventures, Community Economic Development Corporations (CEDCs), and trusts. Corporations and joint ventures function in a similar manner as limited partnerships, while CEDCs and trusts are tailored specifically for community-led investment projects and aim to facilitate employment and economic development within indigenous communities. Trusts are deployed in situations where indigenous communities want to ensure that the future benefits of energy and infrastructure projects flow to future generations.

Data and Methodology

The Canada Energy Regulator (CER) is the federal agency in Canada which regulates, pipelines, energy product development, and energy supply. Before a company can build a federally regulated pipeline, power line, or offshore renewable energy project, it must first apply to the CER. The CER reviews and assesses proposed projects, including potential impacts to people, property, and the environment, as well as examines a company's engagement activities. Before construction or installation can begin, the CER reviews and must give its approval for all major energy and infrastructure projects to ensure that they comply with safety standards, economic considerations, and social and environmental assessments. A critical component of this review process is the crown consultation process, which is triggered when there is a potential for infringement of the rights and interests of Canada's Indigenous population. The CER provides an online database of project applications on which it has issued decisions and projects which it is currently reviewing. The database includes detailed project documentation, records of hearings, crown consultations, and regulatory documents. This chapter will utilize content analysis and a case study approach to identify applications which involve Indigenous participation and crown consultation. It will present an analysis of these cases to empirically demonstrate the trend in increased ownership of Indigenous groups in Canada's energy infrastructure.

The Indigenous Resource Network (IRN) is an advocacy group of indigenous leaders and organizations involved in indigenous participation in resource development. As part of its Ownership is Reconciliation campaign, the network collects and publishes data on indigenous equity deals in Canada. This chapter will also take a content analysis and case study approach to analyze the data provided by the IRN.

Trends in Indigenous Ownership of Energy Infrastructure

Over the last 15 years, Canada has announced a total of 135 energy infrastructure projects with indigenous ownership. A significant percentage (28%) of these projects were established by at least 111 Indigenous communities within the last two years (2022–2024). The dominant energy sector for these indigenous equity investments made in Canada between 2022 and 2024 is wind (35%), followed by oil and gas (15%) with solar and transmission having an equal share of 13% of the announced investments (Carruthers & McKlusky, 2024). As indicated in Table 10.1, data from the CER and IRN databases indicate that the number of Indigenous equity deals has increased significantly in the last ten years.

The trends in Indigenous equity ownership of energy projects and infrastructure in Canada have evolved significantly over the past two decades. Historically, Indigenous communities were excluded from the benefits of natural resource development, despite bearing significant environmental and social costs (Angell & Parkins, 2011). However, a shift in legal and regulatory frameworks has enabled Indigenous communities to engage in resource development projects more meaningfully, particularly through equity ownership.

The primary catalyst of the shift in the legal and regulatory environment has been the provision of provincial and federal loan guarantee programmes that have enabled Indigenous groups to take equity stakes in energy and natural resource projects across Canada. The first such programme was the Ontario Aboriginal Loan Guarantee Program, established in 2009, that made it easier for Indigenous communities to acquire equity in clean energy projects (Exner-Pirot, 2023). This programme facilitated the growth of Indigenous equity in wind, solar, hydroelectricity, and transmission projects across the province. The federal government and other provincial governments, such as Alberta

Table 10.1 Indigenous equity deals in energy infrastructure projects listed in CER and IRN databases

Project	*First Nations/ Indigenous community involved*	*Project year*	*Project value*	*Indigenous equity dollar amount*	*Indigenous equity percentage*	*Sector*
Harrowby Potash	Gambler Nation	2023	$12M	$2.4M	20%	Mining
Mount Hayes LNG Terminal	Stz'uminus First Nation, Cowichan Tribes	2012	$200M	$5.7M	3%	LNG Terminal
Cedar LNG	Haisla Nation	2023	$3M	$1.5M	50%	LNG Terminal
East Tank Farm	Fort Mckay First Nation, Cree First Nation	2017	$1B	$503M	49%	Oil Sands
Access NGL Pipeline	Buffalo Lake Metis Settlement, Kikino Metis Settlement, Heart Lake First Nation, Whitefish Lake Fist Nation, Saddle Lake Cree Nation	2023	$103M	$240M	43%	Pipeline
Coastal Gaslink	Haisla Nation, Kitselas First Nation, Lax Kw'alaams Band, Lheidleh T'enneh First Nation, Metlakatla First Nation, McLeod Lake Indian Band, Witset First Nation, Nadleh Whut'en First Nation, Nak'azdli First Nation, Saik'uz First Nation, Skin Tyee First Nation, Stellat'en First Nation, Ts'il kaz Koh First Nation, West Moberly First Nations, Wet'suwet'en First Nation, Nee Tahi Buhn Indian Band	2022	$14.5B	$145M	10%	Pipeline

(continued)

Table 10.1 (continued)

Project	*First Nations/ Indigenous community involved*	*Project year*	*Project value*	*Indigenous equity dollar amount*	*Indigenous equity percentage*	*Sector*
Athabasca Pipeline	Beaver Lake Cree Nation; Cold Lake First Nation; Frog Lake First Nation; Heart Lake First Nation; Athabasca Chipewyan First Nation; Chard Métis Nation; Chipewyan Prairie First Nation; Conklin Métis Local 193; Kehewin Cree Nation; Buffalo Lake Métis Settlement; Onion Lake Cree Nation; Saddle Lake Cree Nation; Whitefish Lake First Nation (Goodfish); Fort Chipewyan Métis Local 125; Fort McKay First Nation; Fort McKay Métis Nation; Fort McMurray 468 First Nation; Kikino Métis Settlement; Elizabeth Métis Settlement; Fishing Lake Métis Settlement; McMurray Métis Local 1935; Mikisew Cree First Nation; and Willow Lake Métis Nation	2023	$10B	$1.1B	11.57%	Pipeline
Deerfoot and Barlow Solar	Chiniki First Nation and Goodstoney First Nation	2023	$120M	$60M	51%	Solar

(continued)

Table 10.1 (continued)

Project	*First Nations/ Indigenous community involved*	*Project year*	*Project value*	*Indigenous equity dollar amount*	*Indigenous equity percentage*	*Sector*
Northern Courier Pipeline	Athabasca Chipewyan First Nation, Chipewyan Prairie Dene First Nation, Ft. McMurray #468 First Nation, Fort Chipewyan Metis Local 125, Fort McKay Metis Nation, Fort McMurray Local Council 1935, Willow Lake Metis Association, Conklin Metis Local 193	2021	$1.3B	$195M	15%	Pipeline
Peter Sutherland Sr. Generating Station	Taykwa Tagamou Nation	2017	$300M	$100M	33%	Hydro
Keeyask Generating Station	Tataskweyak Cree Nation, War Lake First Nation, York Factory First Nation, Fox Lake Cree Nation	2022	$3B	$750M	25%	Hydro
Northwest British Columbia Hydro	Tahltan Nation	2019	$2.5B	$152M	5%	Hydro
Lower Mattagami Project	Moose Cree First Nation	2015	$2.6B	$650M	25%	Hydro
Kokish River Project	Namgis First Nation	2014	$52M	$13M	25%	Hydro
Wuskwatim Project	Nisichawayasihk Cree Nation	2012	$1.6B	$528M	33%	Hydro
Lac Seul Project	Lac Seul Nation	2012	$47M	$11.75M	25%	Hydro
Awasis	Cowessess First Nation	2022	$21M	$20M	95%	Solar

(continued)

Table 10.1 (continued)

Project	*First Nations/ Indigenous community involved*	*Project year*	*Project value*	*Indigenous equity dollar amount*	*Indigenous equity percentage*	*Sector*
Nanticoke Solar	Mohawk First Nation, Cayuga First Nation, Onondaga First Nation, Oneida First Nation, Seneca First Nation, Tuscarora First Nation, Mississaugas First Nation	2019	$100M	$20M	20%	Solar
Waasigan Transmission Line	Couchiching First Nation, Eagle Lake First Nation, Fort William First Nation, Mitaanjigamiing First Nation, Nigigoonsiminikaaning First Nation, Ojibway Nation of Saugeen, Lac La Croix First Nation, Lac des Mille Lacs First Nation, Lac Seul First Nation, Seine River First Nation	2022	$1.2B	$600M	50%	Transmission
McMurray West Transmission Line	Athabasca Chipewyan First Nation, Bigstone Cree Nation, Lac Ste. Anne Métis Community, Mikisew Cree First Nation, Paul First Nation, Sawridge First Nation, Sucker Creek First Nation	2019	$1.7B	$680M	40%	Transmission

(continued)

Table 10.1 (continued)

Project	*First Nations/ Indigenous community involved*	*Project year*	*Project value*	*Indigenous equity dollar amount*	*Indigenous equity percentage*	*Sector*
NorthShore Transmission Line	Batchewana First Nation, Garden River First Nation, Thessalon First Nation, Mississauga First Nation, Serpent River First Nation, Sagamok Anishnawbek First Nation, Atikameksheng Anishnawbek First Nation, Wahnapitae First Nation	2023	1.25B	$625M	50%	Transmission
Bruce to Milton Transmission Line	Saugeen Ojibway Nation	2013	$240M	$72M	30%	Transmission
Henvey Inlet Wind	Henvey Inlet First Nation	2017	$1B	$500M	50%	Wind
Mesgi'g Ugju's'n Wind Farm	Mi'gmaq communities	2016	$330M	$165M	50%	Wind
Bow Lake Wind Farm	Batchewana First Nation	2015	$240M	$120M	50%	Wind
Adelaide Wind Farm	Aamjiwnaang First Nation	2015	$200M	$50M	25%	Wind
McLean's Mountain Wind Farm	United Chiefs and Councils of Mnidoo Mnising	2014	$135M	$67.5M	50%	Wind
Grand Renewable Energy Park	Grand River First Nations	2012	$364M	$36.4M	10%	Wind

and Saskatchewan, have followed suit with their own loan guarantee programs. Alberta's Indigenous Opportunities Corporation (AIOC) has been a notable success, providing over $500 million in loan guarantees since 2020. The loan guarantees work by allowing Indigenous communities to access loans from financial institutions without the need for traditional collateral, as the government acts as a guarantor for the loan and makes a commitment for its repayment should there be a default or inability to pay. With this greater ability to access capital without the hinderance of collateral, which Indigenous communities often lack due to legal restrictions on reservation lands and assets, these communities are able to put up the cash required to purchase stakes in major projects.

A direct outcome of the provincial and federal loan guarantee programmes has been a relatively rapid change from the basic revenue-sharing agreements which governed the previous relationships between the energy sector and Indigenous communities in Canada to more active engagement through co-ownership of projects. Indigenous equity deals, particularly in the energy sector, have grown significantly, with Indigenous communities acquiring or negotiating options for nearly $10 billion in energy and resource projects since 2012 (Exner-Pirot, 2023). This shift towards equity ownership creates a strategic alignment of interests between Indigenous communities and project developers, as both parties share incentives for long-term project success. By granting Indigenous communities a direct financial stake in projects, equity models foster more responsible and sustainable management practices, ensuring that projects prioritize environmental stewardship and community concerns alongside economic returns. It also provides project developers with social license, enabling smoother regulatory approvals and reducing the likelihood of protests or opposition.

Several high-profile deals have marked this evolution, including the Enbridge-Athabasca pipeline deal in 2022, where 23 First Nations and Métis communities in Alberta acquired an 11.57% stake in the pipeline system for $1.12 billion. This deal, the largest of its kind, exemplifies the growing trend of Indigenous ownership in major energy infrastructure. Similar transactions have taken place in the oil sands, renewable energy, and other resource sectors.

The role of government-backed loan guarantees has been pivotal, but these programs are not without limitations. Many Indigenous communities still face challenges in securing affordable financing, particularly for high-risk projects like mining or upstream oil and gas development. Additionally, most government-backed programs focus on low-risk projects with established revenue streams, limiting their applicability to more speculative ventures. Energy companies are exploring other mechanisms, such as investment tax credits, royalty trusts, and corporate share ownership, to provide more flexible and scalable tools for Indigenous engagement (Exner-Pirot, 2023).

In the federal budget for 2024, the Canadian government announced up to CAD $5 billion in Indigenous loan guarantees and outlined a financial strategy for the guarantees. The loans eligible for guarantee should range from CAD$20 million to CAD$1 billion and, once guaranteed by the federal government, are offered by banks and financial institutions at lower interest rates. Applicants must be Section 35 rights holders (Indigenous groups or subsidiaries) impacted by the proposed project, and projects are required to demonstrate financial viability. A notable feature of the financial strategy of the loan guarantees is that the government provides funding for legal, financial, and technical analysis. Stacking limits for federal-provincial guarantees cover up to one hundred percent of equity while federal stacking is capped at 75% of project costs.

An important development driving the sustainable energy transition in Canada is the fact that Indigenous communities are increasingly involved in environmental regulation and oversight of energy projects. The Squamish Nation, for example, developed an independent environmental review process for the Woodfibre LNG project, becoming both a partner and a regulator. This unique arrangement allowed the Squamish to implement their own environmental standards, ensuring that the project aligned with their cultural and environmental priorities. This approach sets a precedent for Indigenous-led environmental regulation in future energy projects (Exner-Pirot & Authier, 2023). In addition to regulatory involvement, Indigenous communities have also become involved in CER approval of proposed energy projects through the crown consultation process (von der Porten, Nilsen et al., 2024). Table 10.2 details CER-approved projects that have been through the crown consultation process and the Indigenous communities that have been involved.

Case Study of Capital Mobilization for Cascade Power Plant

Indigenous communities, historically excluded by both government and industry from meaningful participation in economic development on their lands, are now leveraging equity ownership as a key mechanism for securing financial capital. The Cascade Power Plant Project, which

Table 10.2 Energy infrastructure projects approved by CER which underwent Crown Consultation Process

Project	*Project description*	*Project application date*	*CER decision date*	*Crown Consultation Process (CCP)*	*Indigenous groups involved in CCP*	*Information request rounds*
Westcoast Energy Inc.—Upper South Taylor Hill HDD Pipe Replacement Project	Replacement of 2,700 m of gas pipeline	18 October 2022	5 April 2022	No	None	3
Enbridge Pipelines Inc. Line 3 Replacement Programme	Replacement of 863.6 mm of crude oil pipeline	18 November 2014	1 December 2016	Yes	33	6
Enbridge Pipelines Inc.—Line 9B Reversal and Line 9 Capacity Expansion Project	Reversal in flow direction and expansion of oil pipeline	29 November 2012	6 March 2014	No	None	1
Enbridge Line 10 Westover Segment Replacement Project	Replacement of 32 km of gas pipeline	4 December 2015	26 January 2017	No	None	1
Enbridge Pipelines (NW) Inc.—Line 21 Segment Replacement Project	Installation of 2.5 km of new gas pipeline	10 March 2017	25 January 2018	No	None	1
Hydro-Québec TransÉnergie—Appalaches—Maine Interconnection Power Line Project	Construction of power line	30 September 2019	20 May 2021	No	None	4
Manitoba Hydro—Manitoba-Minnesota Transmission Project	Construction of power line	16 December 2016	15 November 2018	Yes	12	1

Project	*Project description*	*Project application date*	*CER decision date*	*Crown Consultation Process (CCP)*	*Indigenous groups involved in CCP*	*Information request rounds*
ITC Lake Erie Connector LLC—Lake Erie Connector Project	Construction of power line	22 May 2015	19 January 2017	Yes	3	1
NOVA Gas Transmission Ltd. System 2017 Expansion Project	Construction of new pipeline	31 March 2015	28 October 2016	Yes	10	1
NOVA Gas Transmission Ltd. System 2021 Expansion Project	Construction of new pipeline	20 June 2018	19 February 2020	Yes	15	1
NOVA Gas Transmission Ltd.—Edson Mainline Expansion Project	Construction of new pipeline	13 April 2019	23 June 2021	Yes	13	1
Trans Mountain Pipeline ULC—Trans Mountain Expansion Project	Construction of new pipeline	16 December 2013	18 June 2019	Yes	7	2
Westcoast Energy Inc.—Wyndwood Pipeline Expansion Project	Construction of new pipeline	21 October 2016	10 August 2017	No	None	1

secured a $93 million loan guarantee from the AIOC, demonstrates how Indigenous groups can harness financial instruments to gain a significant stake in large infrastructure projects (von der Porten & Podlasly, 2021).

This case illustrates that Indigenous communities can raise capital through innovative partnerships that include government-backed loan guarantees. The six First Nations involved in this project—Alexis Nakota Sioux Nation, Enoch Cree Nation, Kehewin Cree Nation, O'Chiese First Nation, Paul First Nation, and Whitefish Lake First Nation—formed a consortium, the Indigenous Communities Syndicate LP, to hold their equity stake (von der Porten & Podlasly, 2021). This structure allowed them to pool their resources and secure financing that would have otherwise been difficult to obtain due to the restrictions imposed by the Indian Act and the historical denial of Indigenous property rights (von der Porten, Edwards et al., 2022).

The role of the AIOC, a Crown corporation, was crucial in this process. By providing loan guarantees, the AIOC reduced the perceived financial risk associated with Indigenous investment, allowing the First Nations to secure more favourable financing terms from lenders and investors (von der Porten & Podlasly, 2021). This not only provided the necessary capital but also ensured that the First Nations could maintain control over their investment and decisions regarding the project.

One of the central challenges for Indigenous communities in accessing financial capital is the colonial legacy embedded in the Indian Act, which prevents First Nations from using their lands or other assets as collateral for loans (von der Porten, Savic et al., 2024). This systemic barrier has long limited the ability of Indigenous communities to participate in economic ventures on their own lands. In the case of the Cascade Power Plant Project, the involvement of the AIOC and its loan guarantee addressed this issue by providing a workaround to the limitations of the Indian Act. Instead of relying on traditional forms of collateral, the First Nations were able to leverage government support to secure financing. This model demonstrates that with the right financial tools and institutional backing, Indigenous communities can overcome historical barriers and participate as equity owners in major projects (von der Porten & Podlasly, 2021; von der Porten et al., 2022).

Furthermore, the participation of multiple First Nations in a consortium allowed them to collectively access capital and share both the risks and rewards of the investment. This collaborative approach not only strengthened their bargaining position but also enabled them to undertake a project of this scale, which would have been challenging for any single community to finance independently.

Another key factor in the success of this project was the partnership between Indigenous communities and private investors. The Cascade Power Plant Project involved major private sector partners, including OPTrust, Axium Infrastructure, and DIF Capital Partners, who contributed to its financing and development (von der Porten & Podlasly, 2021). These partnerships were mutually beneficial. For the private sector, having Indigenous equity ownership in the project reduced the risk of opposition from Indigenous communities, which has historically been a significant concern in infrastructure projects on Indigenous lands. By securing the consent and participation of the First Nations, the project minimized the likelihood of legal challenges or protests, making it a more attractive investment for institutional investors (von der Porten & Podlasly, 2021). For the Indigenous communities, partnering with private investors provided access to additional capital and expertise, which were crucial for the successful development of the project. It also allowed them to participate in a revenue-generating venture that will contribute to their long-term economic resilience. By holding an equity stake, the First Nations stand to benefit not only from the immediate financial returns but also from the broader economic opportunities created by the project, such as procurement, employment, and skills development (von der Porten & Podlasly, 2021).

The government, at both the provincial and federal levels, plays a pivotal role in enabling Indigenous communities to access the financial sector and raise capital. The creation of the AIOC by the Government of Alberta is an example of how provincial governments can facilitate Indigenous investment by providing financial tools like loan guarantees. This type of government support is essential for levelling the playing field for Indigenous communities, who often face higher borrowing costs due to the perception of increased financial risk (von der Porten, Nilsen et al., 2024). By offering loan guarantees, the government not only reduces the cost of capital for Indigenous investors, but also signals to the private sector that Indigenous participation in projects is both feasible and desirable (von der Porten, Savic et al., 2024).

In addition to financial support, government policies and frameworks such as the Truth and Reconciliation Commission's Call to Action #92 and the UNDRIP have created a legal and moral imperative for governments and corporations to include Indigenous communities in the decision-making processes for projects on their lands (von der Porten & Podlasly, 2021). The Cascade Power Plant Project is an example of how energy companies and Indigenous communities have operationalized these frameworks, with Indigenous equity ownership becoming a standard feature of major infrastructure projects (von der Porten & Podlasly, 2021).

The Cascade Power Plant Project showcases how Indigenous communities in Canada can mobilize the financial sector to raise the capital needed for large-scale, long-term investments in environmental sustainability and economic resilience. By leveraging government-backed financial instruments, forming partnerships with the private sector, and asserting their rights as equity owners, Indigenous communities are overcoming historical barriers and shaping the future of development on their lands.

This model has the potential to be replicated in other sectors and regions, providing a blueprint for how Indigenous communities can take control of their economic futures while contributing to broader goals of sustainability and reconciliation. As the demand for ESG-conscious investments continues to grow, projects like the Cascade Power Plant will likely become more common, setting a new standard for Indigenous participation in the financial and economic life of Canada.

Economic and Social Impacts

Indigenous ownership in the energy sector has led to transformative economic and social benefits. From financial returns and job creation to improvements in infrastructure, healthcare, and education, these projects are helping to revitalize Indigenous communities. Furthermore, the control that Indigenous nations now have over energy projects empowers them to make decisions that align with their cultural values and environmental priorities, ensuring that development occurs in a sustainable and inclusive manner. As Indigenous participation in energy continues to grow, it will play an increasingly central role in shaping both local and global energy futures.

A key advantage of ownership is the long-term financial returns from projects. As equity holders, Indigenous nations are no longer merely passive recipients of royalties; they have a stake in the ongoing profitability of energy ventures (von der Porten et al., 2022). This allows them to generate substantial revenue that can be reinvested into local economies. For example, in the natural gas sector, Indigenous nations are gaining control over critical infrastructure like pipelines and LNG terminals, which offer significant returns over the long term, especially as global demand for natural gas grows (Exner-Pirot & Authier, 2023).

The economic benefits also extend beyond direct revenue generation. Indigenous-owned businesses provide services to energy projects, such as construction and logistical support, creating employment opportunities for local residents. Indigenous workers, who are often employed at higher wages in the energy sector compared to other industries, experience economic upliftment (Exner-Pirot & Authier, 2023). In this way, the financial returns from energy ownership go hand in hand with broader economic development in Indigenous communities.

Indigenous communities are increasingly reinvesting the economic benefits derived from their ownership in energy projects into essential community needs, such as healthcare, education, and infrastructure. This reinvestment is critical for many Indigenous communities, which have historically suffered from underfunding and inadequate services (Exner-Pirot & Authier, 2023). For example, several Indigenous nations have used profits from energy projects to fund community infrastructure, including new housing developments, healthcare facilities, and schools. These investments not only improve the quality of life for current residents but also help attract future generations to stay and contribute to community development (von der Porten et al., 2022). Moreover, the availability of capital allows these nations to pursue other entrepreneurial ventures, further diversifying their local economies and reducing dependence on external sources of funding. In addition to physical infrastructure, these nations also invest financial returns from energy projects in social services. Indigenous communities, particularly those in remote areas, often struggle with inadequate healthcare and revenue from energy projects has been instrumental in providing better healthcare access and facilities, addressing long-standing health disparities (von der Porten et al., 2022). Similarly, investments in education ensure that younger generations are better equipped to participate in economic activities, including energy-related sectors, thereby ensuring the sustainability of these projects.

One of the most significant social impacts of Indigenous ownership in the energy sector is the sense of empowerment and self-determination it brings to Indigenous nations. Historically marginalized by both government and industry in resource development, Indigenous communities are now gaining control over their lands and resources. This shift in ownership and governance allows them to make decisions aligned with their values, particularly regarding environmental stewardship (Exner-Pirot & Authier, 2023).

The ability to exert control over resource projects helps Indigenous nations protect their cultural heritage and environmental interests. For instance, the FNMPC has emphasized the importance of sustainable development that respects Indigenous rights and the environment. Indigenous ownership ensures that these projects meet high environmental standards, thereby minimizing harm to the land and promoting long-term ecological sustainability (von der Porten et al., 2022). This is particularly important as many energy projects take place on or near Indigenous lands, and their potential environmental impacts could affect the cultural and spiritual practices tied to these territories. In addition to environmental stewardship, Indigenous ownership also enhances governance capacity. Indigenous nations are increasingly taking on roles as regulators, project proponents, and business leaders. This involvement in governance ensures that energy projects serve the interests of Indigenous communities and adhere to social and environmental principles that reflect Indigenous values (von der Porten, Podlasly et al., 2022).

Conclusion and Recommendations

While the policy landscape is increasingly supportive of Indigenous ownership, several challenges remain that require targeted solutions. One of the persistent challenges is the difficulty Indigenous communities face in accessing affordable, competitive capital. The high cost of borrowing often undermines the potential financial benefits of equity ownership. The establishment of a national benefits-sharing framework could be instrumental in reducing these costs. The inconsistent application of UNDRIP and Free Prior and Informed Consent (FPIC) across Canadian provinces creates legal uncertainties for Indigenous groups and project developers alike. Harmonizing regulatory standards across jurisdictions is essential for streamlining Indigenous participation and ownership of clean energy projects.

Policy makers have made several recommendations for enhancing Indigenous ownership of energy infrastructure. While programs like

AIOC and Aboriginal Loan Guarantee Program (ALGP) are effective, they are limited in scope. Expanding these programs nationally or developing a national loan guarantee fund would provide broader financial support for Indigenous communities. Indigenous ownership should be coupled with governance rights, ensuring that Indigenous nations can shape the operational and environmental outcomes of energy projects. This aligns with the principles of self-determination and UNDRIP. Governments and the private sector should collaborate to fund long-term capacity-building programs that provide Indigenous communities with the skills and knowledge necessary to lead large-scale infrastructure projects. These initiatives should focus on both the technical aspects of project management as well as the broader regulatory and financial landscape of energy projects. Private sector partnerships with Indigenous nations can be highly beneficial for all parties involved and incentivizing these partnerships through tax credits or other financial mechanisms could increase the number of Indigenous-led clean energy projects.

The path towards increased Indigenous ownership of energy infrastructure in Canada is both a legal and economic imperative. As Canada transitions to a net-zero energy system, Indigenous nations must be positioned not only as participants but as leaders and owners of clean energy projects. This requires sustained policy reforms that address the financial, regulatory, and capacity-related challenges that Indigenous communities face.

References

Angell, A. C., & Parkins, J. R. (2011). Resource development and aboriginal culture in the Canadian north. *Polar Record, 47*(240), 67–79.

Berger, T. (1977). *Northern frontier, northern homeland: The report of the Mackenzie Valley Pipeline Inquiry Volume One*. Minister of Supply and Services Canada. https://publications.gc.ca/collections/collection_2015/bcp-pco/CP32-25-1977-1-eng.pdf

Canada Revenue Agency, Interpretation Bulletin IT-2016-064503117, "Indian Act Bands" (27 July 2016) [IT 2016].

Carruthers, A., & McKlusky, E. (2024). Update on trends in indigenous equity investments in Canada. *Indigenous Law Bulletin*. Fasken.

Coates, K. (2016). *First Nations engagement in the energy sector in the energy sector in Western Canada*. Indian Resource Council.

DesBrisay, D. (1994). *The impact of major resource development projects on aboriginal communities: A review of the literature.* Royal Commission on Aboriginal Peoples.

Exner-Pirot, H. (2023). *Indigenous equity and its growing role in Canadian energy and resource development.* Macdonald-Laurier Institute.

Exner-Pirot, H., & Authier, Z. (2023). *Ownership and beyond: How indigenous peoples ae shaping a more secure energy future.* Energy for a Secure Future.

Isaac, T., & Hoekstra, A. J. A. (2020). Canadian laws and realpolitik regarding indigenous-industry agreements. In I. Odumosu-Ayanu & D. Newman (Eds.), *Indigenous-Industry agreements, Natural resources and the law* (pp. 1–19). Routledge.

Munzar, A. (2021a). Infrastructure policy trends: Critical minerals mining in Canada. *University of Calgary School of Public Policy Publications, 14*(1).

Munzar, A. (2021b). Mackenzie Valley Gas Project Pipeline in retrospect. *University of Calgary School of Public Policy Publications, 14*(33).

Nelson, R. (2019). Beyond dependency: Economic development, capacity building and generational sustainability for indigenous people in Canada. *SAGE Open, 9*(3). https://doi.org/10.1177/2158244019879137

von der Porten, S., & Podlasly, M. (2021). *ESG-Indigenous case study: Cascade Power Plant project.* First Nations Major Project Coalition.

von der Porten, S., Podlasly, M., & Csicsai, P. (2022). *Indigenous leadership and opportunities in the net zero transition.* First Nations Major Project Coalition.

von der Porten, S., Edwards, N., Vickers, B., Mullen, D., & Podlasly, M. (2022). *National roundtable on Indigenous access to capital in Canada: Roundtable primer.* First Nations Major Project Coalition.

von der Porten, S., Nilsen, E., Podlasly, M., & Ransom, A. (2024). *Clean energy project types and opportunities.* First Nations Major Project Coalition.

von der Porten, S., Savic, K., Edwards, N., McCormick, J., & Cole Sayers, K. (2024). *National Indigenous electrification strategy: Strategy to accelerate Indigenous ownership of net zero infrastructure in Canada.* First Nations Major Project Coalition.

Warrier, V., Morrison, L., White, A., & Buffalo, S. (2021). Indigenous ownership of natural resource projects: A framework for partnership and economic development. *Alberta Law Review, 59*(2), 394–422.

CHAPTER 11

Gendered Aspects of Electrification

Caroline Zimm and Natalia de Assis Brasil Weber

The Gendered Dimensions of Electrification and Energy Access

This chapter will introduce a gender perspective on electrification and a broader perspective on energy services. First, we provide an introduction to gendered analyses of electrification and energy access. Next, we assess the overall status of modern energy access and its gaps. To this end, understanding how energy access is measured and how this is linked to gender roles is useful. Introducing broader concepts of energy services for wellbeing is likewise beneficial from a gender perspective in overcoming various gendered outcomes of electrification. The chapter introduces a mentoring program for women as well as case studies of how to increase empowerment and gender-sensitivity in energy projects.

C. Zimm (✉)
International Institute for Applied Systems Analysis, Laxenburg, Austria
e-mail: zimmc@iiasa.ac.at

N. de A. B. Weber
GWNET–Global Women's Network for the Energy Transition, Vienna, Austria
e-mail: weber.nati@gmail.com

T. Walker et al. (eds.), *Wiring the Future*, Palgrave Studies in Emerging Risk Management and Sustainable Finance,
https://doi.org/10.1007/978-3-032-14717-2_11

Significance of Gender Analysis in Advancing Electrification

Electrification is a transformative force that has the potential to significantly improve lives across diverse communities and households—but men and women do not experience its benefits uniformly (Kohlin et al., 2011). Research that applies a gender lens to energy studies highlights the problem of women's exclusion from energy services and from influence in decision-making in energy-related matters (Clancy et al., 2020; Listo, 2018; Malakar & Day, 2020; Michael & Ahlborg, 2024).

Women and girls often face distinct challenges in energy-poor settings, where traditional roles and gender norms limit their access to modern energy services. For example, women disproportionately shoulder time—and energy–intensive tasks like cooking and fuel collection, exposing them to health risks, environmental hazards, and even threats of violence when collecting wood. The availability of modern energy, therefore, is more than a technical upgrade—it is a pathway to enhancing women's rights, safety, and economic opportunities (Standal & Winther, 2016; Upadhayay, 2009; Winther, 2008, 2014). By understanding how electrification affects all genders differently, we can create energy systems that promote equitable development and empowerment.

Historically, electrification projects have focused primarily on technical and economic factors, often overlooking the diverse needs of end-users. Yet, as more recent studies and practical experiences have revealed, gender-sensitive approaches to planning and implementing energy infrastructure can shape the reach and efficacy of these projects (Orlando et al., 2018). A gender-informed approach can clarify who benefits from electricity, who participates in its supply, and who gains from its expansion. This is especially relevant for addressing inequalities that may inadvertently arise in areas such as land use, forced displacement, and access to energy-related employment—issues that can have particularly acute impacts on women.

Planning that includes gender perspectives can also address social challenges such as unequal property rights, limited access to technical training, and risks of gender-based violence associated with electrification projects (Orlando et al., 2018). By anticipating these challenges, planners can work on more than providing energy; they can actively support women's participation in energy decision-making and employment, creating more inclusive energy systems that benefit entire communities equitably (Cecelski & Dutta, 2011).

However, existing infrastructure policies, financial mechanisms, and pricing structures rarely consider gender equity. For instance, policies on subsidies, tariffs, and energy reforms frequently fail to account for gendered differences in income, making female-headed households, which are often financially vulnerable, especially susceptible to tariff hikes. In many settings, male-dominated household decision-making further complicates energy access for women, who may lack financial autonomy. For energy policies to be genuinely inclusive, they must integrate the perspectives of all genders, ensuring that all voices contribute to shaping strategies that address the energy needs of everyone in the household and community.

A significant barrier to achieving this inclusivity is the lack of reliable data on gender differences in energy access and use. Traditional household electricity surveys rarely collect sex-disaggregated data, leaving policymakers in the dark about specific inequalities that affect how different genders benefit from electrification. Collecting and analyzing gender-disaggregated data is essential for assessing the impact of electrification on women's economic opportunities, health, and educational outcomes, as are more pluralistic methods such as participatory or observational approaches (Das et al., 2023). This data gap extends to the supply side as well, where more research is needed to quantify women's participation and the challenges they face in energy industry and employment.

Gender-sensitive policy design in electrification programs is only slowly evolving. From 2019 to 2021, gender sensitivity in electrification plans showed signs of improvement, particularly in Latin America and the Caribbean, with Sub-Saharan Africa making strides as well (Orlando et al., 2018). However, according to the World Bank's Regulatory Indicators for Sustainable Energy (RISE) framework, most regions scored below 29 out of 100 on addressing the specific needs of female-headed households (Orlando et al., 2018). By integrating gender analysis into the design, implementation, and evaluation of electrification efforts, policymakers and practitioners can develop more inclusive and equitable energy systems. These systems will not only bring light and power to underserved communities and households but also promote social progress by expanding opportunities and improving the quality of life for all genders, especially in times of a changing climate (Aguilar et al., 2015).

Global Progress in Electrification and Access to Modern Energy and Current Gaps

In the 2030 Agenda, the United Nations established ambitious goals to eradicate energy poverty and promote gender equality—fundamental steps toward sustainable and inclusive global development. Sustainable Development Goal (SDG) 7, which focuses on affordable and clean energy (electricity, modern energy for cooking, and energy efficiency), and SDG 5, which is dedicated to gender equality, are central pillars supporting broader improvements in well-being, economic opportunity, and environmental resilience. Access to energy enhances quality of life and fuels economic growth by opening up services, jobs, and market opportunities (UNDESA, 2018).

Despite significant advancements, however, the journey to universal energy access remains challenging. By 2022, 91% of the world's population had electricity access (Target 7.1, indicator 7.1.1.), yet the remaining 9% translates to 685 million people still living without it—an increase of 10 million from the previous year. This concerning reversal signals the ongoing barriers to progress, with estimates suggesting that only 92% will be connected by 2030. An even more urgent issue lies with access to clean cooking fuels and technologies (indicator 7.1.2). Although 74% of the global population had access to clean cooking solutions in 2022, 2.1 billion people still relied on traditional fuels like wood, charcoal, coal, and kerosene for daily cooking, with only modest recent improvements (IEA, 2024). This predominantly affects the lives of women and girls, as they are the primary cooks—some studies report >99% of female respondents as primary cooks (Balakrishnan et al., 2019; Government of India, 2019)—and responsible for cooking fuel.

Actors in the energy sector treat these two targets of SDG 7 on access to electricity and clean cooking as distinct. In India, for example, since 2000, more than 900 million citizens have received electricity connections (IEA, 2021) thanks to ambitious government programs. Nearly 97% of households are now connected to the grid at high daily service hours (Agrawal et al., 2020). Yet, still 300 million people across the country do not use clean cooking fuelsPracticioners and policymakers tackle these two targets often tackled separately, as in the majority of unelectrified areas (except for some parts of southern Africa) electricity is not used for cooking. Efforts to achieve clean cooking have traditionally centered on enhancing traditional cookstoves, biomass and biogas stoves, and Liquified Petroleum Gas (LPG)-based cooking. However, with recent advancements, electric cooking is emerging as a viable option for an increasing segment of the population. It is now crucial to explore the advantages and challenges associated with various forms of electric cooking and to identify strategies for expanding access. Electric cooking opens up new possibilities for improving access to clean cooking solutions, offering a sustainable and scalable pathway to address health (see Section "Gendered Outcomes of Electrification"), environmental, and energy access challenges (see Section "Advancing Women's Inclusion in the Energy Sector for Just Transitions"). This will require improving the capacity and reliability of electricity supply infrastructure in these parts of the world, which is challenging given that many still lack access to electricity altogether. Addressing these disparities and accelerating progress in energy access will be crucial for reaching the SDGs by 2030.

Achieving Universal, Reliable, and Sustainable Access for Decent Living Services

Multi-tier Framework for Measuring Electricity Access

It is important to recognize that the access figures mentioned above do not reflect access to stable, 24-hour grid electricity for all. The World Bank has shifted from a simple binary "yes or no" approach to a more comprehensive Multi-Tier Framework (ESMAP, 2015) for measuring electricity access, which comes with several implications from a gender perspective. This framework evaluates electricity service across seven key attributes—capacity, availability, reliability, quality, affordability, formality, and health and safety—organizing them into six tiers of access, each with specific minimum requirements. To determine a household's overall access level, each attribute is assessed individually, and the household's tier is set by the lowest level scored across these attributes (ESMAP, 2015). Basic solar lights and home systems usually provide Tier 1 or Tier 2 levels of access, offering limited power, while mini-grids and grid connections generally support higher tiers, from Tier 3 up to Tier 5, with more robust and diverse energy services. Higher tiers allow for different productive uses of electricity and the transition to, for example, electric cooking, predominantly benefitting women and girls.

Electricity supply reliability remains a significant challenge, even in regions with high access rates like India, where frequent outages, load shedding, and brownouts increase the cost of electricity due to the need for back-up equipment, which may be unaffordable or unusable (Sedai et al., 2021). Despite substantial progress in providing connections, there has been little improvement in electricity reliability in many countries (Sedai et al., 2021). For grid-connected households, limited supply often results from lack of capacity, while off-grid households face constraints due to fuel availability and affordability (Aklin et al., 2017). These issues can hinder post-electrification decisions, like purchasing time-saving appliances, which are crucial for women's labor force participation and efficient time allocation for household tasks (Ferrant & Thim, 2019; Klasen, 2019). Since women typically spend more time at home and are tasked with home production due to social norms, they are disproportionately affected by unreliable electricity (Fletcher et al., 2019).

The Gendered Nature of Lacking Access to Clean Cooking

The Multi-Tier Framework survey data shows a positive link between women's empowerment and household energy access in many countries (Chandrasekaran et al., 2023). However, the study only looks at household-level electricity and does not include community-level energy services (e.g., education, health facilities) or energy for other services in the household (e.g., mobility, cooking if done with other energy carriers such as biomass or LPG). Given the dominant gender norms with regards to household chores such as cooking, a sole focus on providing access to electricity falls short of the challenge.

Decent Living Standards and Energy Services

A *Decent Living Standards* (DLS) (Rao & Min, 2017) approach offers a more holistic framework for understanding the essential energy needs that support wellbeing at both household and community levels. It moves beyond minimal energy access, focusing instead on energy sufficiency that meets basic human needs and enables human development. Rather than simply focusing on energy access or consumption levels at the household level, DLS emphasizes the range of services necessary to ensure a decent quality of life for all. This includes energy for crucial daily activities like food preparation and safe storage, which are essential for nutrition and food security, as well as reliable energy for educational environments that support learning, including lighting for studying after dark. Health services also benefit immensely from dependable power, enabling medical equipment operation, vaccine refrigeration, and hygiene practices that are vital for public health. Energy for sanitation systems further contributes to community health, supporting clean water access, which is crucial for women who are often responsible for supplying households with water. Similarly, energy for mobility is needed to access diverse services and participate in society. Finally, household and community lighting enhance safety, extend productive hours, and improve overall quality of life. By focusing on these interconnected services, the DLS framework emphasizes energy as a means for well-being, creating the conditions for individuals and communities to thrive and driving social equality by ensuring equitable access to essential services, which support education, health, and economic opportunities for all genders. A drawback of DLS is that the framework looks at the household and community levels but does not account for power dynamics within the household or conduct a deeper dive into individuals' needs.

Globally, there are significant gaps in access to DLS. These gaps are larger than the gaps in electricity access discussed above, and affect women's wellbeing. The highest rates of populations living below DLS are found in Sub-Saharan Africa, where over 60% of people lack basic services like electricity, clean cooking, decent housing, sanitation, transport, and water. South and Pacific Asia also face large gaps, especially in clean cooking and heating, often relying on biomass with negative health effects. North Africa, the Middle East, Latin America, and parts of Asia and Eastern Europe have moderate gaps in sanitation, water, heating, and clean cooking (Kikstra et al., 2021).

Gendered Outcomes of Electrification

The pathways to women's empowerment through electrification are multifaceted (Winther et al., 2017). While there is a need for more systematized knowledge on the electricity-gender nexus to disentangle the causal chain further, several insights are available.

Norms and Values, Agency and Gender Dynamics

The meaningful use of electricity goes beyond simply providing light—it is about enabling access to a range of transformative technologies like mobile phones, televisions, and household appliances. Television, in particular, has played a pivotal role in shifting gender norms, offering new perspectives on issues like domestic violence and the value of girls compared to boys. Studies have shown that access to television can significantly alter societal attitudes by exposing viewers to alternative gender roles and empowering women with knowledge they might not otherwise have had (Belmin et al., 2021; Hoque, 1988; Jensen & Oster, 2009; La Ferrara et al., 2012). For example, women who watch TV may gain awareness of family planning options, which can influence their reproductive choices, leading to smaller family sizes (Belmin et al., 2021). This exposure to different cultural norms and ideas allows women to challenge traditional expectations and adopt more progressive views on gender equality and family dynamics (Jensen & Oster, 2009; La Ferrara et al., 2012) (see Section "Advancing Women's Inclusion in the Energy Sector for Just Transitions"). With technological change, mobile phone and internet access are becoming increasingly important for providing people with access to information on reproductive health (Rotondi et al., 2020). A combination of energy access and education contributes to lower fertility rates, particularly in regions with high initial fertility, thereby supporting both gender equality and sustainable development goals (Belmin et al., 2021).

Access to electricity has significantly improved women's agency in household management (Grabher et al, 2024). In many contexts, entrenched cultural norms continue to dictate control over household finances and wealth-building decisions. For example, in Afghanistan and India, men generally control expenditure decisions, with women's income primarily allocated to family needs rather than investments that could build wealth (Standal & Winther, 2016). This power imbalance in financial decision-making persists despite increased access to energy. In contrast, in South Africa, targeted social protection policies and electrification subsidies for female-headed households have allowed some women to gain greater control over electricity-related decisions, showing that policy interventions can help shift gendered power dynamics (Annecke, 2005; Matinga, 2010).

Electrification projects that integrate women into energy supply chains, such as the installation of solar home systems, have demonstrated that gender-sensitive approaches can enhance women's agency and involvement in the energy sector (see Section "Advancing Women's Inclusion in the Energy Sector for Just Transitions"). In Zanzibar, women's participation in these projects provided them with new skills and greater involvement in decision-making, contributing to a shift in their social status and empowerment (Upadhayay, 2009) (see Section "Advancing Women's Inclusion in the Energy Sector for Just Transitions"). However, cultural norms and societal expectations can still restrict women's full participation, highlighting the need for interventions that not only provide technical training but also address the local gender norms that shape women's roles and opportunities. This underscores the importance of context-specific solutions that consider both the practical and cultural barriers women face in engaging with new technologies.

The usage patterns of electricity further reveal gendered dynamics, particularly in households with low levels of empowerment. Women in such households tend to use electricity primarily during the morning hours, between 6 a.m. and 12 p.m., with limited access or control over energy usage during the rest of the day. This finding reflects broader trends where men in the household often have greater access to and control over electricity use (Winther et al., 2018). Power dynamics within the household—where women in more empowered households experience less of a gap in electricity awareness and usage—play a significant role in shaping these disparities. The material and sociocultural barriers women face, such as limited investment in household appliances and the persistence of unequal power structures, reinforce gender inequalities in energy access and usage. These barriers highlight that while electrification can improve women's day-to-day management of household tasks, more fundamental shifts in gender relations and societal norms are needed for true empowerment.

Health and Wellbeing

Access to clean energy is a fundamental factor for improving human health, economic opportunities, and gender equity. The lack of clean energy significantly affects various aspects of development, For example, household air pollution caused by the use of biomass, kerosene, or coal for cooking leads to approximately 3.2 million premature deaths annually, with women and children shouldering the greatest health burden due to their roles in collecting firewood and cooking (Sagar et al., 2023) (see Section "Advancing Women's Inclusion in the Energy Sector for Just Transitions"). Access to electricity and modern cooking fuels can mitigate these risks by reducing exposure to indoor air pollution, improving respiratory health, and contributing to maternal health improvements.

Electrification supports broader socio-economic development by enabling essential health services, such as use of health diagnostic technologies and refrigerated medicine. Other social impacts, particularly for women and girls are related to security: for instance, 42% of households reported incidents of gender-based violence during firewood collection in Chad over a six-month period (CCA, 2021).

Changes in Time Use

Electrification has reshaped domestic labor for all genders, although women typically see greater impacts. Access to electrical appliances and lighting enables households to reallocate time spent on traditional domestic tasks, facilitating women's participation in education, income-generating activities, and community involvement. Women and children in developing countries can spend up to 10 hours a week gathering fuel (CCA, 2021). In an example from India, electricity access reduced the time spent collecting fuel, saving women and men about 3.3 hours monthly (Khandker et al., 2014). In Zanzibar, electric water pumps saved women up to 25 hours per week, freeing girls to go to school (Winther, 2008). In South Africa, the shift from firewood to electric stoves eased the domestic burden, particularly for women, and encouraged some men to participate in cooking and ironing, tasks traditionally assigned to women (Annecke, 2005; Matinga, 2010). Electrification has enabled a rescheduling of household activities, illustrating how electricity facilitates more equitable domestic workloads, sometimes challenging entrenched gender roles.

Social Implications and Resources

Access to electricity influences social interactions within households and communities. With electricity, women have more time for socializing, leisure, and personal agency. Television and mobile phones have become powerful tools for social connection and self-expression, especially for women. Television, in particular, has created new social spaces in homes; in South Africa, it has become a hub for social gatherings among youth, fostering community bonds (Matinga, 2010). Electrification also allows people to stay up later, though studies in the Philippines and Laos found that it sometimes reduced overall rest (IEG, 2008).

Electricity improves women's mobility and safety around the home, enhancing their confidence and security. Mobile phones can provide women with a crucial connection to family, offering support in emergencies (Tenhunen, 2014). These technologies not only strengthen women's social ties but also support their independence and assertiveness.

Education

Electrification's positive impact on children's education, particularly for girls, is well-documented. In many cases, access to electricity has allowed girls to increase school attendance and study hours, narrowing educational disparities (see Section "Advancing Women's Inclusion in the Energy Sector for Just Transitions"). In Bangladesh, electricity access significantly improved school completion rates for girls, particularly from low-income households (World Bank, 2009). In rural India, electrification boosted girls' school enrollment by 14%, while boys' enrollment remained unchanged (World Bank, 2013). Access to cable television further reinforced gender-equitable attitudes, contributing to increased school attendance for girls by challenging restrictive norms (Jensen & Oster, 2009).

Studies from Zambia and India show that electricity-enabled flexible scheduling allows girls to manage both study and household duties. This facilitates their academic success, since they can continue their domestic contributions according to prevailing gender norms. In many cases, access to electric lighting allows mothers to assist daughters with homework after dark, fostering an environment conducive to educational attainment (Winther et al., 2017).

Employment and Economic Empowerment

Access to electricity has the potential to reduce gender disparities, particularly by improving women's employment opportunities, providing avenues for entrepreneurship (Ojong, 2022), and reducing the time spent on domestic tasks (Clark, 2021). Studies in India have shown that reliable electrification helps improve women's status relative to men by increasing their labor force participation and providing more time for paid work rather than household chores (Kumar, 2022). However, while electricity connections are associated with increased female labor force participation in some contexts (Rathi & Vermaak, 2018; Samad & Zhang, 2019), the economic benefits are highly contextual and not universally experienced. In some cases, women's labor force participation has increased, but this does not always result in significant income gains or control over financial resources (Standal & Winther, 2016; Winther, 2014). Yet, common examples are electricity-powered technologies such as solar-powered cold storage and food processing equipment that can create new income-generating opportunities, thereby contributing to women's economic empowerment (Sagar et al., 2023).

The mixed impact of electrification on women's economic and empowerment outcomes is evident across various countries (Chandrasekaran et al., 2023) and case studies, often linked to lacking standardized measures (Das et al., 2023). In countries like South Africa, Mali, and Afghanistan, women have used electricity to generate income through small businesses, such as running shops or processing food, though these gains are modest (Sovacool et al., 2013). In Afghanistan, although women received salaries for maintaining solar home systems, they often lacked control over their income (Standal, 2008). These findings highlight that while electrification can facilitate income-generating activities, its impact on financial empowerment is shaped by local economic conditions, gender norms, and control over resources.

Moreover, studies suggest that the role of financial support mechanisms, such as subsidies, is crucial in enhancing women's access to electricity. For instance, in Lao PDR, subsidies increased the connection rates for female-headed households (World Bank, 2017). However, the benefits of electricity extend beyond economic gains; involvement in energy supply processes can provide women with valuable skills and increase their social status. Despite these potential benefits, institutional support remains essential to ensure that women can fully benefit from electrification.

Conditional Factors and Challenges

Overall, electrification has shown potential in enhancing women's welfare, time management, and agency within households. The benefits of electrification on gender relations are shaped by various sociocultural factors, such as traditional norms, control over resources by men, and project-specific factors across contexts. South Africa's experience, where men began participating in household tasks after electrification, suggests that social shifts can sometimes challenge gender norms (Annecke, 2005; Matinga, 2010).

Gender-targeted interventions, typically led by Non-Governmental Organizations (NGOs) or international donors, have been more effective in off-grid and decentralized electrification projects than grid-based initiatives, which historically lack gender considerations (Ilskog & Kjellström, 2008; Sovacool et al., 2013; Winther et al., 2018). Targeted public-sector interventions and policies encouraging women's involvement in electrification processes, however, have shown promise in reshaping social norms and increasing women's influence in household and community decision-making (Winther et al., 2018). This also includes targeted information and learning campaigns (Zhang et al., 2022) and capacity building initiatives (Leduchowicz-Municio et al, 2023).

While electrification can empower women, it sometimes reinforces social and economic disparities. In certain regions, electrification deepens financial and gender inequalities. In Zanzibar, for example, men's control over wealth-building assets, such as home appliances, continued to restrict women's financial security, complicating empowerment efforts (Winther, 2008). Similarly, electrification in India has led to increased expectations for dowries, including modern appliances, which has imposed new financial burdens on families and contributed to social pressures like sex-selective abortions (John, 2011; Wilhite, 2008).

Advancing Women's Inclusion in the Energy Sector for Just Transitions

In this section, we will present a mentoring program for women in the energy sector, as well as case studies and approaches on how to increase gender-sensitivity in energy projects. Realizing electrification's full potential for gender equality depends on addressing local gender norms and ensuring that electrification interventions explicitly target women's empowerment within the social and cultural context. Policymakers can achieve this by including women in these projects from the onset and getting more women into the energy sector in general.

Mentoring for Empowering Women in the Energy Transition—Clean Cooking

Expanding access to modern energy is a vital, immediate approach to address gender equality, mitigate climate change, and enhance women's health and safety. Women's roles and networks also make them pivotal in the adoption of clean cooking technologies serve as key advocates and facilitators within their communities. The emerging clean cooking value chain offers entrepreneurial opportunities that contribute to both gender equality and the growth of a sustainable global clean cooking industry. Women entrepreneurs sold three times more stoves than their male counterparts following an empowerment training for women and men (CCA, n.d.)

Sustainable Energy for All (SEforALL), the Global Women's Network for the Energy Transition (GWNET), and the Clean Cooking Alliance have been collaborating since 2021 on a program that empowers women to build sustainable careers in the clean cooking sector through targeted mentorship. The Women in Clean Cooking (WiCC) mentorship program provides tailored mentorship and professional development for women in early to mid-management roles, entrepreneurs, and other professionals engaged in the clean cooking ecosystem (GWNET, 2024).

Since its inaugural cohort in 2021, the WiCC program has supported 180 women in the clean cooking sector from across Asia, Africa, and Haiti, with 30 mentees (from 17 nationalities) in the first cohort, 60 (19 nationalities) in the second, and 90 (21 nationalities) in the third cohort in 2023 (Fig. 11.1). Each cohort spans nine months, during which the program pairs participants with a mentor who, through virtual engagement, provides guidance toward collaboratively defined professional development targets and goals. Alongside monthly mentor-mentee meetings, participants gain exclusive access to targeted knowledge-transfer webinars and virtual networking sessions.

Evaluation of Mentees' Experience

To assess the impact of the mentoring program at the end of each cohort, the organization sends a survey to both mentors and mentees to provide their feedback of the mentoring experience as well as the impact of the mentoring program on their personal and professional life.

The testimonials from participants of the Women in Clean Cooking Mentoring Program highlight the transformative and impactful nature

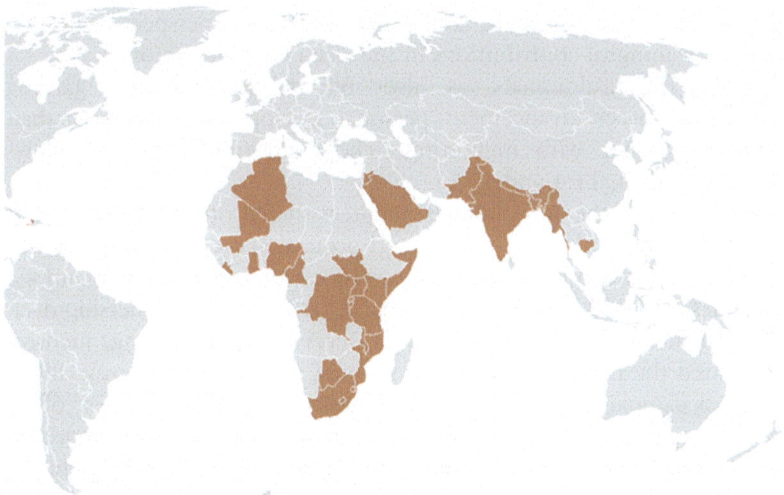

Fig. 11.1 Illustrative map showing the national diversity of the WiCC mentees. In total, 30 different nationalities were represented in the last three cohorts (2021–2023) (*Source* Author's own elaboration)

of the initiative. Mentees shared how their mentors' expertise in clean cooking and renewable energy technologies provided them with valuable knowledge, including the design and sizing of solar-powered electric pressure cookers. One participant described the experience as both transformational and informative, noting how it upgraded her technical skills and broadened her perspective by learning from her Ugandan mentor about local clean cooking approaches, prevalent fuels, and challenges related to electricity and renewable energy adoption. Many emphasized the empowerment they gained through discovering their potential to inspire others, which boosted their confidence, enhanced their sense of purpose, and enriched both their personal and professional lives (Ojwang, 2024). The program also provided insights into female leadership, fostering growth through shared experiences and stories. A standout success was the achievement of a grant for a participant's company, made possible by her mentor's guidance throughout the application process (Waraga, 2024). These testimonials underscore the program's ability to equip women with the skills, confidence, and global perspective needed to drive meaningful contributions in the clean cooking sector.

Besides the testimonials, we also collected information regarding participants' career advancement, personal development, and about their overall experience with mentoring. The main results are as follows (Figs. 11.2, 11.3, 11.4 and 11.5).

In addition to the core findings, mentees shared other relevant results related to their work environment and professional development needs. 44% of mentees observed increased awareness of gender barriers among colleagues, and 33% noted greater female participation in leadership decisions. To further accelerate women's careers in clean cooking, mentees highlighted the need for specialized training in renewable energy (89%), awareness campaigns (76%), and more networking opportunities (68%). They also emphasized the importance of mentoring programs, improved access to education, and better support for balancing work and family life.

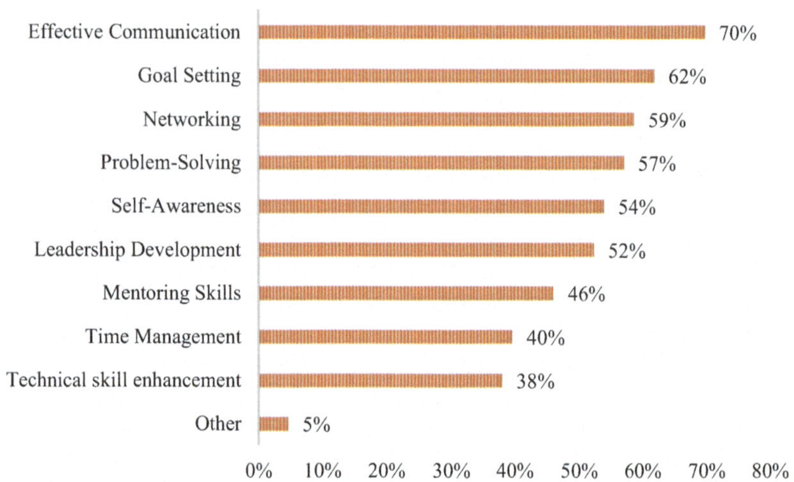

Fig. 11.2 Results Mentees' responses to the question: What key concepts or skills did you gain from your mentoring experience? Communication, goal setting, and networking emerged as the most frequently cited benefits, followed by problem-solving, self-awareness, and leadership development. Mentees mentioned skills such as mentoring, time management, and technical knowledge to a lesser extent (*Source* Author's own elaboration)

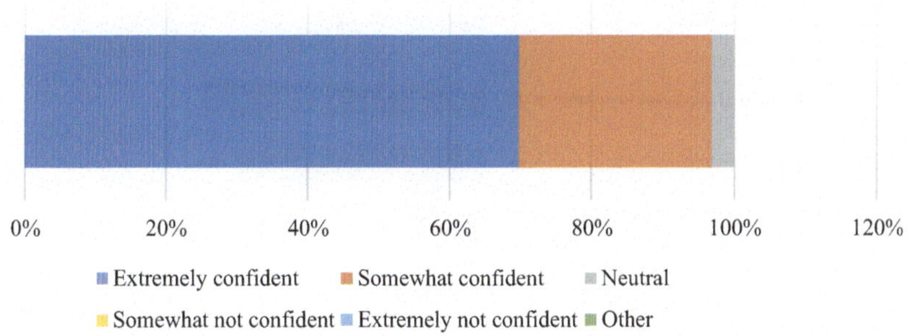

Fig. 11.3 Results Mentees' responses to the question: How confident do you feel applying these new skills or knowledge at work? The majority of mentees (70%) reported feeling extremely confident, while 27% felt somewhat confident in using the skills gained through the program (*Source* Author's own elaboration)

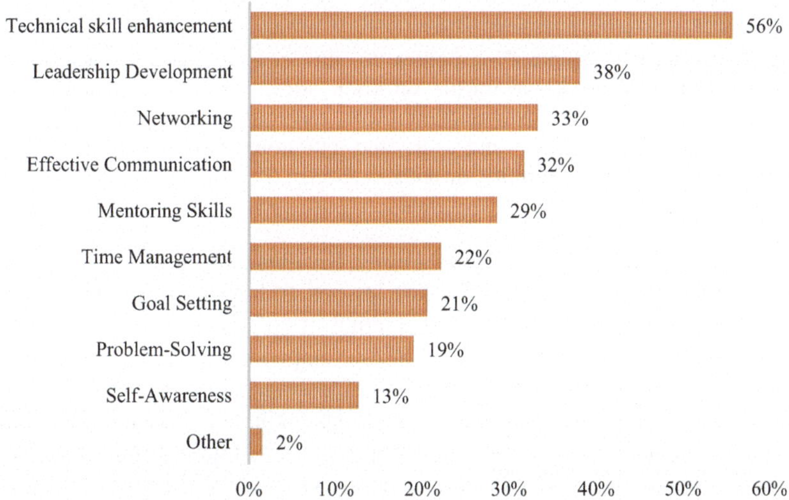

Fig. 11.4 Results Mentees' responses to the question: In what areas do you feel you need further instruction or practice? Mentees cited technical skills emerged as the most significant need (56%), followed by leadership development (38%), networking (33%), and communication (32%) (*Source* Author's own elaboration)

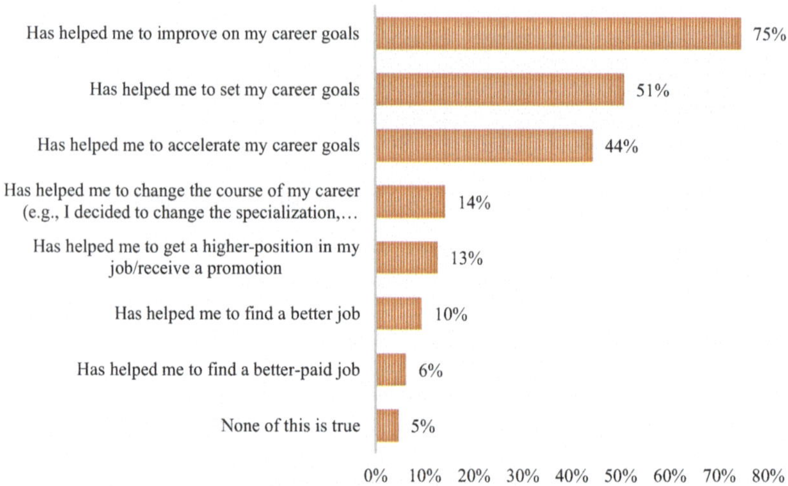

Fig. 11.5 Results Mentees' responses to the question: If the mentoring program has had a positive impact on your career, which of the following statements are true? Most respondents (75%) reported improved career goals, 51% identified support in setting goals, and 44% noted faster progress toward achieving them (*Source* Author's own elaboration)

Lessons Learned, Limitations, and Future Recommendations

The survey results reveal several key lessons and conclusions. First, the mentoring program has been instrumental in enhancing career development, as most mentees report significant improvements in their career goals, confidence, and leadership skills. However, there remains a clear demand for additional support in technical skill enhancement, leadership development, and networking opportunities, highlighting areas for future program focus. Furthermore, mentees recognize the importance of creating awareness about gender barriers and increasing women's representation in leadership roles, suggesting that targeted campaigns and advocacy are crucial. The strong call for specialized training in renewable energies underscores the need for tailored professional development to empower women in the clean cooking sector. Overall, these insights emphasize the program's positive impact while identifying opportunities to expand its scope and deepen its influence.

Mentoring programs such as the WiCC program play a critical role in advancing gender analysis in the context of electrification, particularly within the clean cooking and renewable energy sectors. By empowering women with the skills, knowledge, and networks necessary to lead in these fields, the programs help address the gender disparities that often hinder progress in electrification efforts. Women's involvement in energy decision-making is essential to ensure that electrification strategies are inclusive, equitable, and effective in reaching underserved populations. Moreover, these programs highlight the importance of integrating gender considerations into energy planning and policy development, ensuring that solutions are tailored to the needs of all genders. When women are equipped with the tools to drive change, they contribute to more sustainable and socially just energy transitions, making gender targeted programs/actions a fundamental component of advancing global electrification goals.

Successful Case Studies with a Gender Focus

Combining Rural Electrification and Education for Remote Communities, Guajira, Venezuela

A study conducted across eight previously unelectrified communities in La Guajira in Venezuela on the impact of electrification on educational access among indigenous communities has shown interesting gendered outcomes (López-González et al., n.d.). This study assessed changes in educational access since the 2010 electrification project, drawing comparisons to 2006 baseline data. With electrification and the introduction of the national "Misión Robinson" program—a TV and video-assisted literacy initiative—access to basic education in the studied communities has expanded significantly. Among younger generations, gender disparities in primary education have largely disappeared, although they persist among older community members. Secondary education has especially benefited young women, who now use nighttime lighting to study after completing domestic tasks. Electrification, coupled with the education program, has fostered higher female participation in tertiary education, while many men delay or bypass further schooling. Beyond individual access, electrification has stabilized community life and introduced a bilingual educational model that leverages technology to support indigenous cultural preservation and learning.

Electrification, when combined with targeted educational programs, can significantly elevate educational access and opportunities, especially for women. Electricity has not only enabled women to balance household responsibilities with studies but has also strengthened their roles within families and the community. The findings suggest that while electricity access is critical, it is the integration of complementary programs that maximizes educational and social benefits, particularly in addressing gender inequalities linked to poverty. The study further advocates for decentralized, solar-based electrification as a sustainable model for Venezuela and similar contexts in the Andean region, where centralization has led to reliability issues. This decentralized approach is seen as a path toward more resilient energy access that can support long-term social and educational development.

Empowering Female Solar Entrepreneurs: Solar Mamas, Zanzibar, Tanzania

The "Solar Mamas" project (Michael & Ahlborg, 2024), an initiative led by Barefoot College in Zanzibar, trains rural women with little or no formal education to become solar engineers. Through intensive six-month courses, these women learned to install, repair, and maintain solar panels, which would then provide renewable energy to their communities. This project primarily targeted women, as they tend to stay in and contribute to their communities rather than migrating for jobs. By focusing on women, the Solar Mamas initiative aimed to empower female participants not only with technical skills but also with the confidence and agency to become community leaders in a traditionally male-dominated field.

From a gender perspective, the Solar Mamas project has achieved substantial gains in empowering women and reshaping gender norms. It has created employment opportunities in sustainable energy for women who were otherwise limited to domestic and subsistence roles. This new skill set has elevated women's social status, as they are now recognized in their communities for their technical expertise and contribution to reliable, clean energy access. Beyond personal economic empowerment, the project has had a ripple effect, challenging perceptions about women's capabilities and roles within the community. Men, too, have gained respect for these women's abilities, which helps to shift broader gender expectations. In doing so, the Solar Mamas project is not just addressing energy poverty but also promoting gender equality and inspiring future generations of women to pursue skills and leadership roles in sustainable development.

Beyond Access: The Impacts of Reliable Electricity for Women in India
Sedai et al. (2021) explore how reliable electrification impacts gender equality in labor and household roles. Unlike prior research, their study examines the effect of both electricity access and reliability on a multidimensional framework of gender outcomes, using household data from India. Reliable electricity increases women's labor force participation by reducing time spent on unpaid labor, such as fuel and water collection. This shift enables women to pursue employment and make gains in earnings, health, and decision-making, especially in areas like economic agency, reproductive choices, and mobility.

The analysis shows that adding hours of reliable electricity benefits women more than men in terms of reduced domestic workload, healthier fuel choices, and access to basic amenities. Improved reliability also supports time allocation for employment, suggesting that reliable electricity can enhance women's agency and overall productivity. The findings advocate for treating reliable electrification as a right, emphasizing its potential to mitigate gender disparities and align with broader goals of economic development and gender equality.

Gender-Responsive Electric Cooking in Nepal
A study conducted by ENERGIA (2021) approached cooking from a gender perspective in Nepal, covering all services provided by electric cooking appliances, such as preparing main meals, tea, and snacks. The study also examined the use of other stoves in target households and explored time use and time savings associated with adopting electric cooking. Findings revealed that while stove usage time is difficult to quantify through interviews, the more relevant metric is the total time spent preparing a meal, which often involves multiple stoves and additional tasks like chopping vegetables. Key topics for further research on clean cooking from a gender perspective include understanding how cooking practices evolve with different fuels and technologies, assessing the value of time saved through electric cooking, and exploring its potential for productive uses. Research is also needed on women's roles in the electric cooking supply chain, the impact of flexible payment options on adoption, and the relationship between electric cooking and gender-based violence.

Additionally, the study calls for analysis of gender differences in access to electricity and cooking appliances, the substitution of prior fuels by electric cooking, and the design of pots and stoves to improve usability. These insights aim to support inclusive and gender-sensitive approaches to clean cooking solutions.

Conclusion: Toward an Inclusive Approach to Electrification

This chapter has explored the multifaceted relationship between gender and energy access, demonstrating that electrification can have profound social, economic, and cultural impacts, particularly on women. While there is consensus on the broad socio-economic benefits of improved energy access, the outcomes are deeply gendered and highly context-dependent. Women are uniquely positioned to benefit from electrification in terms of time savings, access to information, and improvements in health and education. However, these benefits are not guaranteed and depend on the design of energy interventions, local gender norms, and the broader socio-political context. Therefore, a gender-sensitive approach to energy projects is critical in ensuring equitable outcomes and advancing gender equality within the energy sector.

The reviewed literature highlights the gendered dimensions of electrification, noting that women often bear the brunt of domestic energy provision, with energy poverty exacerbating their time burdens and limiting opportunities for paid employment, education, and social participation. Studies consistently point to the positive impacts of modern energy on women's ability to engage in paid work, access information, and participate in public life, as well as call for providing access to modern and clean cooking.

However, as the chapter has shown, these benefits are not always realized. This is particularly evident in contexts where women's increased economic participation does not correspond to greater control over household income or decision-making. The complexity of these outcomes underscores the importance of adopting a nuanced, context-specific approach to energy policy and implementation that considers the intersecting factors of gender, class, and culture.

Furthermore, the chapter emphasized the need to rethink traditional frameworks around energy access and empowerment. While much of the discourse around women and energy has focused on increasing women's labor productivity and economic outcomes, it is equally important to address the relational dimensions of empowerment. Women's empowerment must be understood in terms of their relative rights, control over resources, and decision-making power in relation to men. As pointed out by Winther et al. (2017), energy access alone is not a panacea for gender inequality; rather, the distribution of power and resources within the household and society must be considered when evaluating the impacts of electrification. This calls for research that goes beyond the binary of access versus non-access, examining how energy interventions shape gender relations and challenge or reinforce existing power structures.

One of the key recommendations emerging from this chapter is the need for more inclusive and gender-sensitive energy policies, projects, and initiatives. Governments, development partners, and international organizations must prioritize gender equality in energy planning and policy formulation. This includes ensuring women's representation in decision-making processes, designing policies that address the specific energy needs of women, and developing monitoring and evaluation frameworks that incorporate gender-disaggregated data (Kooijman et al., 2023). Importantly, energy access efforts must go beyond simply increasing the number of connections to focus on ensuring reliable, sustainable, and socially just energy services that benefit all genders and groups.

Case studies that highlight successful gender-focused interventions, such as the Barefoot College's Solar Mamas program or the mentoring program by Women in Clean Cooking, demonstrate the transformative potential of decentralized, locally managed energy solutions. These initiatives not only provide women with access to clean energy but also equip them with the skills and knowledge to become energy entrepreneurs, thereby disrupting traditional gender roles and fostering a sense of agency and empowerment (Michael & Ahlborg, 2024). Such programs underline the importance of valuing women's contributions to energy systems as skilled labor, rather than simply as beneficiaries of aid or services.

In conclusion, achieving gender equality in the context of electrification requires a multi-faceted approach that integrates gender considerations into all aspects of energy policy, project design, and implementation. The evidence reviewed in this chapter highlights the importance of ensuring that policymakers design energy access initiatives with a clear understanding of the gendered dynamics at play. For electrification to contribute to the broader goal of social sustainability, it must empower women and challenge historically gendered power imbalances.

This requires not only the involvement of women in decision-making but also a critical examination of the social, economic, and cultural factors that shape their energy needs and access. As we move toward achieving universal energy access (SDG 7) and gender equality (SDG 5), it is imperative that energy transitions are inclusive, just, and transformative for all.

Acknowledgements The authors thank Camille Belmin, Shonali Pachauri, and the editors Simone Donders and Hanna Murray for valuable inputs and feedback during the preparation of this chapter.

References

Agrawal, S., Mani, S., Jain, A., & Ganesan, K. (2020). *State of electricity access in India: Insights from the India Residential Energy Survey*. Indian Revenue Service. https://www.ceew.in/sites/default/files/ceew-research-on-state-of-electricty-access-and-coverage-in-india.pdf

Aguilar, L., Granat, M., & Owren, C. (2015). *Roots for the future: The landscape and way forward on gender and climate change*. IUCN & GGCA. https://wedo.org/wp-content/uploads/2015/12/Roots-for-the-future-final-1.pdf

Aklin, M., Byer, P., Harish, S. P., & Ureplainen, J. (2017). Does basic energy access generate socioeconomic benefits? A field experiment with off-grid solar power in India. *Science Advances, 3*(5), e1602153. https://doi.org/10.1126/sciadv.1602153

Annecke, W. J. (2005). *'Whose Turn Is It to Cook Tonight?' Changing gender relations in a South African township* (Report of the Collaborative Research Group on Gender and Energy (CRGGE), the ENERGIA International Network on Gender and Sustainable Energy and the Department for International Development (DFID), UK). Published by independent researcher/consultancy firm: Gender and Energy Research and Training, Cape Town, South Africa. https://assets.publishing.service.gov.uk/media/57a08ca4e5274a27b2001317/R8346_finrep_annecke.pdf

Balakrishnan, K., et al. (2019). The impact of air pollution on deaths, disease burden, and life expectancy across the states of India: The Global Burden of Disease Study 2017. *The Lancet Planetary Health, 3*(1), e26–e39. https://doi.org/10.1016/S2542-5196(18)30261-4

Belmin, C., Hoffmann, R., Pichler, P.-P., & Weisz, H. (2021). Fertility transition powered by women's access to electricity and modern cooking fuels. *Nature Sustainability, 5*, 245–253. https://doi.org/10.1038/s41893-021-00830-3

Chandrasekaran, M., et al. (2023). *Environmental Research Letters, 18*, 045003. https://doi.org/10.1088/1748-9326/acc2d3. Available at: https://iopscience.iop.org/article/10.1088/1748-9326/acc2d3

Clark, L. (2021). Powering households and empowering women: The gendered effects of electrification in sub-Saharan Africa. *Journal of Public and International Affairs*. https://jpia.princeton.edu/news/powering-households-and-empowering-women-gendered-effects-electrification-sub-saharan-africa.

Clean Cooking Alliance s.a.: *Women and clean cooking*. https://cleancooking.org/the-issues/women-and-clean-cooking/

Clean Cooking Alliance. (2021). *Gender and clean cooking*. https://cleancooking.org/wp-content/uploads/2021/07/CCA-gender-sheet-ENGLISH.pdf

Clancy, J., Özerol, G., Mohlakoana, N., Feenstra, M., & Sol Cueva, L. (2020). Engendering the energy transition: Setting the scene. In J. Clancy, G. Özerol, N. Mohlakoana, M. Feenstra, & L. Sol Cueva (Eds.), *Engendering the energy transition* (pp. 3–9). Springer. https://doi.org/10.1007/978-3-030-43513-4_1

Cecelski, E., & Dutta, S. (2011). *A practical handbook mainstreaming gender in energy projects*. ENERGIA International Network on Gender & Sustainable Energy (Eds.) https://ppp.worldbank.org/public-private-partnership/sites/ppp.worldbank.org/files/documents/Energia_Mainstreaming_gender_in_energy_projects_A_practical_Hand_book.pdf

Das, I., Klug, T., Krishnapriya, P. P., Plutshack, V., Saparapa, R., Scott, S., Sills, E., Kara, N., Pattanayak, S. K., & Jeuland, M. (2023). Frameworks, methods and evidence connecting modern domestic energy services and gender empowerment. *Nature Energy, 8*, 435–449 (2023). https://doi.org/10.1038/s41560-023-01234-7

ENERGIA. (2021). *Gender-responsive electric cooking in Nepal*. https://mecs.org.uk/wp-content/uploads/2021/05/Gender-responsive-electric-cooking-in-Nepal_Full-report_FINAL-vs-06052021.pdf

ESMAP. (2013). *Integrating gender considerations into energy operations*. Knowledge Series 014/13. http://www.esmap.org/sites/esmap.org/files/ESMAP_Interating_Gender_Into_Energy_Operations.pdf

ESMAP. (2015). *Beyond connections. Energy access redefined* (ESMAP/Sustainable Energy 4 All (SE4ALL). Technical Report 008/15). https://openknowledge.worldbank.org/bitstream/handle/10986/24368/Beyond0connect0d000technical0report.pdf?sequence=1&isAllowed=y

ESMAP. (2018). *Getting to gender equality in energy infrastructure. Lessons from electricity generation, transmission, and distribution projects* (ESMAP Technical Report 012/18). https://documents1.worldbank.org/curated/en/639571516604624407/pdf/122887-REVISED-GenderEquality-Report-WEB-2-2-18.pdf

Ferrant, G., & Thim, A. (2019). *Measuring women's economic empowerment: Time use data and gender inequality* (OECD Development Policy Papers 16). OECD Publishing. https://ideas.repec.org/p/oec/dcdaab/16-en.html

Fletcher, E. K., Pande, R., & Moore, C. T. (2019). Women and work in India: Descriptive evidence and a review of potential policies. *India Policy Forum, National Council of Applied Economic Research, 15*(1), 149–216. https://ideas.repec.org/a/nca/ncaerj/v15y2019i2019-1p149-216.html

Grabher, H. F., Pichler, M., & Rau, H. (2024). Social practices, gender, and materiality: Women's agency in bioenergy-dependent households and communities in the central Ethiopian Rift Valley. *Energy Research & Social Science, 117*, 103736, https://doi.org/10.1016/j.erss.2024.103736

GWNET. (2024). *Annual Report.* https://www.globalwomennet.org/the-gwnet-2023-annual-report/

Hoque, N. (1988). *Rural electrification and its impact on fertility: Evidence from Bangladesh, 200*. University Microfilms International.

Ilskog, E., & Kjellström, B. (2008). And then they lived sustainably ever after? Assessment of rural electrification cases by means of indicators. *Energy Policy, 36*, 2674–2684. https://doi.org/10.1016/j.enpol.2008.03.022

Klasen, S. (2019, August). What explains uneven female labor force participation levels and trends in developing countries?*The World Bank Research Observer, 34*(2), 161–197. https://doi.org/10.1093/wbro/lkz005

IEA. (2021). *India Energy Outlook 2021.* https://www.iea.org/reports/india-energy-outlook-2021

IEA. (2024). *Energy access: Delivering electricity and clean cooking for all.* Available at: https://www.iea.org/topics/access-andaffordability#key-findings

IEG. (2008). *The welfare impact of rural electrification: A reassessment of the costs and benefits. An IEG impact evaluation.* World Bank Independent Evaluation Group. https://documents.worldbank.org/en/publication/documents-reports/documentdetail/317791468156262106/the-welfare-impact-of-rural-electrification-a-reassessment-of-the-costs-and-benefits-an-ieg-impact-evaluation

IRENA. (2024). *Tracking SDG 7.* https://www.irena.org/-/media/Files/IRENA/Agency/Publication/2024/Jun/IRENA_Tracking_SDG7_energy_progress_2024.pdf

Jensen, R., & Oster, E. (2009). The power of TV: Cable television and women's status in India. *The Quarterly Journal of Economics, 124*(3), 1057–1094. https://doi.org/10.1162/qjec.2009.124.3.1057

John, M. E. (2011). Census 2011: Governing populations and the girl child. *Economical Political Weekly, 46*(16), 10–12.

Khalifa, Y. (2024). *Gender equality, equity and women's empowerment framework a proposed framework to assess and monitor gender equality, equity and women's empowerment in modern energy cooking services (MECS)*. Loughborough University. https://mecs.org.uk/wp-content/uploads/2024/07/Gender-Framework-Report.pdf

Khandker, S. R., Samad, H. A., Ali, R., & Barnes, D. F. (2014). Who benefits most from rural electrification? Evidence in India. *The Energy Journal, 35*(2), 75–96. https://doi.org/10.5547/ISSN0195-6574-EJ.

Kikstra, J., Mastrucci, A., Min, J., Riahi, K., & Rao, N. D. (2021). Decent living gaps and energy needs around the world. *Environmental Research Letters, 16*(9), 095006. https://doi.org/10.1088/1748-9326/ac1c27.

Kohlin, G., Sills, E. O., Pattanayak, S. K., & Wilfong, C. (2011). *Energy, gender and development: What are the linkages? Where is the evidence?* (World Bank Policy Research Working Paper No. 5800). Available at SSRN: https://ssrn.com/abstract=1931364

Kooijman, A., Clancy, J., & Cloke, J. (2023). Extending energy access assessment: The added value of taking a gender perspective. *ERSS, 96*, 102923. https://doi.org/10.1016/j.erss.2022.102923.

Kumar, P. (2022). Energy access and women's empowerment. *Nature Energy, 7*, 684–685. https://doi.org/10.1038/s41560-022-01080-z

La Ferrara, E., Chong, A., & Duryea, S. (2012). Soap operas and fertility: Evidence from Brazil. *American Economic Journal: Applied Economics, 4*(4), 1–31. https://doi.org/10.1257/app.4.4.1

Leduchowicz-Municio, A., Domenech, B., Ferrer-Martí, L., Udaeta, M. E. M., & Gimenes, A. L. V. (2023). Women, equality, and energy access: Emerging lessons for last-mile rural electrification in Brazil. *Energy Research & Social Science, 102*, 103181. https://doi.org/10.1016/j.erss.2023.103181

Listo, R. (2018). Gender myths in energy poverty literature: A critical discourse analysis. *Energy Research & Social Science, 38*, 9–18. https://doi.org/10.1016/j.erss.2018.01.010

López-González, A., Domenech, B., & Ferrer-Martí, L. (2020). The gendered politics of rural electrification: Education, Indigenous communities, and impacts for the Venezuelan Guajira. *Energy Research & Social Science, 70*, 101776. https://doi.org/10.1016/j.erss.2020.101776

Malakar, Y., & Day, R. (2020). Differences in firewood users' and LPG users' perceived relationships between cooking fuels and women's multidimensional well-being in rural India. *Nature Energy, 5*, 1022–1031. https://doi.org/10.1038/s41560-020-00722-4

Matinga, M. N. (2010). *'We grow up with it': An ethnographic study of the experiences, perceptions and responses to the health impacts of energy acquisition and use in rural South Africa* (PhD diss.). University of Twente/CSTM, Enschede, The Netherlands.

Matinga, M. N., & Annegarn, H. J. (2013). Paradoxical impacts of electrification on life in a rural South African village. *Energy Policy, 58*, 295–302. https://doi.org/10.1016/j.enpol.2013.03.016

Michael, K., & Ahlborg, H. A. (2024). A conceptual analysis of gendered energy care work and epistemic injustice through a case study of Zanzibar's Solar Mamas. *Nature Energy, 9*, 947–954. https://doi.org/10.1038/s41560-024-01539-1

Ojwang, C. A. (2024, April). *Reflections on the women in clean cooking mentoring experience* [LinkedIn post]. LinkedIn. https://www.linkedin.com/posts/concepta-atieno-ojwang-0b1179128_womenincleancooking-activity-7221184054740602880-fGBA

Orlando, M. B., Lopes Janik, V., Vaidya, P., Angelou, N., Zumbyte, I., & Adams, N. (2018). Getting to gender equality in energy infrastructure: Lessons from electricity generation, transmission, and distribution projects (Energy Sector Management Assistance Program (ESMAP) Technical Report 012/18). World Bank. https://documents1.worldbank.org/curated/ar/639571516604624407/pdf/122887-REVISED-GenderEquality-Report-WEB-2-2-18.pdf

Ojong, N. (2022). Energizing entrepreneurship. *Nature Energy, 7*, 392–393. https://doi.org/10.1038/s41560-022-01014-9

Rathi, S. S., & Vermaak, C. (2018). Rural electrification, gender and the labor market: A cross-country study of India and South Africa. *World Development, 109*(C), 346–359. https://ideas.repec.org/a/eee/wdevel/v109y2018icp346-359.html

Rao, N., & Min, J. (2017). Decent living standards: material prerequisites for human wellbeing. *Social Indicators Research*, 1–20. https://doi.org/10.1007/s11205-017-1650-0.

Rotondi, V., Kashyap, R., Pesando, L. M., Spinelli, S., & Billari, F. C. (2020). Leveraging mobile phones to attain sustainable development. *Proceedings of the National Academy of Sciences of the United States of America, 117*(24), 13413–13420. https://doi.org/10.1073/pnas.1909326117

Sagar, A. D., Mathur, A., Birol, F., Mulugetta, Y., Ogunbiyi, D., Sokona, Y., & Steiner, A. (2023). Mission energy access for a just and sustainable future for all. *Nature Energy, 8*, 1171–1173. https://doi.org/10.1038/s41560-023-01380-y

Samad, H. A., & Zhang, F. (2019). *Electrification and women's empowerment: Evidence from rural India (March 28, 2019)* (World Bank Policy Research Working Paper No. 8796). https://ssrn.com/abstract=3362640

Government of India. (2019). *Saubhagya dashboard*. Ministry of Power. https://saubhagya.gov.in/.

Sedai, K. A., Vasudevan, R., Alves Pena, A., & Miller, R. (2021). Does reliable electrification reduce gender differences? Evidence from India, *Journal of Economic Behavior & Organization, 185*, 580–601. https://doi.org/10.1016/j.jebo.2021.03.015

Standal, K., & Winther, T. (2016). Empowerment through energy? Impact of electricity on care work practices and gender relations. *Forum for Development, 43*(1), 27–45. https://doi.org/10.1080/08039410.2015.1134642

Standal, K. (2008). *Giving light and hope in Rural Afghanistan: Enlightening women's lives with solar energy. Studies, 43*(1), 27–45. Master Thesis, University of Oslo, Norway. https://doi.org/10.1080/08039410.2015.1134642.

Sovacool, B. K., Clarke, S., Johnson, K., Crafton, M., Eidsness, J., & Zoppo, D. (2013). The energy-enterprise-gender nexus: Lessons from the Multifunctional Platform (MFP) in Mali. *Renewable Energy, 50*, 115–125. https://doi.org/10.1016/j.renene.2012.06.024

Tenhunen, S. (2014). Gender, intersectionality and smartphones in Rural West Bengal. In K. B. Nielsen & A. Waldrop (Eds.), *Women, gender and everyday social transformation in India* (pp. 33–46). Anthem Press.

Upadhayay, S. (2009). *Evaluating the effectiveness of micro-hydropower projects in Nepal* (Master Thesis San Jose State University, CA). SHSU Scholar Works.

UNDESA. (2018). *Policy Brief #12 global progress of SDG7—Energy and gender*. United Nations.

Waraga, M. (2024, April). *Clean cooking fundraising success with mentor support* [LinkedIn post]. LinkedIn. https://www.linkedin.com/posts/martha-waraga-b75614195_cleancooking-cleancooking-activity-7213440536844615680-UHH8

Wilhite, H. (2008). New thinking on the agentive relationship between end-use technologies and energy-using practices. *Energy Efficiency, 1*, 121–130. https://doi.org/10.1007/s12053-008-9006-x

Winther, T. (2008). *The impact of electricity: Development, desires and dilemmas*. Berghahn Books.

Winther, T. (2014). The introduction of electricity in the Sundarban Islands: Conserving or transforming gender relations? In K. B. Nielsen & A. Waldrop (Eds.), *Women, gender and everyday social transformation in India* (pp. 47–61). Anthem Press.

Winther, T., Matinga, M. N., Ulsrud, K., & Standal, K. (2017). Women's empowerment through electricity access: Scoping study and proposal for a framework of analysis. *Journal of Development Effectiveness, 9*(3), 389–417. https://doi.org/10.1080/19439342.2017.1343368

Winther, T., Ulsrud, K., & Saini, A. (2018). Solar powered electricity access: Implications for women's empowerment in rural Kenya. *Energy Research & Social Science, 44*, 61–74. https://doi.org/10.1016/j.erss.2018.04.017

World Bank. (2013). *Breaking barriers, improving futures. Challenges and solutions for girls' education in Pakistan*. https://documents1.worldbank.org/curated/en/099907205312429006/pdf/IDU141cd1e1516588142c119f741ba6fd510021d.pdf

World Bank. (2009). *Bangladesh: IDCOL solar home systems project*. Project Information Document (PID). https://documents1.worldbank.org/curated/en/122961467997888816/pdf/405010PID01IDCOL1.pdf

World Bank. (2017). *Why measuring energy access for women makes smart business. The case of Lao PDR. A lessons learned note from the EAP gender and energy facility*. https://documents1.worldbank.org/curated/en/547301506533359927/pdf/120104-BRI-PUBLIC-P151262-M-E-Laos-Energy-and-Gender-Note.pdf

Zhang, A. T., Patnaik, S., Jha, S., Agrawal, S., Gould, C. F., & Urpelainen, J. (2022). Evidence of multidimensional gender inequality in energy services from a large-scale household survey in India. *Nature Energy, 7*, 698–707. https://doi.org/10.1038/s41560-022-01044-3

PART III

Technological Innovations and Regional Case Studies

CHAPTER 12

A Decentralized eRoaming Business Service Ecosystem to Promote Citizen-Centric Energy Communities

Anthony Jnr. Bokolo

Introduction

In 2023, close to one in five cars sold globally was electric, increasing Electric vehicles (EVs) sales to about 14 million in 2023, with 95% of the sales happening in China, Europe, and within the United States of America (USA). In Europe, the first part of 2024 recorded a growth in sales of over 5%, a number marginally higher than the total car sales, highlighting the increase of EV sales compared to other years.[1] In energy communities, EVs are now being adopted by residents as an alternative to traditional combustion engine vehicles as a more environmentally friendly mode of transportation (Yaqub et al., 2020). However,

[1] iea.org/reports/global-ev-outlook-2024/trends-in-electric-cars

A. Jnr. Bokolo (✉)
Department of Digital Sovereignty, Institute for Energy Technology, Halden, Norway
e-mail: anthony.bokolo@ife.no

T. Walker et al. (eds.), *Wiring the Future*, Palgrave Studies in Emerging Risk Management and Sustainable Finance,
https://doi.org/10.1007/978-3-032-14717-2_12

with the continuous increase in the number of EVs in energy communities, there is a need for distributed, unified, and integrated charging infrastructure for vehicle batteries that supports seamless charging which enables interoperable EV charging across cities and communities (Adam, 2016). Within these energy communities, citizens generate renewable energy sources (RES) in their residential areas. They do so, for example, by using Photovoltaics (PV) solar panels to charge their EVs. Moreover, some citizens use commercial public charging stations for vehicle charging, though this practice depends on the accessibility and availability of such facilities (Adam, 2016; Anthony Jnr, 2020). To address this shortcoming, practitioners in the domain of EV charging have developed EV roaming,[2] something analogous to roaming in mobile telecommunications. EV Roaming also allows the use of data-driven services such as the locating of charging stations, quick access to information on fees, and the availability of charging stations for its use based on the deployed communication protocols (Roth et al., 2022; van der Kam et al., 2020). EV Roaming further enables the use of publicly shared charging infrastructure, independent of the vendor or technology, without legal and economic obstacles (Visser, 2019).

However, eRoaming development is still in its early phase and like other technological innovations, the widespread adoption of eRoaming faces challenges such as interoperability, incompatible communication protocols, standardization, etc. Hence, eRoaming is still in its development as several actors such as charge point operators (CPOs), who operate the actual charging stations or the electric mobility service providers (eMSPs), who offer eRoaming services for mobility services and payment (Adam, 2016; van der Kam et al., 2020; Anthony Jnr, 2021). Consequently, citizens can only access a limited number of charging stations and must subscribe to many different services provided by CPOs across different territories (van der Kam et al., 2020). The ineffectiveness of this system hinders the widespread adoption of EVs, thus slowing energy communities' transition towards sustainable mobility (Jnr, 2024a). Therefore, we aim to address the following research questions:

[2] Please note that eRoaming and EV Roaming are used interchangeable throughout this paper

- What is the state-of-the-art of eRoaming services in energy communities?
- Which requirements and stakeholders are needed to achieve a decentralized eRoaming business model ecosystem?
- Which existing proprietary protocols and standards that are applicable to citizens, CPOs, and eMSPs, and which key barriers impact decentralized eRoaming in energy communities?

Accordingly, our article explores the barriers that impact eRoaming services. More importantly, we develop a decentralized eRoaming business service ecosystem to facilitate eRoaming services for charging EVs and further support the provision of transparent information to citizens about the source of the energy, its availability, and the accessibility of charging station locations and the fees for charging. This will ultimately allow EV drivers and users to charge at any charging station across a designated region, thereby fostering the creation of one regional market ecosystem for EV owners. Findings from this article further identify the requirements for achieving decentralized eRoaming in energy communities. Specifically, this study adopts a narrative literature review as a research method, which helps to address the research questions this paper explores. This study then concludes with recommendations to pave the way for future developments, thus facilitating a seamless, citizen-centric, eRoaming services in energy communities. This study contributes towards maximizing connectivity and integration in the EV charging market, particularly in regard to the ability of different charging infrastructures to seamlessly communicate with each other effectively and contribute to achieving a harmonized EV charging ecosystem. The remainder of this chapter is arranged as follows: Sect. 2 outlines the research method employed, where a narrative review is adopted to explore our research questions. Sect. 3 presents the key findings followed by a discussion in Sect. 4, and Sect. 5 presents the conclusion.

Research Method

We employ a narrative literature review on the aspects of eRoaming in energy communities. Accordingly, this study adopted a narrative literature review to outline relevant findings. A narrative literature review was opted for as it outlines findings of relevant studies to conceptualize and present key arguments in specific areas. Narrative literature reviews are well-suited for research that seeks to examine the current state of the field, synthesize knowledge, and specify future research directions (Snyder, 2019). Additionally, narrative literature reviews are specifically suited to areas that have a multidisciplinary theoretical background (Snyder, 2019). Narrative literature review is pertinent for this study, where two broad sets of sources from different disciplines are being examined concurrently, i.e. eRoaming and EV charging. Although findings from eRoaming and EV charging are presented in the results section, the findings are more aligned with eRoaming as this research area of decentralized EV charging has not been well researched in the literature. Therefore, this narrative review analyses existing sources with a focus on eRoaming to explain its importance for the adoption of EVs for the progression towards sustainable mobility in energy communities.

The scope of our article is aligned with the implementation of eRoaming services from a user-centric perspective about citizen-centric energy communities. This specificity is captured in the keywords utilized for the search strategy employed to identify relevant sources. Therefore, the search was conducted by the corresponding author using the following terms: eRoaming, EV roaming, decentralized eRoaming, eRoaming business model, eRoaming ecosystem, citizen-centric, energy communities, electric vehicle charging, and electric vehicle adoption. Using online databases (Google Scholar, Web of Science, and Scopus) (Jnr, 2024c), we used these keywords to identify appropriate sources such as journals, conference proceedings, book chapters, and technical reports written in the English language. We analyzed more than 200 sources in 2023 to identify relevant studies for inclusion in the literature review. Also, three duplicates (same sources), were excluded, resulting in 197 sources. We assessed the abstracts of the remaining sources to determine their relevance to eRoaming. This step led to the selection of 24 useful contributions that were added to the final literature selection as shown in the reference section of the paper. Finally, we employed narrative analysis to present the data extracted from the selected sources.

Findings

Background of eRoaming in Energy Communities

EVs are indispensable for communities to achieve more sustainable transportation. However, the rate of adoption of EVs is still low in cities and communities (Jnr, 2024b). This low rate is due to challenges like the limited accessibility and availability of charging stations. Additionally, public charging infrastructures often require users to have an existing contract with the charging service provider as well as a dedicated platform or Radio Frequency Identification (RFID) cards that grant access to the charging stations. As a result, eRoaming was introduced by charging point providers to provide users with access to available charging points operated by different providers using a single contract (Maheshwari & Nair, 2021). EV Roaming allows for the aggregation of charging stations from different charge point operators (CPOs) in one ecosystem, making public and private charging available to EV owners regardless of regional boundaries or platform used (Maheshwari & Nair, 2021). EV Roaming service will help EV owners to easily locate chargers, give them access to the stations with a single card, offer similar authentication applications or devices, provide a unified map of available charging stations, give a fair and transparent pricing, etc. Enabling citizens to charge their EVs at different CPOs in turn enhances accessibility and (gets rid of?) hidden charging costs (Ferwerda et al., 2018). EV Roaming typically involves four major processes, as seen in Figure 12.1.

12.1 depicts the eRoaming process in energy communities where before accessing the EV charging point, the EV owners and users must register for the service and sign a digital contract with a CPO (as seen in Fig. 12.2). The CPO then assigns the user with a unique user identification number as part of the identity management process. The user

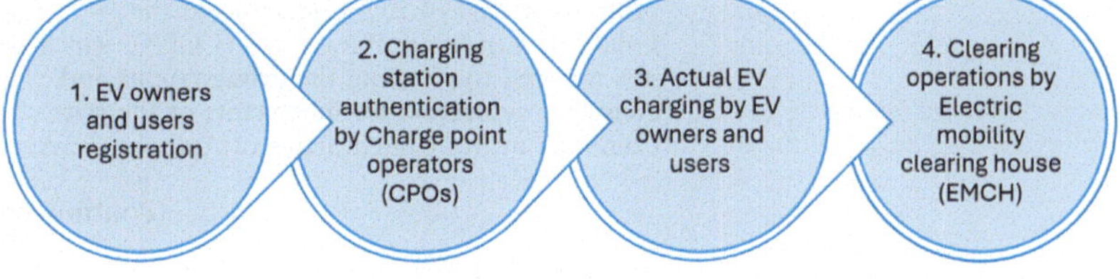

Fig. 12.1 eRoaming process in energy communities

Table 12.1 Requirements for decentralized eRoaming in energy communities

Requirements	*Description*
Autonomy	Existing eRoaming platforms provided by EMCHs are deployed based on proprietary communication protocols. As such EMCH(s) have an oligopolistic or monopolistic position within the eRoaming service ecosystem. To reduce this effect there is a need to limit market power concentration by implementing decentralized eRoaming systems that are autonomous of intermediaries or not self-governed by actors such as the EMCH(s) (Hoess et al., 2022).
Cyber security and data privacy	For the eRoaming ecosystem to be efficiently operational, there is a need to ensure sufficient protection of EV owners' personal information. As personal information of EV owners, such as the geographic information of charging stations, can profile and define users' charging patterns. To safeguard data security and privacy, CPOs, EMCHs, and eMSPs should limit correlating personal identifiable EV charging data beyond the information needed for eRoaming. Thus, only needed information should be collected and processed. For example, data related to charging events should only be made accessible to the EV owner and the respective eMSP and EMCHs for billing purposes. Similarly, confidential or sensitive business data of CPOs, EMCHs, and eMSPs should be safeguarded. Data minimization procedures should be employed based on a need-to-know basis. If possible, eRoaming services should limit the centralized storage of data to mitigate a single point of failure, thereby minimizing unauthorized access to data (Hoess et al., 2022).To mitigate such privacy breaches, decentralized eRoaming services should adhere to data protection regulations such as the General Data Protection Regulation (GDPR), which is a European Union (EU) regulation put in place to protect the data of EU residents, and the California Consumer Privacy Act (CCPA), which was initiated to promote the privacy rights and consumer protection for inhabitants of the state of California in the United States of America (USA).

(continued)

Table 12.1 (continued)

Requirements	*Description*
Identity management	To enable trustworthy eRoaming services, CPOs, EMCHs, and possibly electric mobility service providers (eMSPs) who are actors who rent EVs should be able to mitigate costs and potential losses such as EV owner charging with an illegal contract. Thus, all actors participating in the charging ecosystem must have the correct credential authorization and authentication. Hence, access to charging-related information should be provided only to actors with the right authorization and authentication. CPOs can facilitate this process by deploying user verifications to check the validity of EV owners' contracts to guarantee that they are authorized to charge their EVs on behalf of EMCHs and eMSP (Hoess et al., 2022). Also, a standardized ID schema can simplify the verification of EV owners' identity.
Scalability	Due to the increase in the use of EVs in society, there is a need for the decentralized eRoaming service to be able to cope with an increasing demand of EV charging and corresponding eRoaming activities. Thus, decentralized eRoaming services should be able to process many energy sharing transactions per minute without long response times, blackouts, or power outages (Hoess et al., 2022; Jnr, 2024b).
Adeptness	For a dependable decentralized eRoaming service, there is a need for a seamless data exchange between all actors involved. Actors involved in eRoaming service also need to support micro payment processes for EV charging. Accordingly, ensuring information availability and transferring machine-readable and standardized data exchange between all the actors is essential. Data from public registries can be employed to provide data such as coordinates on the available, accessible, and authorized charging stations (Hoess et al., 2022).

(continued)

Table 12.1 (continued)

Requirements	*Description*
Ease of use	To foster the uptake of EV, the deployed decentralized eRoaming service should be user-friendly for EV owners using a single contract, an application, or a RFID card. For CPOs, EMCHs, and eMSPs, the decentralized eRoaming service should require few implementation efforts compared to exiting ad hoc charging solutions.
Fair pricing	The business model for the decentralized eRoaming service should be based on an equitable pricing scheme with a transparent pricing communication medium for EV owners (Hoess et al., 2022).

identification number is then used at the charging point by EV owners and users to authenticate if the user is an authorized subscriber using a mobile application or via a specific RFID card (Hoess et al., 2022).

Figure 12.2 depicts a centralized eRoaming service ecosystem in energy communities where an electric mobility clearing house (EMCH) requires the payment of transferred energy processed using the contract for the EV user between CPO1 and CPO2 (Noyen et al., 2013). The ecosystem also provides data-driven services based on data collected from CPOs, EMCH, EV owners, and other actors (Anthony Jnr et al., 2020; Zelle et al., 2018). An EV owner utilizing the eRoaming service shown in Fig. 12.2 registers and subscribes to their local CPO1. At some point, the EV owner embarks

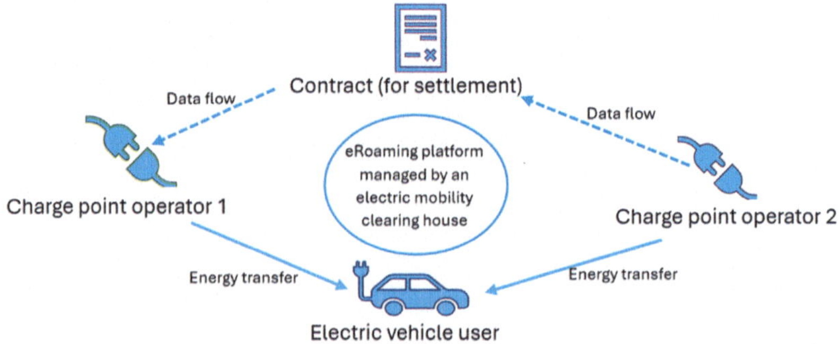

Fig. 12.2 Ecosystem of a centralized eRoaming service

on a journey and becomes a tenant for another charge point operator (CPO2) for a certain period of time in a different region managed by CPO2 (Farao et al., 2021). Then, a contract is established among the CPOs based on a definite service level agreement (SLA) between CPO1 and CPO2 that defines how CPO2 will be reimbursed for the energy consumed to charge the EV (Farao et al., 2021).

Furthermore, in energy communities, most EV owners charge their vehicles at their place of residence using a privately installed charge station. Public charging stations (mostly highway fast DC, direct current fast charging (DCFC)) accommodate their charging needs when they drive for longer distances (van der Kam & Bekkers, 2022; Höfer et al., 2013). Also, in energy communities, publicly available curbside charging facilities across the neighborhood, for example, in garages in multi-dwelling building or apartments, in offices, or public parking can be shared and used by the public when available. Today, existing eRoaming platforms managed by an electric mobility clearing house (EMCH) are based on a centralized architectural design that supports multiple CPOs to connect and share their charging stations (Hoess et al., 2022; Gonder et al., 2007). Many users and stakeholders find centralized eRoaming platforms problematic due to data security, privacy concerns, and the consolidated storage of data. These platforms often lead to the concentration of market autonomy in favor of EMCHs. As advocated by Hoess et al. (2022) this study suggests a decentralized eRoaming approach.

EV Roaming Business Models in Energy Communities

The process of EV adoption in energy communities is developing rapidly, but there are still issues to address in regard to the most efficient way to connect charging infrastructure. EV Roaming requires the collaboration between market competitors that will otherwise compete against one another (Maheshwari & Nair, 2021). For improved eRoaming services in energy communities, we need to explore how to achieve a form of alignment among different participating actors while still having a high degree of flexibility to incorporate new digital services enabled by business models. A business model specifies how an organization creates, captures, and delivers value in social, economic, cultural, environmental, or other contexts (Costa et al., 2022). The business models developed for EV charging are still evolving as the deployment of eRoaming services by CPOs, EMCH, and eMSP are typically based on different business models. Although, findings from Costa et al. (2022) revealed that the development of a business model needed to manage the interface between electricity supply system and EVs charging is faced with several challenges because eRoaming involves an ecosystem of multiple actors (see Fig. 12.4) that collaborate on charging-related services. And yet, getting competitors to collaborate is challenging without clear economic benefit. Hence, it is important to outline a viable business model that is based on the number of participants, openness, existing market structure, degree of collaboration, etc. (Maheshwari & Nair, 2021).

Moreover, there is a need for a business model that offers full-service packages such as charging stations, energy contract, a roaming contract, and other additional services including an EV if needed by citizens. The business model should aim to provide a better charging experience for EV owners and users (Maheshwari & Nair, 2021). In addition, the business model should enable the flexibility of CPOs, EMCH, and eMSP by combining new services across the eRoaming business ecosystem via communication infrastructure (van der Kam & Bekkers, 2020). Business models for eRoaming in energy communities include a peer-to-peer market structure, an eRoaming hub, and a hybrid model as seen in Fig. 12.3.

In a *peer-to-peer (P2P)* business model, the CPOs bilaterally offer eRoaming across each other's charging networks. This decentralized business structure fosters the open exchange of data such as charging station information, availability statuses, reservations, and payments. The P2P approach is mostly suitable when there are fewer actors connected within the eRoaming ecosystem (van der Kam & Bekkers, 2022) because it supports EV market incumbents by helping them to strengthen their market share. However, the P2P model requires a substantial number of resources and time depending on the customization level of the infrastructures. In the P2P model, the associated costs incurred with deploying, maintaining, and regularly upgrading charging infrastructures is the responsibility of all participants involved (van der Kam & Bekkers, 2022). The P2P business model is like the business-to-customer (B2C) approach as distributed charging stations are provided to EV owners and users (Maheshwari & Nair, 2021).

In comparison, an *eRoaming hub* offers an intermediary platform that serves as a clearing house for data exchange. The *eRoaming hub* offers a centralized clearing house which connects several nodes (Maheshwari & Nair, 2021). While an *eRoaming hub* enables eRoaming services for EV

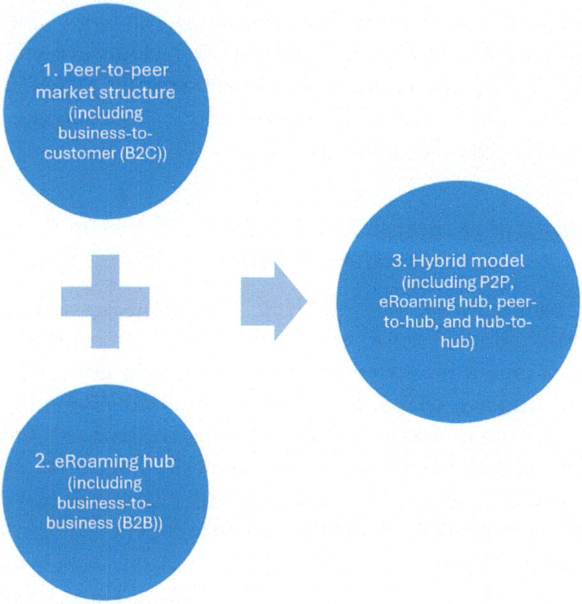

Fig. 12.3 Business models for eRoaming in energy communities

owners and users, it serves as a central node and not fully as a market contributor. An eRoaming hub mostly employs a business-to-business (B2B) approach offering standardization in technology and business to all participants across the business ecosystem. The eRoaming hub may be a non-profit or profit-based entity (Maheshwari & Nair, 2021).

The last business model is the *hybrid approach,* which is the intersection between a P2P and eRoaming hub. The current eRoaming *approach* that supports business decisions for eRoaming market structure is mostly determined by the participants within the ecosystem. As seen in most European countries, there are different models adopted, such as P2P, peer-to-hub, or hub-to-hub market structure and the choice of market topology is based mostly on the existing market structure in the region (Maheshwari & Nair, 2021). But, irrespective of the business model employed to improve eRoaming services in energy communities, the deployed business model must evolve in parallel with the development of EV charging such as vehicle-to-building (V2B), vehicle-to-grid (V2G), vehicle-to-vehicle (V2V), and vehicle-to-everything (V2X) technologies. Thus, eRoaming services will provide actors within the ecosystem with ample flexibility to co-create a user-centric, inclusive, open, and distributed eRoaming service.

Decentralized eRoaming Business Service Ecosystem in Energy Communities

In energy communities, decentralized eRoaming services consist of different stakeholders or actors that impact their current business model, technologies adopted for operation, and policies that govern the business of energy sharing for charging EVs. Decentralized eRoaming typically involves the sharing of energy and the exchange of data and funds within a regulated sphere, ecosystem, or actors (Maheshwari & Nair, 2021). One of the main actors involved in decentralized eRoaming services is the *CPOs* that are responsible for the maintenance/operation (technical) and the management (administrative) of charging stations. CPOs mostly own the charging stations (Maheshwari & Nair, 2021). Thus, CPO manages the charging stations and maintain the charging process while the *electric mobility service provider(s?) (eMSP)* lease EV to the public. By subscribing to an eMSP, EV owners and users can access all charging stations provided by CPOs. This is because the CPOs and eMSPs establish a mutual B2B eRoaming agreement that facilitates such services (Hoess et al., 2022). In some regions, the CPOs and eMSP can be assumed by a single entity who is responsible for installing, monitoring, and maintaining charging stations while also providing charging as a commercial service to EV drivers and owners (Maheshwari & Nair, 2021).

Another actor is the *electric mobility clearing house (EMCH)*, which is situated within the eRoaming clearing house or eHub. The EMCH is the actor that orchestrates eRoaming transactions between multiple CPOs, EV owners and users, and eMSPs (Maheshwari & Nair, 2021). *The Electric vehicle supply equipment (EVSE)* is the actors that provide the physical hardware equipment that transfers/supplies electricity to EVs. EVSE can also be referred to as chargers (Maheshwari & Nair, 2021). Furthermore, regulatory, governmental, and standardization bodies at different levels are also actors that regulate and eRoaming services in energy communities. The energy suppliers such as Transmission System Operators/Distribution System Operators (TSOs/DSOs) are also actors that connect with the smart grid. Charging infrastructure providers are another actor that uses communication and interface protocols employed by CPOs. Data and platform providers are (also/in turn) responsible for the data exchange via dedicated protocols (van der Kam & Bekkers, 2022). Then, there are the automotive suppliers that implement software systems compatible with the communication protocols employed by CPOs, EMCH, and eMSP. There are also IT system and platform developers that develop and implement applications and protocols to govern the entire decentralized eRoaming business ecosystem. The IT system and platform developers develop eRoaming platforms and systems used by all actors to support eRoaming transactions. In summary, the key actors involved in decentralized eRoaming services can be seen in Fig. 12.4:

Figure 12.4 depicts a simplified eRoaming business service ecosystem in energy communities that captures the overall market and the stakeholders involved in the whole value chain for eRoaming services based on findings from prior studies (van der Kam & Bekkers, 2020).

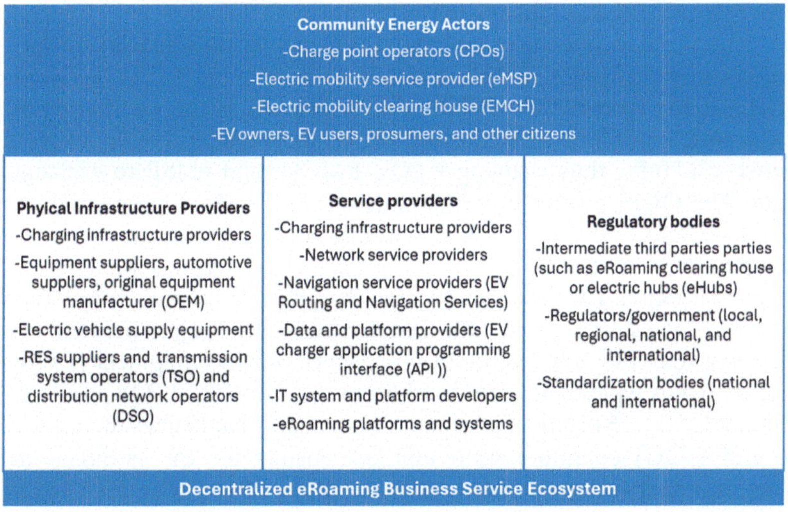

Fig. 12.4 Decentralized eRoaming business service ecosystem in energy communities

Requirements for Decentralized eRoaming in Energy Communities

The following are plausible requirements that support the development of decentralized eRoaming in energy communities.

Existing Proprietary Protocols and Standards Deployed for eRoaming

Today, several actors responsible for eRoaming deploy gateway technologies that are systems that provide interfaces via two or more different communication protocols. These communication protocols differ by functionality, market penetration, openness, ownership, maturity, and adoption (van der Kam & Bekkers, 2022). These communication protocols deployed for eRoaming operations are either based on a centralized or decentralized architecture. The centralized architecture has been widely employed to connect multiple actors in a single platform by employing proprietary protocols (*Open Charge Point Information* (OCPI), *Open Clearing House Protocol* (OCHP), *Open InterCharge Protocol* (OICP), *eMobility Inter-Operation Protocol* (eMIP), *and Open Charge Point Interface Protocol* (OCPI)), that enable authentication and clearing. However, the centralized architectures are limited with significant scalability, privacy, and security challenges, as well as with market monopolization (Ferwerda et al., 2018; Hoess et al., 2022). Protocols such as the OCPI support eRoaming and information sharing. OCPI also supports communication between eHubs, thus enabling CPOs, EMCH, and eMSP to exchange data. The OCPI further supports functionalities such as reservations, to remotely start and stop charging stations, and the exchange of data on charging fees and the availability and location of charging stations.

Furthermore, EVs connect to the charging station to charge their batteries by using a cable. The interface communication between the charging stations known as the Electric vehicle supply equipment (EVSE*)* and the EV is typically implemented using the *ISO/IEC 15118* protocols. The ISO/IEC 15118 is an international standard that defines the vehicle-to-grid (V2G) communication interface mainly for the bi-directional charging/discharging of EVs (Ferwerda et al., 2018). The ISO 15118 supports car-based authentication, for instance. By using the plug & charge feature that enables EVs to automatically identify and authorize themselves to a particular charging station on behalf of the EV owner or user for recharging their battery (van der Kam & Bekkers, 2020). The

ISO 15118 supports data collection functionalities relevant to key business actors (CPOs, EMCH, and eMSP) (Ferwerda et al., 2018). It also provides information on the capacity, availability of charge station, and related services surrounding charging stations such as contextual information on close-by facilities available near the charging station (van der Kam & Bekkers, 2020).

Also, communication between the charging stations such as the Supply Equipment Communication Controller (SECC) and the eRoaming gateway router (e.g. R-GW/R gateway) are implemented by the ISO/IEC 61850, enabling communication connection to the EVSE (Yaqub et al., 2020). Another protocol is the IEC 63119–2. Developed by the International Electrotechnical Commission (IEC), it is a protocol that supports information exchange for EV charging eRoaming service between CPOs, EMCH, and eMSP platforms via roaming endpoints (IEC, 2019). Overall, each of the reviewed protocols provides similar functionalities or features in relation to CPO name, charging stations availability, connector type, location, etc. Countries introducing eRoaming can decide to adopt one of the existing protocols or to develop new protocols that are more tailored to their current business needs and infrastructure requirements. However, deploying a new protocol is time-consuming, complicated, and a resource-intensive process. A typical eRoaming protocol should adhere to certain design principles such as quality control, openness, scalability, and business model flexibility (van der Kam & Bekkers, 2020).

Barriers Impending Decentralized eRoaming in Energy Communities

The increasing adoption of EVs requires the provision of a charging infrastructure to facilitate eRoaming services in energy communities. However, citizens, as EV owners and users, still face challenges regarding the availability and accessibility of these charging stations across different regions (Anthony Jnr, 2021). This difficulty is due to barriers that impact eRoaming operations. Currently, multiple eHub for EV exist in Europe and in the USA. Also, some eRoaming operations are more mature than others depending on the level of EV adoption across various regions (Ferwerda et al., 2018). The identified barriers that influence decentralized eRoaming in energy communities are shown in Fig. 12.5.

Each of the identified barriers that impact decentralized eRoaming in energy communities are discussed below.

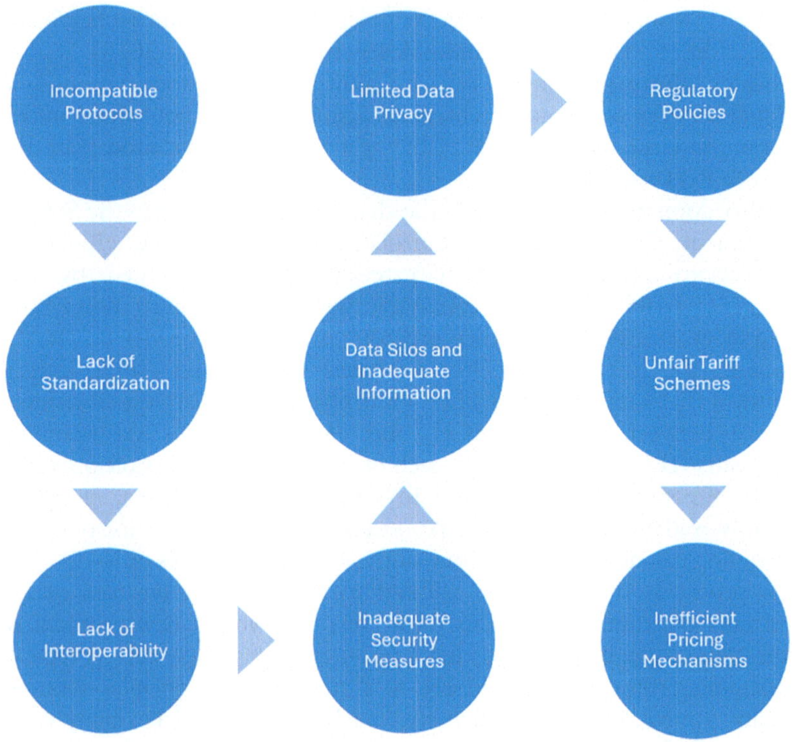

Fig. 12.5 Barriers that impact decentralized eRoaming in energy communities

Incompatible Protocols

EV Roaming requires actors from different regions to communicate with each other. Protocols are vital in ensuring an effective communication between all stakeholders involved (Ferwerda et al., 2018). From a technical perspective, the protocol enables two or more system components to communicate with each other by exchanging data (Maheshwari & Nair, 2021). Most CPOs implemented their own legacy protocols which result in incompatibility (van der Kam & Bekkers, 2020). Hence, the existence of several communication protocols with similar functionalities has resulted in a fragmented eRoaming landscape (van der Kam & Bekkers, 2022). Most eHub managed by CPOs, EMCH, and eMSP typically develop their own proprietary communication protocols with different levels of openness that enable newer functionalities. Thus, the absence of a common protocol for eRoaming in cities and communities has severely hindered the actualization of cross-border EV charging (Ferwerda et al., 2018). Researchers such as van der Kam and Bekkers (2020) mentioned that gateway technologies, which are systems that create communication interface with two or more different protocols, can be employed to support eRoaming.

Lack of Standardization
The technologies and infrastructures deployed to support eRoaming require some level of standardization to enable compatibility among various components, e.g., hardware connectors (Maheshwari & Nair, 2021). This is primarily due to the lack of a single (open) standard that enables information exchange between the different actors involved. (Ferwerda et al., 2018). Also, existing standards will need to be reviewed and/or created to enable technological advancements such as V2B, V2G, V2V, and V2X technologies (Anthony Jnr, 2021; Brown et al., 2010).

Unfair Schemes for Fees
The payment schemes that CPOs employ for eRoaming services are not always fair as most fees are not based on energy charged but on how many minutes the EV is parked, thus necessitating a time-based charging mechanism (van der Kam & Bekkers, 2020). Moreover, there are issues related to high capital and operational costs. This is because in energy communities, EV owners and users charge their EV at public charging stations that are made available as a commercial service to EV owners and users through mobile applications that require them to register and subscribe to charging services and carryout Ad-hoc payment that makes charging highly tedious for citizens (Maheshwari & Nair, 2021).

Inadequate Security Measures
Security involves protecting physical equipment from physical attacks and implementing data security guidelines that are important to future-proof the eRoaming ecosystem. As such, there is a need to define a set of standard security policies for the entire network operations, both for legacy systems and electricity systems (Farao et al., 2021). Accordingly, it is important to address physical protection, data confidentiality, integrity and authenticity, non-repudiation, access control, authorization, and accountability. Physical protection involves protection mechanisms

against adversaries that can cause physical damage or access the charging stations (Farao et al., 2021). When it comes to accountability, EV owners/users should be held responsible for their actions and for how they use their personal data. As such, data confidentiality ensures that energy consumption data should be available only to the responsible energy utility and the EV owners/users. No other actors should have access to or be able to track EV owners/users' activity. For data integrity and authenticity, all data that is to be exchanged between the actors' entities should be safeguarded against any alteration and replication. Also, each actor should be able to verify the source of the data. Regarding non-repudiation, the EV owners/users should be able to deny or accept data-related actions. Finally, access control and the authorization to access the eRoaming service should only be granted to legitimate EV owners/users registered by a CPO (Farao et al., 2021).

Inefficient Pricing Mechanisms
The price of EV charging is another important determinant as the cost of charging differs in the different jurisdictions (Maheshwari & Nair, 2021). The CPO implements a variety of pricing schemes for charging fees, whether it be a flat fee or a pay per charge, based on a "per electricity unit" pricing. NOK/KWh, for example, (NOK is the local currency in Norway), is the most common pricing method based on per-time and per-session pricing. Once the EV owners and users roam, a standard fee is usually included by the inbound actor and is charged to EV owners and users' total charging cost. This charge is needed for setting up, managing, and maintaining the entire eRoaming infrastructures. Although, evidence from the literature (Maheshwari & Nair, 2021) suggests that a standard scheme does not exist for determining this price, as each CPO carryout their own calculation for the fee payable by the users of eRoaming services.

Lack of Interoperability
Interoperability is associated with the compatibility of key software and hardware system components needed to enable an ecosystem to function seamlessly (Maheshwari & Nair, 2021). Interoperability reduces fragmentations and silos, thereby providing a more accessible network of charging stations for EV owners and users in energy communities (Maheshwari & Nair, 2021). Interoperability enables the use of a standardized infrastructure that supports different market participants and allows them

to seamlessly work together. Thus, there are issues related to selective compatibility due to different charging connectors used by EVs. Hence, full interoperability has not been achieved yet (van der Kam & Bekkers, 2020).

Data Silos and Inadequate Information

There is an inadequate localization data on the availability and accessibility of charging stations, and the applications used by EV owners and users are not compatible with legacy systems, limiting the user access to charging stations. There is a need for information availability about the state of charging stations and their future availability, real-time price information for charging, and the availability of local energy sources, RES, and more (van der Kam & Bekkers, 2022). There is also a lack of contextual information on the type of charging plugs and the number of operational plugs. Not every charging station has the same type/specification of plug, and it is important for EV owners and users to know if their EV can be charged at that location. The specified number of plugs can indicate the availability of charging stations. As the charging stations require the use of real-time data on availability, data on planned availability, and historical data on availability (van der Kam & Bekkers, 2020). Likewise, within the eRoaming ecosystem, there exists limited information about local pricing schemes and transparent pricing for charging stations, resulting in non-transparent business models and a possibly high cost due to the complexity of the eRoaming value chain. Because rates may differ among CPOs, it will be useful for users to have an idea of how much they are willing to pay for a charging session.

Limited Data Privacy

Data privacy is important to achieve a citizen-centric energy community, mainly for eRoaming services. Thus, there is a need to deploy privacy preserving data aggregation mechanisms. Evidence from the literature (Farao et al., 2021) presented a few strategies that are to be addressed such as the protection of identity privacy, location privacy, "unlinkability," minimum data disclosure, and privacy preserving data aggregation practices. To ensure identity privacy, the real identity of EV owners and users should only be known by their registered CPOs. The location privacy should be hidden via pseudonyms to prevent trajectory tracking of EV owners and users' details. The use of unlinkability schemes can ensure that different charging sessions from the same EV owners and users cannot be associated with each other. Also, data disclosure should be kept to a minimum, guaranteeing the control to the access of EV owners and users' information. Privacy preserving data aggregation must be deployed in a secure manner that protects EV owners and users' consumptions from being revealed or altered by unauthorized actors (Farao et al., 2021).

Regulatory Policies

Regulatory policies mostly refer to the rules regarding the responsibilities and compliance of actors in the eRoaming ecosystem. Because eRoaming involves stakeholders from multiple sectors (see Fig. 12.4) such as electricity, automobiles, software, fintech, public, etc., there is a need to formulate regulations that govern how eRoaming business service ecosystem can be operated in energy communities. Therefore, regulatory policies can be seen as barriers that mitigate against the upscale of eRoaming services in energy communities. These include policies such as energy market regulation, GDPR, CCP, etc. While compliance with these regulations appears to be much manageable for local eRoaming, a substantial challenge such as unclear regulatory policies remains, especially for regional eRoaming (Maheshwari & Nair, 2021).

In summary this section presents the barriers that influence the adoption of decentralized eRoaming in energy communities. Hence, to foster the widespread deployment of eRoaming service by citizens, CPOs, and eMSPs there is a need to address these aforementioned barriers in energy communities.

Discussion

In energy communities, various municipalities are promoting the adoption of EV by installing public charging infrastructures to support their regional EV fleet (van der Kam et al., 2020). Municipalities are also initiating different policies and strategies, which include information campaigns, reducing taxing, and subsidizing public charging stations (van der Kam & Bekkers, 2020). Findings from this study identify proprietary protocols such as OCHP, OICP, eMIP, OCPI, and gateway technologies that support eRoaming operations in energy communities. Using eRoaming, EV owners and users can charge their EVs with renewable energy sources (RES) such as solar, wind power, hydropower, etc. (Ferwerda et al., 2018). EV Roaming services should be open, decentralized, user-centric, and inclusive allowing participants such as CPOs, EMCH, eMSP to have the flexibility to innovate (Maheshwari & Nair, 2021). Though several eRoaming initiatives exist, there is no common protocol yet. However, OCPI is seen as the most attractive and is a widely adopted protocol by the EV market (Ferwerda et al., 2018).

This study has an important contribution to energy communities by providing knowledge in improving decentralized eRoaming service to promote citizen-centric energy communities and providing evidence on the state-of-the-art of eRoaming services in citizen-centric energy communities. This study also identifies the stakeholders needed to achieve a decentralized eRoaming business model ecosystem and further specifies which key barriers impact decentralized eRoaming in energy communities. A decentralized eRoaming business service ecosystem for energy communities is developed as seen in Fig. 12.4. It aims to capture and provide additional or new services to all key stakeholders involved in eRoaming services. Implications from this study provide insight on the current state of protocol development, the potential future of seamless EV charging, and on how to achieve infrastructure and system interoperability. Furthermore, this article explores requirements and proprietary standards such as ISO/IEC 15118, ISO/IEC 61850, and IEC 63119–2 deployed to achieve interoperability to support decentralized eRoaming. The findings offer recommendations as a roadmap for the future of cross-border charging for sustainable regional eRoaming operations.

Conclusion

EV Roaming can foster access to charging infrastructure for EV owners and users. EV Roaming provides a medium to access, use, share, and pay for public charging infrastructure. However, existing eRoaming systems are based on a centralized approach and are faced with several limitations. Accordingly, this research provides a more practical understanding of the decentralized eRoaming business service ecosystem to promote citizen-centric energy communities. A business service ecosystem is developed for decentralized eRoaming services to support an interoperable and standardized use of public and private charging infrastructures. The findings mainly provide practical guidance for decentralized eRoaming services based on secondary data. For future research, this study will be extended to employ a qualitative research approach based on case studies using real-life applications of eRoaming services. This study provides evidence on the current practice on eRoaming services in energy communities, requirements, and stakeholders needed to achieve a decentralized eRoaming business model ecosystem. The findings further provide information to key stakeholders, such as local governments, energy policymakers, or industry practitioners on existing proprietary protocols and standards that are required by citizens, CPOs, and eMSPs, and inform policymakers on which key barriers impact decentralized eRoaming in energy communities.

References

Anthony Jnr, B. (2020). Smart city data architecture for energy prosumption in municipalities: Concepts, requirements, and future directions. *International Journal of Green Energy, 17*(13), 827–845.

Anthony Jnr, B. (2021). Integrating electric vehicles to achieve sustainable energy as a service business model in smart cities. *Frontiers in Sustainable Cities, 3*, 685716.

Anthony Jnr, B., Abbas Petersen, S., Ahlers, D., & Krogstie, J. (2020). API deployment for big data management towards sustainable energy prosumption in smart cities-a layered architecture perspective. *International Journal of Sustainable Energy, 39*(3), 263–289.

Brown, S., Pyke, D., & Steenhof, P. (2010). Electric vehicles: The role and importance of standards in an emerging market. *Energy Policy, 38*(7), 3797–3806.

Costa, E., Wells, P., Wang, L., & Costa, G. (2022). The electric vehicle and renewable energy: Changes in boundary conditions that enhance business model innovations. *Journal of Cleaner Production, 333*, 130034.

Farao, A., Veroni, E., Ntantogian, C., & Xenakis, C. (2021). P4G2Go: A privacy-preserving scheme for roaming energy consumers of the smart grid-to-go. *Sensors, 21*(8), 2686.

Ferwerda, R., Bayings, M., Van der Kam, M., & Bekkers, R. (2018). Advancing E-roaming in Europe: Towards a single "language" for the European charging infrastructure. *World Electric Vehicle Journal, 9*(4), 50.

Gonder, J., Markel, T., Simpson, A., & Thornton, M. (2007). *Using GPS travel data to assess the real world driving energy use of plug-in hybrid electric vehicles (PHEVs)* (no. NREL/CP-540-40858). National Renewable Energy lab.(NREL), Golden, CO (United States).

Hoess, A., Roth, T., Sedlmeir, J., Fridgen, G., & Rieger, A. (2022). With or without Blockchain? Towards a decentralized, SSI-based eRoaming architecture. In *proceedings of the 55th Hawaii international conference on system sciences (HICSS)*.

Höfer, C., Petit, J., Schmidt, R., & Kargl, F. (2013). POPCORN: Privacy-preserving charging for eMobility. In *proceedings of the 2013 ACM workshop on security, privacy & dependability for cyber vehicles* (pp. 37-48).

IEC. (2019). Information exchange for electric vehicle charging roaming service— Part 1: General. *IEC Standard, 63119-1*(2019), 2019.

Jnr, B. A. (2024a). Enabling sustainable energy sharing and tracking for rural energy communities in emerging economies. *Renewable Energy Focus, 100633.*

Jnr, B. A. (2024b). Decentralized AIoT based intelligence for sustainable energy prosumption in local energy communities: A citizen-centric prosumer approach. *Cities, 152*, 105198.

Jnr, B. A. (2024c). Artificial intelligence of things and distributed technologies as enablers for intelligent mobility Services in Smart Cities-a Survey. *Internet of Things, 101399.*

Maheshwari, S., & Nair, M. (2021). Laying the groundwork for electric vehicle roaming in India. https://www.ceew.in/cef/solutions-factory/publications/CEEW-CEF-EDRV_Laying_the_Groundwork_for_EV_Interoperability.pdf.

Noyen, K., Baumann, M., & Michahelles, F. (2013). Electric mobility roaming for extending range limitations. *2013 international conference on Mobile business. 13.*

Roth, T., Utz, M., Baumgarte, F., Rieger, A., Sedlmeir, J., & Strüker, J. (2022). Electricity powered by blockchain: A review with a European perspective. *Applied Energy, 325*, 119799.

Snyder, H. (2019). Literature review as a research methodology: An overview and guidelines. *Journal of Business Research, 104*, 333–339.

van der Kam, M., & Bekkers, R. (2022). Mobility in the smart grid: Roaming protocols for EV charging. *IEEE Transactions on Smart Grid, 14*(1), 810–822.

van der Kam, M., & Bekkers, R. N. (2020). Design principles for an 'IDEAL'EV roaming protocol: Report D6. 3 for the evRoaming4EU project. https://pure.tue.nl/ws/portalfiles/portal/151710081/D6.3_Design_principles_for_an_ideal_EV_roaming_protocol.pdf

van der Kam, M., Ferweda, R., & Bekkers, R. N. (2020). Developing roaming protocols for EV charging: Insights from the field. *Proceedings of the 8th transport research arena (TRA'20).*

Visser, W. (2019). Electric vehicle charging-definitions and explanation.

Yaqub, R., Ahmad, S., Ali, H., & Asar, A. U. (2020). AI and blockchain integrated billing architecture for charging the roaming electric vehicles. *IoT, 1*(2), 382–397.

Zelle, D., Springer, M., Zhdanova, M., & Krauß, C. (2018). Anonymous charging and billing of electric vehicles. In *proceedings of the 13th international conference on availability, reliability and security* (pp. 1-10).

CHAPTER 13

Imagining Energy Futures: Disrupt, Downsize, Scaleup, or Conserve?

Johan Kask, Per Carlborg, Nina Hasche, Vojtěch Klézl, and Gabriel Linton

Introduction

The transition to a carbon neutral energy system presents a complex challenge that intertwines technological advancements with social and behavioral changes. This challenge arises because of the rapid evolution in technological innovations in renewable energy, storage solutions,

J. Kask (✉) · G. Linton
CREDS—Center for Research on Digitalization and Sustainability, University of Inland Norway, Kongsvinger, Norway
e-mail: johan.kask@inn.no

G. Linton
e-mail: gabriel.linton@inn.no

P. Carlborg · N. Hasche · V. Klézl
Örebro University, Örebro, Sweden
e-mail: per.carlborg@oru.se

N. Hasche
e-mail: nina.hasche@oru.se

T. Walker et al. (eds.), *Wiring the Future*, Palgrave Studies in Emerging Risk Management and Sustainable Finance,
https://doi.org/10.1007/978-3-032-14717-2_13

and smart grids, placing the energy sector at a crossroads (IEA, 2021b; IRENA, 2020). At the same time, there is a growing recognition of the role that consumer behavior, policy frameworks, and societal values play in shaping energy consumption and production (Shove & Walker, 2010; Sovacool et al., 2015). Consequently, current discussions focus on how technological and social dimensions can be integrated to create a sustainable and efficient future (Geels, 2011; Markard et al., 2012). For example, community energy initiatives often combine local energy generation, storage and smart grid technologies with participatory decision making, demonstrating the benefits of technological innovations and active engagement from local communities (Smith et al., 2016).

However, even though the amount of data and the variety of data sources are greater now than ever before (IEA, 2021a), predicting the future of energy systems remains full of uncertainty. Although the abundance of data allows us to weave together a large set of parameters—technological development, political orientation, societal behavior, climate change, etc.—to estimate and predict the future with accuracy, unpredictability of technological breakthroughs, shifting political landscapes, and evolving societal values make it difficult to chart a definitive course. This uncertainty raises critical questions: How can we plan effectively for an energy future amid such volatility? What strategies and policies can help balance technological innovation with social and behavioral change? Addressing these questions is essential for policymakers, industry leaders, and communities striving to meet climate targets and ensure energy security.

Our chapter explores these challenges by presenting an analysis of four scenarios along technological and socio-behavioral change dimensions toward fulfilled climate targets. Our four scenarios—*Disrupt, Downsize, Scaleup*, and *Conserve*—offer a comprehensive framework for understanding how the interaction between technology, society, and individuals can influence the future of energy systems. By examining these scenarios, we want to initiate a debate about how we could work with scenario analysis and discuss potential futures and their implications for stakeholders, including utilities, investors, consumers, and policymakers. In this chapter, we use data from a Swedish research project to explore alternative futures

---

V. Klézl
e-mail: vojtech.klezl@oru.se

of the electricity market. Historically, Sweden's electricity market has been characterized by centralized energy generation from hydro and nuclear sources, with long-distance transmission to passive consumers. Our analysis is informed by case studies, a Delphi study involving experts from academia, industry, and government, and scenario analysis.

Our chapter is structured as follows: First, we outline our methodological approach, and thereafter, we discuss the technological and socio-behavioral dimensions that form the foundation of our scenarios. We then present each scenario in detail, analyzing its potential impact on industry dynamics and stakeholder roles. Finally, we address the broader implications for the energy transition and offer practical recommendations for navigating the complexities of electrification.

Methods to Imagine Futures

Actively shaping the future, rather than being passively influenced by it, is an essential aspect of scenario analysis. Traditionally, scenarios are constructed using historical data and trends, which are then evaluated to determine whether these trends will continue or change. As Schot observes, "history is not only a mirror, it is also a set of scenarios. It teaches us the path-dependencies which shape who we are today, roads not taken, and hidden alternatives, which still might have a future, never in a similar way as they may have in the past, yet in an unmistakable way may shape what is yet to come." (2016, p. 446) Nonetheless, forecasting the future has proven extremely difficult, often due to underestimations of pace and scope of technological advances and an inability to foresee the unknown and disruptive. Instead, traditional models tend to extrapolate from existing technology to determine how these could be optimized and improved. Developing scenarios that consider social and behavioral shifts in addition to technology—such as changes in consumer behavior, values, and policy frameworks—presents another layer of complexity for the forecasting process, challenging the assumption that future developments can be captured through linear models. Scenarios involving societal development are particularly difficult in turbulent times marked by transformations in socio-economic, behavioral, and technical systems, alongside heightened political systems instability. These factors make it difficult to predict the interaction between societal changes and technological advancements. Despite these challenges, constructing scenarios remains a valuable exercise. Imagined futures can guide policymakers, managers, and investors in tackling uncertainties because decisions made today are typically anchored in expectations about economic and technological trajectories (Beckert, 2013; Gümüsay & Reinecke, 2022). Scenarios that explore diverse pathways to carbon neutrality can support stakeholders by anticipating potential developments, identifying opportunities and risks, and making informed strategic decisions. They offer us a way to visualize possible futures, enabling proactive planning rather than reactive responses.

In the context of energy systems, a critical global issue is mitigating the rapid rise in Earth's temperature. The Paris Agreement sets a clear target: limiting the global average temperature increase to well below 2 °C above pre-industrial levels, striving for a 1.5 °C, achieving net-zero emissions globally by 2050 (UNFCCC, 2015). This sets a defined goal for energy system scenarios, while acknowledging that multiple technological, societal, and policy pathways may lead to the same outcome. Achieving this goal requires, anyhow, a multitude of strategies, including the electrification of industries and transportation systems, enhancing energy efficiency, and the restoration of natural ecosystems. While this future may seem distant, studies like those conducted by the MIT Sloan Sustainability Initiative (2024) demonstrate that change is possible. Analyzing nearly one thousand scenarios, the initiative explored equifinal pathways to zero emissions by considering factors such as energy supply, transportation, land use, technological progress, and policy measures. These scenarios highlight the interaction between variables, showing that significant temperature reductions can be achieved using current technologies in sound combinations. Methodologically, these scenarios are not predictions or forecasts of the most probable future. Instead, they are tools for exploring and imagining a range of possible futures to support long-term planning and foresight. Depending on dominating political discourse and social change, different scenarios will at different speed shape the future.

In a research project conducted in Sweden from 2018 to 2023, we employed diverse study designs and data collection methods—including case studies, a Delphi study, and scenario analysis—to develop transition scenarios for the Swedish energy and electricity market. Our goal was to understand possible futures of the Swedish market by incorporating insights from various stakeholders, such as industry professionals, academics, and government representatives. Through case studies, we gained an in-depth understanding of companies involved in the Swedish electricity market, collecting data through 24 in-depth interviews across four different companies. The participants that we interviewed held key strategic and operational positions within their companies and provided valuable insights into industry changes, market trends, business models, and transition processes. We employed an open-ended interview approach to encourage detailed responses, with interviews lasting between 30 and 100 minutes. All sessions were recorded and transcribed to ensure accuracy. In addition to interviews, we spent six days gathering additional context in the field through observations at the company sites. We took detailed field notes during these visits and later compiled these as part of the company reports. Additionally, we reviewed extensive secondary data, including strategic plans, reports, and industry publications. This combination of primary and secondary data allowed us to develop a comprehensive understanding of the companies and their operations within the broader ecosystem.

We used the Delphi method (see, e.g., Dalkey & Helmer, 1963; Linstone & Turoff, 1975) to gather expert insights on the future of the Swedish electricity market and power system. This method is well-suited for exploring future developments and harnessing expert opinion in areas characterized by uncertainty and complexity (Okoli & Pawlowski, 2004; Schmidt, 1997). Using iterative rounds of anonymous feedback helps build consensus among experts while reducing the influence of dominant voices, making it effective for shaping informed scenarios in areas where future developments cannot be determined from empirical data alone. Through a web-based, three-round process, Swedish and international experts from academia, industry, and government—including those from the electric utility sector and related industries—anonymously estimated future developments through structured cycles of questions and answers. This anonymity minimized the influence of dominant individuals and encourages independent thinking (Keough & Shanahan, 2008). The first round of estimations featured nine participants who responded to open-ended questions regarding the market in 2030, with particular attention on their thoughts and visualization of the future. Subsequent questions addressed trends, themes, challenges, game changers, and enablers. In the second round, 34 experts evaluated 28 theses regarding future developments, assessing the probability, impact, and desirability of each. We considered themes like the rise of renewable energy, decentralization of power production, technological advancements, and new consumer preferences. By the third round, 29 experts focused on 13 critical theses which we believe have high impact and/or highlighted high disagreement in previous rounds. This iterative process provided a robust picture of potential transitions in the researched market, highlighting areas where experts agreed or differed on key issues.

Finally, we used scenario analysis to synthesize insights from the case studies and the Delphi study. Hence, the cases and Delphi study informed the scenarios with information, inspiration, and restrictions. However, the scenarios do not represent preferable nor probable outcomes, instead they all represent possible futures. An explorative scenario analysis approach was employed to generate knowledge about both the present and the future (Kosow & Gaßner, 2008). The scenario analysis followed five phases: scenario field identification, key factor identification, key factor analysis, scenario generation, and scenario transfer (Amer et al., 2013). This process created detailed hypothetical constructs of possible futures for the Swedish electricity market in 2030. By thematically organizing these scenarios along two dimensions—technological change and socio-behavioral change—we were able to highlight central elements that could shape the future of the industry, such as technological advancements, shifts in consumer behavior, and the integration of renewable energy. These two dimensions represent key mechanisms for change in the industry; the change driven by technology and social behavior, respectively. Both these dimensions are informed by the empirical results from the case studies and the Delphi panel. When all the scenarios had been developed, we arranged workshops together with all the industry partners to test and validate them. This gave us the opportunity to reflect on the result and to discuss mechanisms for change.

Although our research is based on the Swedish electricity market specifically, the insights gained from our study have broader applicability. Sweden is embedded in the Nordic (Nord Pool, n.d.) and European (European Commission, n.d.) electricity systems, meaning that market structure and policy development are influenced by EU-wide directives and global industry trends. Moreover, the topics discussed across our study, such as decentralization, the need for grid development, customer participation and effects of digitalization, are common to current discussion in energy markets globally. As such, we believe the scenarios presented in the following sections are not limited to Sweden or even Europe specifically but instead aim to present possible futures for global stakeholders addressing these challenges.

Technological and Socio-Behavioral Change

The future of energy systems is influenced by two interconnected but distinct dimensions: technological and socio-behavioral change. These dimensions both influence the implementation of energy innovations and influence the energy systems' pathways to decarbonization. The technology dimension reflects the advancement in renewable energy and energy storage technologies, and grid infrastructure that continue advancing in cost-effectiveness and integration (Fouquet, 2016). Conversely, the socio-behavioral dimension captures how shifts in societal norms—for example, from individual behaviors to community values—influence energy usage and demand, as well as new roles for consumers in a decentralized energy future (Stirling, 2014). It is important to note that these two dimensions are not confined to national settings, as they reflect on global dynamics that are occurring in energy markets. While local markets might be departing from different starting points (e.g., from more integrated systems in Europe and North America or decentralized systems in the Global South), the interplay between technological advancement and social change is shaping energy transitions across regional contexts. In our study, we explore these two dimensions by drawing on current academic literature as well as insights from real-world examples.

Technological Change

The technological dimension is characterized by rapid advancements in generating, storing, and distributing energy. As renewable energy sources such as solar and wind become more economically feasible, reliance on traditional fossil-fuel sources is expected to decrease, provided that supportive policies and adoption follow. Solar photovoltaics, for example, have benefited from economies of scale, technological advancements, and sharp learning rates—i.e., price decline associated with cumulative number of panels deployed. These factors collectively drive down costs, making photovoltaics a leading solution in clean energy deployment. Studies show that learning rates in PV manufacturing have enabled cost reductions and increased adoption on both small and large scales (see, e.g., Bogdanov et al., 2021; Kavlak et al., 2018; Rubin et al., 2015). Wind energy also plays a large role, with modern turbines achieving higher efficiency and helping to decrease installation costs. Innovations in turbine design, materials, and deployment methods contribute to these gains, making wind a competitive renewable energy source. Regions with favorable wind resources are increasingly integrating wind energy into their grids (Bogdanov et al., 2021).

Another trending technology to consider is energy storage. A grid needs to be able to handle the variability inherent in renewable energy sources like solar and wind. Lithium-ion batteries are popular choices in current storage systems due to their scalability, declining costs, and efficiency in balancing short-term demand fluctuations. Storage solutions enable energy systems to accommodate peak demand periods and smooth out supply inconsistencies. This enhances stability in grids with high renewable energy penetration. Integration of storage solutions with renewable energy enables a more resilient and flexible energy system.

Additionally, grid infrastructure is undergoing a significant transformation to support the decentralization of energy systems. Traditionally, grids are designed to transport electricity from large, centralized power plants to consumers. However, as individuals and communities generating their own power through rooftop solar or micro-wind systems, distributed generation has become more common, and grid architecture must adapt to accommodate bidirectional flows of electricity. Modern smart grids incorporate digital technologies that allow for real-time monitoring and control, making them more adaptable to fluctuating supply and demand conditions. This adaptability enhances reliability, reduces transmission losses, and supports a more efficient use of generated energy.

Socio-Behavioral Change

The socio-behavioral dimension of change in energy transitions highlights the shifts in societal behaviors, values, and policy frameworks that affect how energy systems are adopted, used, and managed. In contrast to purely technical advances, this dimension highlights the important role of consumer behavior, community agency, and the influence of regulatory structures on the progression toward sustainable energy systems. A core component in the socio-behavioral shift toward decarbonization and net-zero is the rise of prosumerism, where individuals and communities who produce as well as consume energy evolve from passive consumers to active participants in energy generation by integrating technologies such as rooftop solar and micro-wind systems (Parag & Sovacool, 2016). This movement reflects an expanding societal preference for autonomy, environmental stewardship, and local control, which aligns with the broader goals of decentralization and sustainability (Vainio et al., 2020). Research by van Klingeren and De Moor (2024) further illustrates that prosumers are often driven by ecological, financial, and social motives, highlighting a commitment to sustainability, resilience, and energy independence.

These changes are influenced by what Geels (2019) calls multi-level transitions in socio-technical systems, where broad societal pressures and cultural shifts meet new, localized innovations like decentralized energy technologies. This interaction leads to a more resilient and inclusive energy model, where people choose sustainable practices not just because the technology is available, but because cultural values and economic incentives are evolving as well. In such an environment, pro-environmental behaviors become part of daily life, gradually shaping how communities collectively use and manage energy. Bandura (2000) describes collective efficacy as a powerful force that emerges when people feel a shared sense of responsibility and believe they can make a difference together. This sense of community empowerment inspires action, making it easier for people to come together around sustainable energy practices. When communities see that their collective efforts can help reach environmental goals, they're more likely to embrace decentralized energy systems and renewable technologies. This shared belief helps communities rally behind local energy projects, advocate for policies that reflect their values, and build resilience within their energy systems (Smith et al., 2016).

Economic incentives, like subsidies and tax breaks for renewable energy installations, also play an important role in encouraging these changes. Policies that lower financial barriers make sustainable choices more attainable for a broader range of people, encouraging adoption across different regions and demographics (Geels, 2019). However, the pace of adoption is not uniform; urban areas often lead the adoption movement due to better infrastructure and access to resources. In essence, the socio-behavioral pathway envisions a future where community values, collective action, and supportive policies come together to create a resilient and decentralized energy system. This approach relies on active engagement from individuals and communities, creating a model where sustainable energy practices are affected by social connection and commitment.

The Scenarios

To explore how change along the technological and socio-behavioral dimensions influence the future of the energy system, we present in this section four different scenarios: *Disrupt, Downsize, Scaleup*, and *Conserve*. Each scenario represents a distinct combination of technological innovation and socio-behavioral change. By examining each scenario, we highlight the opportunities and challenges presented by each one for stakeholders across the energy sector.

Disrupt Scenario

This scenario envisions the most transformative future characterized by high levels of both technological innovation and socio-behavioral change. In this future, decentralized energy generation, driven by prosumerism and renewable sources like solar and wind, creates a disruptive market where traditional utilities must adapt to survive. This means, for instance, that active consumers gain greater control over their energy production and consumption, empowered by technologies that enable peer-to-peer energy trading and dynamic pricing to balance supply and demand. Policymakers thus need to develop regulatory frameworks that support prosumerism and facilitate the integration of decentralized energy resources.

In this scenario, the financial landscape shifts toward low marginal costs of electricity, presenting challenges in grid management and backup services. Traditional revenue models for utilities become less viable, demanding agility and business model innovation. Utilities may evolve into service providers, focusing on managing decentralized assets, integrating storage solutions, and supporting flexibility markets. Investors find significant opportunities in technologies like battery storage, grid software for flexibility services, and decentralized generation systems. However, these investments carry risks such as the need for adaptable infrastructure and regulatory uncertainties. This high-reward, high-risk environment favors stakeholders who prioritize flexibility and customer engagement.

An illustrative example of this scenario is the Brooklyn Microgrid. It can be described as a groundbreaking project aimed at creating a decentralized, community-based energy market, developed in collaboration with the firm LO3 Energy (see, LO3 Energy, 2024). By connecting residents with solar panels to a blockchain-based platform, the microgrid allows participants to trade surplus energy directly with each other. This peer-to-peer (P2P) energy trading model enables residents to become prosumers, giving them control over where their energy comes from, how it is priced, and whom it benefits. This approach gives participants the power to set and negotiate energy prices and as well as build community resilience by providing a local energy network that can operate independently of the main grid in emergencies. Blockchain technology enables transparent transactions in the Brooklyn Microgrid, creating a reliable and community-driven energy market. LO3 Energy's transaction fees support a more sustainable business model that shifts from traditional utilities to centralized production and distribution. It is worth noting that this is still a pilot project, and scaling it will require overcoming regulatory and scalability challenges.

Downsize Scenario

In contrast, the *Downsize* scenario emphasizes socio-behavioral change with limited or very limited technological innovation. In this scenario, energy efficiency and reduced consumption are prioritized, leading to an overall downsized energy system reliant on small-scale, distributed renewables. A more radical off-grid movement also leads to diminished demands for large-scale generation from fossil fuels, hydro, and nuclear sources. Technologically, the focus is on existing renewable technologies, with marginal efficiency improvements rather than major breakthroughs or widespread adoption of new technologies. Socially, there is a shift toward minimalism and reduced consumption, where consumers try to minimize their carbon footprint much more actively than they currently do. Communities prioritize sustainability and self-sufficiency, embracing energy-saving practices and local renewable energy projects.

Utilities adapt by supporting community-based renewable initiatives and adopting localized energy management approaches. Investors still find opportunities in energy efficiency solutions, sustainable building retrofits, and community-driven energy projects. This scenario favors lower-cost, lower-return investments in localized solutions rather than high-tech, high-cost systems. Policymakers play a crucial role by promoting incentives for energy-saving and energy efficiency programs to foster a mindset shift among both consumers and utilities. For local governments and community organizations, this scenario offers a chance to lead in creating resilient, self-sufficient energy systems. However, large-scale investors and traditional utilities may face challenges in terms of scale and profitability, requiring a focus on smaller, diversified portfolios.

As an illustration of this scenario, Earthship Biotecture designs and builds off-grid homes using natural and recycled materials such as earth-packed tires, cans, and bottles. These self-sustaining homes generate their own electricity through solar and wind power, collect and treat water from rain and snow, and regulate indoor temperatures using passive solar design and thermal mass. Communities embracing Earthship principles have emerged worldwide, including one in New Mexico, USA, which is over 600 acres and houses over 130 residents living entirely off-grid. In Australia, an Earthship project near Adelaide demonstrates sustainable building practices. Europe has seen the establishment of a district in Olst, Netherlands, with 23 Earthships. Earthships are also built in the UK, Belgium, France, South Africa, Guatemala, Uruguay, and Argentina. Residents in these communities focus on sustainability, independence from traditional utilities, and minimal environmental impact. The homes are designed to meet all basic needs—shelter, energy, water, and food production—thereby reducing reliance on external resources. These communities embody the Downsize scenario by prioritizing self-sufficiency and minimal consumption, demonstrating how a high degree of socio-behavioral changes and a shift to simpler, more sustainable living can reduce the demand on large-scale energy systems without relying on advanced technological solutions.

Scaleup Scenario

This scenario envisions a future with high technological advancements but low-to-moderate socio-behavioral change. Large-scale generation units and multinational corporations still dominate the energy market, with minimal consumer-side generation or peer-to-peer energy sharing. Technologically, significant investments in utility-scale renewable energy projects, such as large solar farms and offshore wind installations, drive growth. Extensive grid interconnections, long-distance transmission, and advanced infrastructure support the centralized distribution of electricity. Socially, consumer behavior remains largely unchanged, with limited engagement in energy production or management. The emphasis is on delivering affordable, reliable power through centralized infrastructure, keeping the consumer experience similar to the current system.

Investors and utilities benefit from economies of scale and centralization. This scenario is attractive due to its stability, established regulatory frameworks, and access to capital markets. However, it is capital-intensive and requires long-term commitments, making it suitable for institutional investors and established energy companies with resources for large-scale projects. While this approach leverages technological advancements to enhance energy production and distribution, it may miss the benefits of decentralization and prosumer engagement. Potential public resistance could arise against large-scale projects, especially in regions sensitive to land use and environmental impact. Policymakers need to balance corporate interests with environmental regulations and community needs to ensure that sustainability goals are met without alienating local populations.

An example of the Scaleup Scenario is the East Anglia Array, a proposed set of large offshore wind farms in the United Kingdom with a target capacity of 7.2 GW by 2030. It is co-developed in partnership by two multinational energy corporations Sweden's state-owned Vattenfall and the Spanish electric utility giant Iberdrola. The project involves the installation of numerous high-capacity wind turbines, generating significant amounts of electricity to power hundreds of thousands of homes. The scale of projects like East Anglia ONE requires vast capital investment, advanced technology, and extensive infrastructure, aligning with the characteristics of the Scaleup scenario. In this example, the co-founding firms' global reach and financial strength allow them to undertake a massive project that smaller entities cannot.

Conserve Scenario

Lastly, the Conserve scenario is the most conservative one, as the name would suggest, with low technological and socio-behavioral change. The existing energy system remains largely intact, requiring only incremental financial and regulatory adjustments. Technologically, there is minimal adoption of new renewable energy technologies or advancements in grid infrastructure. The focus is on maintaining and slightly upgrading current assets, with modest investments in efficiency improvements rather than radical transformation. Socially, consumers continue to rely on centralized power generation without significant involvement in energy production or management. There is limited change in consumer behavior or societal values regarding energy consumption.

For utilities and investors, this scenario provides the least disruption, preserving traditional business models and infrastructure. Investors may find this scenario appealing for its low volatility and continuity, as returns are more predictable and risks are lower. However, this conservative approach offers limited growth potential and increases exposure to future shocks, especially as global pressures for decarbonization and resilience intensify. Policymakers might view this imagined future as a "safe" option, avoiding the complexities of rapid transition. However, it carries the risk of falling behind in global energy innovation, leaving the system vulnerable to emerging technologies that could render traditional models obsolete. While the status quo approach may satisfy short-term stability, it could pose long-term risks to energy security, environmental targets, and economic competitiveness.

Milestones and Scenario Probability

As our chapter focuses mainly on the effects of two major changes—technological and socio-behavioral—we base the milestones and their effects on these factors as well. We also provide a flowchart (Fig. 13.1) illustrating how these dimensions of change affect one another and how they might result in one of the scenarios presented in the previous section.

The first milestone pertains to changes in the technological landscape. We argue that if the current development of sustainable energy generation and storage continues in the latter half of the 2020s and new technologies are being adopted, further decreases in renewable energy and battery prices will occur. Thus, two of the scenarios (Scaleup and Disrupt) rely on the continued development of sustainable energy generation and storage. After 2030, changes in technology become largely path dependent and will follow this commenced trajectory.

The second milestone is connected to social and behavioral change and a mindset shift. We believe this comes after the first milestone, as the possibility of a changing mindset relies to a certain degree on the results of the first milestone. If this change of mindset occurs, and consumers become more interested in sustainability and their energy use, another subset of scenarios would be favored (Downsize and Disrupt). Moreover, the nature of the mindset change is also important; A more radical change away from the growth paradigm, would lead to the Downsize scenario, regardless of the results of the first milestone, as it is largely not reliant

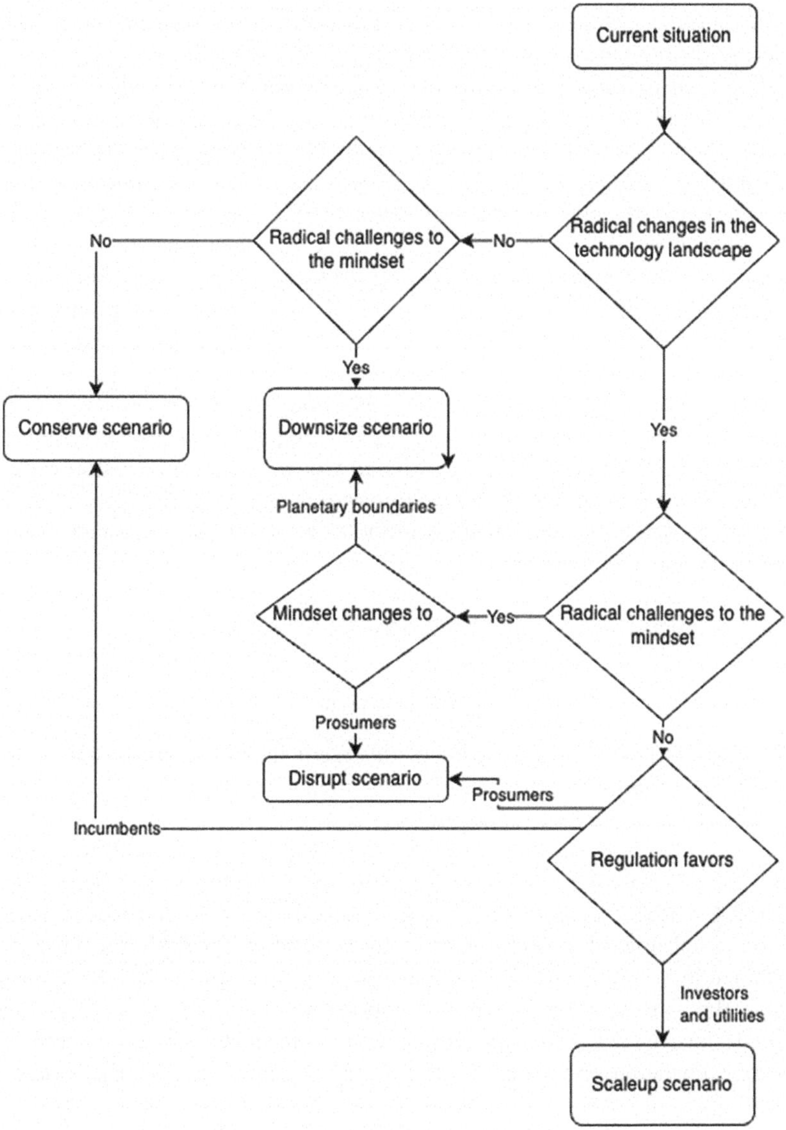

Fig. 13.1 Flow chart with four different endings

on technology. Mindset shifts toward self-production, prosumerism, and microgrid dominance would lead to the Disrupt scenario.

The final milestone is regulatory change, which occurs because of—or in parallel to—the second milestone, thus likely happening in the 2030s. If the mindset changes, there will be strong pressure to modify existing regulations to align with the new perspectives. Adoption is then expected at both international and national levels, as well as with specialized energy regulatory boards such as the Council of European Energy Regulators. At the same time, existing grid and energy companies have traditionally held influence over the regulatory boards and politicians. Since their role is reduced in some scenarios, a struggle can be expected, particularly if existing global conglomerates enter the energy sector and invest in the energy market, as they also hold significant power. Here, three outcomes can emerge. First, if the new regulations favor newcomers with global power generation, grid-based storage, and transmission capabilities, it results in the Scaleup scenario. Second, if the incumbents manage to maintain their business models by establishing further legislative entry barriers for both prosumers and newcomers, the result is the Conserve scenario. Third, if regulations instead favor "smart" microgrids and prosumers, the outcome is the Disrupt scenario.

Discussion

The four outlined scenarios present divergent imagined future pathways toward decarbonization for the future of the energy system, each carrying profound implications for stakeholders across a wide spectrum of stakeholders—e.g., policymakers, utilities, investors, new market entrants, and end users—particularly in relation to business models and user behavior. In the Conserve scenario, the energy landscape remains status quo with incremental upgrades, offering short-term stability but risking long-term sustainability. While utilities and investors might appreciate the continuity, this path may lead to missed opportunities and failure to meet the climate target set by the Paris Agreement. Policymakers would therefore face the challenge of balancing immediate economic interests that potentially hinder progress toward a more sustainable future. Conversely, the Downsize scenario introduces a radical shift driven by societal demands for environmental actions, where consumers transform into active participants, forming microgrids and energy communities that emphasize local control and sustainability. This grassroots movement may compel utilities to adapt to a decentralized model, focusing on localized renewable projects and community engagement, and investors may follow toward supporting small-scale initiatives, redefining traditional notions of profitability. While the Downsize scenario clearly promises strides toward climate mitigation, it challenges traditional energy sector economics and therefore requires a reevaluation of success metrics as well as regulations and other policy incentives to assist it.

Among the two more technology optimistic scenarios, the Scaleup scenario envisions large multinational corporations spearheading technological advancements through large investments in utility-scale projects, enhancing system efficiency but potentially sidelining consumer engagement and smaller market players. Economically, this continuing centralization benefits multi-national corporations and large investors who capitalize on economies of scale, but it may limit opportunities for local businesses and keep consumers as passive participants. In contrast, the Disrupt scenario sees consumers evolving into prosumers, actively generating and managing their own energy, which forces established utilities to reinvent their business models toward supporting decentralized P2P systems. This democratization positively opens diverse business opportunities in emerging technologies and local energy solutions, but introduces complexities in grid management and market regulation. To enhance its feasibility, new policies must support decentralized energy production, invest in smart(er) grid infrastructure and flexibility services, and establish regulatory frameworks that can support these factors. In the Disrupt scenario, policymaking is an act of balancing innovation with stability to ensure a well-functioning and resilient energy system.

When we examine these four potential futures, a central theme emerges around the balance between decentralization and centralization in energy systems. This balance influences how energy is managed and the system's resilience and adaptability. As consumers transition from passive users to active participants, their behaviors can dramatically transform market dynamics, influencing how energy is produced, distributed, and valued. Adapting legal and regulatory frameworks becomes, consequently, important to accommodate new models of energy generation and consumption. Laws and policies must evolve in parallel to support sustainability positive initiatives, protect vulnerable consumer interests, and ensure fair competition and fair pricing. Technological challenges, such as integrating distributed energy resources and maintaining grid stability, will also require innovative policy solutions.

The economic implications vary across the four scenarios. Some approaches may appear cost-effective initially but could lead to higher long-term expenses due to inefficiencies or environmental impacts. Others might demand substantial upfront investments in new technologies and infrastructure but could distribute costs more broadly or yield long-term gains, even if the benefits concentrate among certain entities. Sustainability outcomes differ among these potential paths. While it seems impossible to reach the Paris Agreement in the Conserve scenario, versions of the other three may offer promising prospects for climate mitigation but require significant shifts in societal norms and industry practices. As shown, technological solutions may drive progress, but without also addressing socio-behavioral dimensions like community engagement, they might not achieve enough sustainability outcomes.

Ultimately, the explored scenarios highlight the critical importance of strategic choices in building the future energy system. Stakeholders must balance immediate advantages with long-term responsibilities, creating and spurring innovation while ensuring that transitions are fair and inclusive. Decisions made today will influence the direction and scope of the energy sector of tomorrow, as well as broader social and environmental outcomes for future generations. Engaging in meaningful dialogue and proactive planning is essential to pave the way for an energy future that is sustainable, equitable, and responsive to society's evolving needs. To this end, forecasting and scenarios play an important role even when it is difficult to accurately predict alternative futures.

Concluding Remarks

The four discussed scenarios each present a unique pathway for the future of energy systems to fulfill the Paris Agreement of carbon neutrality in coming decades. While the Conserve scenario might seem tempting for the actors in the short term for their maintained stability, we argue that it is the most unsustainable in the long term and that following that path would exacerbate the situation in the future. The scenarios that introduce an amplified shift in the consumer involvement also require the largest pressure for companies to adapt to new roles. However, whether such shift is feasible and desirable by the public at large remains to be seen. All scenarios demand robust policy interventions and might threaten economic interests in the sector. The Scaleup scenario would leverage the economies of scale, allowing multinational tech companies and investors to take care of the energy systems, however, at a potential price of sidelining the issues of equity. Among the scenarios, the most disruptive pathway appears to balance the technological and socio-behavioral dimension, fostering consumer empowerment and decentralization while encouraging adaptation among existing utilities. However, it requires strong multi-level policy support and collaboration across stakeholders to ensure a balanced and equitable energy transition.

As a final note, we present the four scenarios in this chapter in their rather extreme variants. Perhaps the most likely outcome of the energy transition is some hybrid variant, where some elements of each scenario would emerge, for example, based on prevalent local conditions, regulations or market characteristics. However, adjusting regulatory frameworks and change of behavior of both industry actors and consumers remains necessary to achieve resilient, sustainable, and modern energy systems, locally as well as globally.

References

Amer, M., Daim, T. U., & Jetter, A. (2013). A review of scenario planning. *Futures, 46*, 23–40.

Bandura, A. (2000). Exercise of human agency through collective efficacy. *Current Directions in Psychological Science, 9*(3), 75–78.

Beckert, J. (2013). Imagined futures: Fictional expectations in the economy. *Theory and Society, 42*(3), 219–240.

Bogdanov, D., Ram, M., Aghahosseini, A., Gulagi, A., Oyewo, A. S., Child, M., Caldera, U., Sadovskaia, K., Farfan, J., Barbosa, L. S. N. S., Fasihi, M., Khalili, S., Traber, T., & Breyer, C. (2021). Low-cost renewable electricity as the key driver of the global energy transition towards sustainability. *Energy, 227*, Article 120467.

Dalkey, N., & Helmer, O. (1963). An experimental application of the Delphi method to the use of experts. *Management Science, 9*(3), 458–467.

European Commission. (n.d.). *Electricity market design*. European Commission. Retrieved March 26, 2025, from https://energy.ec.europa.eu/topics/markets-and-consumers/electricity-market-design_en

Fouquet, R. (2016). Path dependence in energy systems and economic development. *Nature Energy, 1*(8), 1–5.

Geels, F. W. (2011). The multi-level perspective on sustainability transitions: Responses to seven criticisms. *Environmental Innovation & Societal Transitions, 1*(1), 24–40.

Geels, F. W. (2019). Socio-technical transitions to sustainability: A review of criticisms and elaborations of the multi-level perspective. *Current Opinion in Environmental Sustainability, 39*, 187–201.

Gümüsay, A. A., Reinecke, J. (2022). Researching for desirable futures: From real utopias to imagining alternatives.*Journal of Management Studies*, *59*(1), 236–242.

IEA. (2021a). *Digitalisation and energy*. International Energy Agency. https://www.iea.org/reports/digitalisation-and-energy

IEA. (2021b). *Net zero by 2050: A roadmap for the global energy sector*. International Energy Agency. https://www.iea.org/reports/net-zero-by-2050

IRENA (2020). *Global renewables outlook: Energy transformation 2050.* International Renewable Energy Agency. https://www.irena.org/publications/2020/Apr/Global-Renewables-Outlook-2020

Kavlak, G., McNerney, J., & Trancik, J. E. (2018). Evaluating the causes of cost reduction in photovoltaic modules. *Energy Policy, 123*, 700–710.

Keough, S. M., & Shanahan, K. J. (2008). Scenario Planning: Toward a more complete model for practice. *Advances in Developing Human Resources, 10*(2), 166–178.

Kosow, H., & Gaßner, R. (2008). *Methods of future and scenario analysis: Overview, assessment, and selection criteria*. Studies 39. German Development Institute.

Linstone, H. A., & M. Turoff (1975). *The Delphi method: Techniques and applications*. Addison-Wesley.

LO3 Energy. (2024). *Brooklyn microgrid: About the project*. Brooklyn Microgrid. https://www.brooklyn.energy/about

Markard, J., Raven, R., & Truffer, B. (2012). Sustainability transitions: An emerging field of research and its prospects. *Research Policy, 41*(6), 955–967.

MIT Sloan Sustainability Initiative. (2024). The en-roads climate solution simulator. https://en-roads.climateinteractive.org/scenario.html

Nord Pool. (n.d.). *The power market*. Retrieved March 26, 2025, from https://www.nordpoolgroup.com/en/the-power-market/

Okoli, S., & Pawlowski, D. (2004). The Delphi method as a research tool: An example, design considerations and applications. *Information & Management, 42*(1), 15–29.

Parag, Y., & Sovacool, B. K. (2016). Electricity market design for the prosumer era. *Nature Energy, 1*, 16032.

Rubin, E. S., Azevedo, I. M. L., Jaramillo, P., & Yeh, S. (2015). A review of learning rates for electricity supply technologies. *Energy Policy, 86*, 198–218.

Schmidt, R. C. (1997). Managing Delphi surveys using nonparametric statistical techniques. *Decision Sciences, 28*(3), 763–774.

Schot, J. (2016). Confronting the second deep transition through the historical imagination. *Technology & Culture, 57*(2), 445–456.

Shove, E., & Walker, G. (2010). Governing transitions in the sustainability of everyday life. *Research Policy, 39*(4), 471–476.

Smith, A., Hargreaves, T., Hielscher, S., Martiskainen, M., & Seyfang, G. (2016). Making the most of community energies: Three perspectives on grassroots innovation. *Environment and Planning A, 48*(2), 407–432.

Sovacool, B. K. (2015). How long will it take? Conceptualizing the temporal dynamics of energy transitions. *Energy Research & Social Science, 13*(March), 202–215.

Stirling, A. (2014). Transforming power: Social science and the politics of energy choices. *Energy Research & Social Science, 1*, 83–95.

Vainio, A., Katainen, A., & Paloniemi, R. (2020). Citizens' sustainable, future-oriented energy behaviors in energy transition. *Energy Research & Social Science, 63*, Article 101404.

van Klingeren, F., & De Moor, T. (2024). Ecological, financial, social and societal motives for cooperative energy prosumerism. *Energy, Sustainability & Society, 14*(1), 13.

UNFCCC. (2015). *Paris Agreement*. United Nations Framework Convention on Climate Change. https://unfccc.int/sites/default/files/english_paris_agreement.pdf

CHAPTER 14

Global Perspectives on Electrification: Diverse Approaches and Solutions

Alberto-Jesus Perea-Moreno,
Quetzalcoatl Hernández-Escobedo,
and David Muñoz-Rodríguez

Introduction

According to the International Energy Agency (IEA) report, the use of fossil fuels has allowed GDP globally to increase, and reversing the consumption of these fuels without resulting in a detrimental impact to

A.-J. Perea-Moreno (✉) · D. Muñoz-Rodríguez
Departamento de Física Aplicada, Radiología y Medicina Física, Universidad de Córdoba, Campus de Rabanales, Córdoba, Spain
e-mail: aperea@uco.es

D. Muñoz-Rodríguez
e-mail: qe2murod@uco.es

Q. Hernández-Escobedo
Escuela Nacional de Estudios Superiores Unidad Juriquilla, UNAM, Querétaro, México
e-mail: qhernandez@unam.mx

T. Walker et al. (eds.), *Wiring the Future*, Palgrave Studies in Emerging Risk Management and Sustainable Finance,
https://doi.org/10.1007/978-3-032-14717-2_14

the world economy is a worthwhile goal (IEA, 2024). For decades, fossil fuels have accounted for 80% of the energy mix, and the target for 2030, according to the Stated Policies Scenario (STEPS), is for fossil fuel use to fall to between 60 and 75% (IEA, 2021). The objective is to achieve a more secure and sustainable energy system by investing in electrical installations that produce low emissions and modernizing the networks and, at the same time, reducing costs through photovoltaic, wind, and electric vehicles and batteries to speed up the transformation (Ram et al., 2018; Schmidt et al., 2017).

In contrast, in many parts of the developing world, access to electricity remains limited. According to data from the IEA, more than 685 million people still lack access to electricity, mainly in Sub-Saharan Africa (World Health Organization, 2024). Initiatives in these contexts often combine innovative solutions such as microgrids and distributed solar power, which offer a more decentralized approach compared to the large, centralized electricity systems prevalent in industrialized nations.

The transition to sustainable electricity systems poses unique challenges for countries, as they must balance universal energy access with climate change mitigation. This tension is evident in emerging countries such as India and China, where the expansion of electricity infrastructure has gone hand in hand with a significant increase in greenhouse gas emissions. Pachauri and Jiang (2008) study the difference in energy consumption between households in India and China. The authors highlight the differences in energy access and energy sources used and conclude that an increase in income favored a shift to more modern energy sources, thus supporting the energy transition. Xu et al. (2014) analyze the factors favoring greenhouse gas emissions, highlighting, among other things, the structure and intensity of energy consumption. They conclude that to reduce greenhouse gas emissions, specific policies are needed to transform energy consumption in industrial sectors into a more efficient consumption. The transition to more sustainable and efficient energy systems requires an analysis of various policies and technologies, including aspects such as the use of electric vehicles, energy storage systems, and their life cycles.

Castillo and Gayme (2014) study the potential of and barriers to the integration of renewable energy specifically through the use of grid-scale energy storage systems. They identify energy storage as a crucial stabilizing factor for the power grid and conclude that this form of integration is essential for supply stability and long-term economic viability, both of which are necessary for the large-scale adoption of renewable energy. In this regard, Akinyele and Rayudu (2014) review various storage system technologies, from electric pumped storage to underwater or adiabatic compressed air energy storage (CAES). The authors conclude that the implementation of these storage systems will enable greater integration of renewables into grids that will allow demand regulation. Zakeri and Syri (2015) focus on studying the economic feasibility of various storage systems, such as pumped hydro storage or CAES, concluding that the life cycle cost of these storage systems is an obstacle to their mass adoption. Gallo et al. (2016) review storage technologies in the context of the transition to more sustainable energy, concluding that, although no single storage system stands out, the future line of research involves technology such as Power-to-Gas and Power-to-Liquids.

The transition from fossil fuel-based to renewable energy-based systems is important for reasons as diverse as reducing greenhouse gas emissions, improving public health, and creating jobs. Mathiesen et al. (2015) examine this transition from a broader perspective, proposing the interconnection of electricity, heating and cooling, and gas grids into a unified energy infrastructure. Their concept of Smart Energy Systems not only integrates these three networks but also incorporates various types of energy storage, such as thermal, gas, and electric vehicles, thereby enhancing the flexibility and efficiency of renewable energy use. The authors conclude that this framework enables more efficient utilization of renewable sources and contributes to cost reductions. Brown et al. (2018) propose to integrate high percentages of renewable energy in pursuit of, on the one hand, the interconnection of electricity with energy from other sectors such as transport and heating and, on the other hand, the reinforcement of the grid at continental level. Using the PyPSA-Eur-Sec-30 model, the authors study the optimal cost structure for the reduction of CO_2 emissions considering the demand for electricity, transport, and heating. In the study, battery electric vehicles, gas-fired power conversion units, and long-term thermal storage play a key role, and the authors conclude that all this is integral for a cost-effective European energy system with a high use of renewable energy.

Child et al. (2019) simulate a 100% renewable energy sector transition with a target year of 2050. The authors contemplate regional scenarios, isolated from each other, and area scenarios, where regions are connected and therefore can share energy resources. Interconnections reduce the need for storage and lower system costs, as well as lower investment in storage infrastructure. The authors conclude that, in Europe, a 100% renewable system is feasible from a technological and economic point of view. Sinsel et al. (2020) explore the technical problems of increasing the proportion of variable renewable energy. These technologies require precise locations and can affect supply stability, power quality, and energy flow in the grid. The authors conclude that there are technical solutions, such as modification of variable renewable energy generators or distributed energy storage, with the ability to solve many of the challenges of renewable energy, and thus they favor the decarbonization of the electricity sector.

Transportation systems are responsible for the growth in energy consumption worldwide. The electric vehicle is an alternative transportation option that does not emit combustion gases while at the same time. Vehicle-to-grid, or V2G, technology not only allows for the charging of the vehicle when connected to the grid, but also allows the electric vehicle to contribute energy to the grid to which it is connected, transforming the electric vehicle into an energy storage system. Tan et al. (2016) review V2G technology for its potential utility as a grid stabilizer by supporting intermittent renewable energy sources such as solar and wind. The authors see V2G technology as having the potential to offer multiple services to the connected grid, as well as a source of revenue for vehicle owners, although improvements in charging infrastructure and battery technology, as well as the participation of governments, businesses, and vehicle owners are necessary for successful implementation of this technology. Li et al. (2019) explore different energy storage systems, management strategies, and charging technologies for pure electric vehicles. While underlining the efficiency of this type of vehicle, they also stress the importance of research in battery technology to overcome problems such as lifetime, capacity, or manufacturing costs. The authors also conclude the need for the involvement of public institutions to make electric vehicles a viable and beneficial alternative for mobility.

While previous studies have examined various aspects of electrification, this chapter offers a comprehensive synthesis of how differing contexts and challenges have shaped electrification globally, underlining the importance of adaptive solutions.

Methodology

This study involved a comprehensive search of the Scopus database, encompassing research published between 2003 and 2024. Figure 14.1 shows the methodology we followed in the bibliometric analysis. The search targeted a range of thematic areas, including but not limited to electrification, smart grids, socioeconomic impacts, and renewable energy. Following the collection of the data, we subjected it to cleaning and preparation through the normalization of terms and the elimination of duplicates.

The analysis used the VOSviewer software[1] to construct a series of keyword co-occurrence maps based on discrete sections of the chapter. For each section, the software generated thematic clusters, which allowed for a comprehensive examination of the structure of research trends. In particular, we used the following keywords for each section:

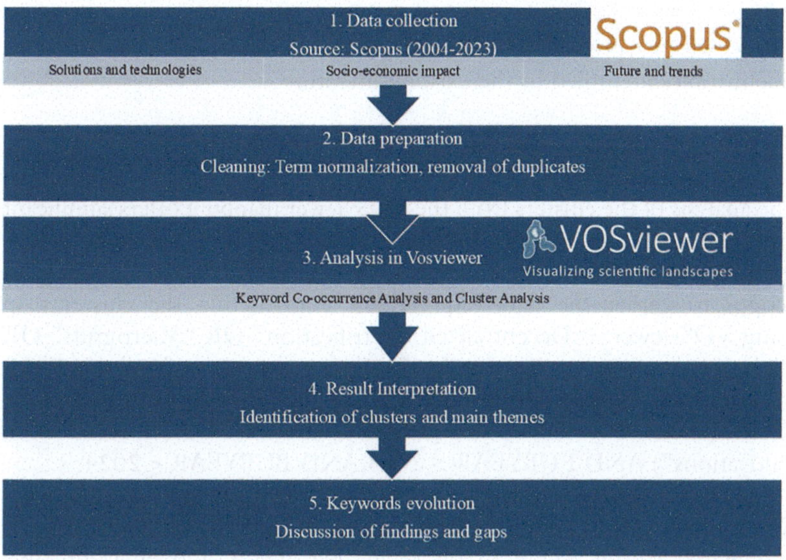

Fig. 14.1 Methodology followed in the bibliometric analysis

[1] https://www.vosviewer.com/

- Solutions and Technologies: To identify studies on electrification technologies, we used terms such as "decentralized electrification," "microgrids," "solar energy for electrification," "energy storage technologies," "smart grids and electrification," "public–private partnerships in electrification," "pay-as-you-go energy models," and "electrification innovations."
- Socioeconomic Impact: This section focused on the social and economic effects of electrification, using keywords like "electrification and economic growth," "electrification and quality of life," "energy poverty," "electrification and inequality reduction," "gender and energy access," and "electrification and social development."
- Future and Trends: To assess emerging trends in electrification, we used keywords such as "future of global electrification," "electrification of transport," "electrification and climate change," "emerging trends in energy access," and "energy transition and electrification."

Finally, we interpreted the clusters and themes by focusing on the most salient topics identified in the existing literature.

Solutions and Technologies

The analysis of the clusters from the VOSviewer mapping offers significant insights into the contemporary landscape of electrification technologies and solutions. The authors applied the following search equation in Scopus to gather the data required for constructing the cluster map using VOSviewer: ("Decentralized electrification" OR "Microgrids" OR "Solar energy for electrification" OR "Energy storage technologies" OR "Smart grids and electrification" OR "Public–private partnerships in electrification" OR "Pay-as-you-go energy models" OR "Electrification innovations") AND PUBYEAR > 2003 AND PUBYEAR < 2024.

Figure 14.2 offers a visual representation of the scientific and thematic clusters associated with electrification technologies and solutions.

The red cluster, centered on renewable energy and environmental sustainability, underscores the pivotal role of renewable sources like solar and wind power in reducing greenhouse gas emissions. These technologies not only provide clean energy alternatives to fossil fuels but also promote sustainable electrification practices, which are essential for achieving global climate goals. The integration of renewable energy into

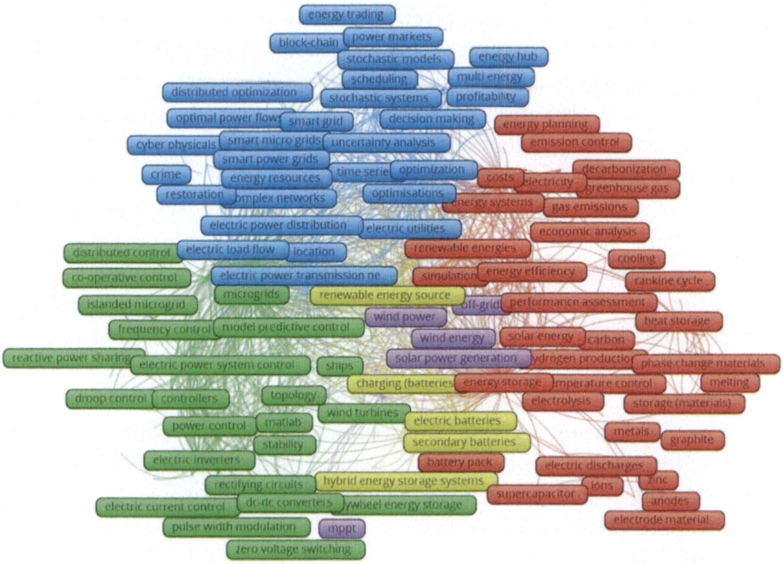

Fig. 14.2 Scientific and thematic communities or clusters associated with the solutions and technologies in electrification

existing power grids provides access to clean energy, thereby stimulating economic growth and addressing environmental challenges.

One of the most frequently referenced articles in this group is the paper by Olivares et al. (2014), which addresses the challenges associated with integrating renewable energy into microgrids and provides a categorization of control strategies into primary, secondary, and tertiary levels. Additionally, the paper explores the key principles of droop control and multi-agent systems for effective grid coordination. In a further pivotal study within the red cluster, Lund et al. (2015) investigate strategies to enhance energy system flexibility for large-scale renewable electricity generation. In particular, they examine measures such as grid expansion, storage, demand-side management, and the integration of electric vehicles. The study highlights the importance of integrating these flexibility solutions into market and utility frameworks to support renewable energy integration. In a further notable contribution to the field, Wang et al. (2019) examine the potential of deep learning techniques to improve the accuracy of renewable energy forecasting. This is a crucial area of research for grid management, given the inherent variability of renewable resources. The authors categorize the approaches into the following types: deep belief networks, stacked auto-encoders, deep recurrent networks, and others. They also discuss data preprocessing and error correction techniques, as well as the challenges and future research directions associated with these approaches.

The blue cluster focuses on optimizing energy systems through advanced data analytics and modeling, aligning with the smart grid and energy management approach. The implementation of smart grids enables the real-time monitoring and management of energy resources, thereby enabling utilities and consumers to respond dynamically to energy demands. Technologies such as smart meters and advanced metering infrastructure facilitate this process, enhancing energy efficiency and reliability while integrating a greater proportion of renewable sources into the grid. The implementation of energy management systems (EMS) serves to enhance the resilience of power networks, particularly in urban areas where demand is high.

The following studies are among the most frequently referenced articles in this cluster. Palensky and Dietrich (2011) discuss energy system management systems, focusing primarily on the demand side. Management systems include various strategies aimed at improving energy efficiency and shaping consumption patterns in a way that balances supply and demand. The authors classify management systems according to energy efficiency, usage-based tariffs, demand, and contingency energy reserve systems, and explain how current communication systems facilitate management. In their paper, Zhong and Weiss (2011) promote, in view of the changing current energy landscape with online or islanded renewable generation systems, the use of synchronous converters, an inverter that simulates the behavior of synchronous generators. Luo et al. (2015) describe the characteristics and operating principles of electrical energy storage systems, classifying them into mechanical or electrochemical, among others, and highlighting their fundamental role in the stabilization of electrical systems. In a similar manner, Cook et al. (2010) highlight the importance of affordable solar energy systems and research for efficiency improvement and cost reduction. In addition, they argue the importance of storage systems such as batteries or hydroelectric storage systems, together with solar production systems, in providing a sustainable and carbon neutral energy source.

The green cluster addresses the field of distributed energy systems, with a particular focus on microgrids and control mechanisms. Such systems provide localized energy solutions that are of paramount importance for the enhancement of energy access in remote and underserved communities. The effective operation of microgrids is ensured by the combination of energy storage systems and distributed generation sources, which are integrated with innovative control strategies. This decentralized approach provides communities with the autonomy to manage their energy needs independently, utilizing local renewable resources and thereby enhancing overall system reliability.

The following most cited studies stand out in this cluster. Lopes et al. (2006) examine the performance of microgrids that are far from the main network. In this situation, control measures are necessary to maintain stability and performance. Within these measures storage systems play a key role in ensuring stable operation. Among the conclusions of the article, the authors propose the use of inverters to control voltage drops. Similarly, Pogaku et al. (2007) also use inverters and highlight the importance of these in the operation of the microgrid and under variable load conditions.

The yellow cluster represents the pivotal role of energy storage technologies in the context of effective electrification, underscoring the significance of innovations in batteries and hybrid systems. The advancement of sophisticated battery technologies, including lithium-ion and flow batteries, is of paramount importance in the storage of surplus energy generated from renewable sources. Such capabilities are of vital importance for maintaining a stable energy supply and supporting the integration of intermittent renewable resources. The promotion of energy storage systems serves to enhance grid stability and to improve the viability of electrification projects, particularly in regions where there is a variable demand for energy.

One of the most frequently referenced articles in the cluster is the study by Cheng et al. (2011), in which the authors review the materials used in rechargeable batteries, including lithium-ion, nickel-metal hydride, and redox batteries, with a particular focus on electrochemical performance. The authors emphasize the necessity for further investigation to develop safe and environmentally sustainable battery technologies. In this context, Wang et al. (2012) examine the potential of redox flow batteries for energy storage. The authors review innovations in redox chemistry, including metal-coordinated redox pairs, hybrid redox flow batteries, and non-aqueous electrolytes, among other developments. In a further investigation, Zhang et al. (2016) examine the cation-deficient $ZnMn_2O_4$ cathode in an aqueous zinc ion battery. The authors introduce an innovative cathode comprising a $Zn(CF_3SO_3)_2$ electrolyte, which enables enhanced energy storage while ensuring safety and efficiency.

The final, purple cluster addresses the technical challenges associated with the management of electric power flows. The implementation of advanced control systems, including model predictive control (MPC) and frequency regulation technologies, is of paramount importance for the optimization of power system performance. These technologies guarantee the reliable delivery of energy, particularly in systems that rely on variable renewable energy sources. The implementation of robust control strategies serves to enhance the overall reliability of the electricity supply, thereby contributing to the broader goals of electrification and sustainable development.

Among the most cited articles in this cluster is Serale et al. (2018), who review the use of MPC for building energy management, enabled by advances in automation and data access. It highlights MPC's benefits for thermal regulation, energy storage, HVAC optimization, and renewable integration, and proposes a unified framework for MPC implementation in building systems. Parisio et al. (2014) investigates the implementation of Model Predictive Control (MPC) for the enhancement of microgrid operations, with a focus on addressing time-varying demands and constraints. The study formulates the problem using mixed-integer linear programming, which allows for the derivation of efficient solutions without the need for complex heuristics. The authors validate their approach through experimental results derived from simulations of a microgrid located in Athens, demonstrating the effectiveness and feasibility of the proposed method.

The study by Olivares et al. (2014) is also a significant and highly cited work within this cluster. It presents a mathematical formulation for energy management in isolated microgrids using model predictive control. The authors divide the problem into Unit Commitment and Optimal Power Flow to simplify the approach, and the study—tested on a CIGRE Benchmark System[2]—emphasizes the importance of detailed

[2] A standardized test system developed by the International Council on Large Electric Systems (CIGRE), an international organization focused on power system expertise.

three-phase models for effective voltage regulation and reactive power support.

Figure 14.3, generated with the VOSviewer software, depicts the progression of keywords associated with electrification-related solutions and technologies. The terms are represented as a function of time, with a color gradient ranging from purple (for words prevalent in 2018) to yellow (for newer terms in 2021).

At the apex of the figure, terms such as "energy trading," "blockchain," and "power markets" reflect the growing integration of emerging technologies like blockchain and energy trading within energy markets. These developments also link to stochastic models and profitability analysis, suggesting a shift towards more sophisticated and optimized planning approaches. In the central part of the figure, terms related to "smart grids," "energy resources," "optimizations," and "energy systems" are displayed, highlighting the focus on energy resource management and optimization and smart grids. These topics have been the focus of numerous studies published in the past, although current studies focus on decarbonization and emissions reduction, demonstrating

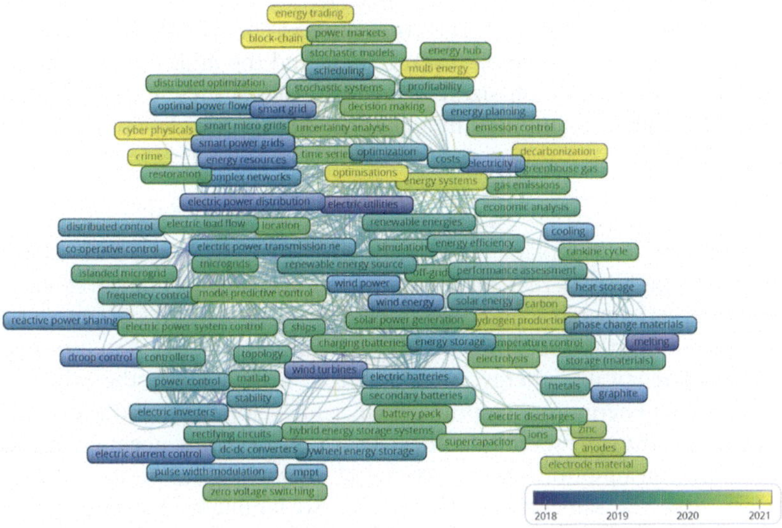

Fig. 14.3 Temporal evolution of keywords in solutions and technologies

a continued concern about the environmental impact of energy technologies. At the bottom of the figure, a range of technical terms related to control and power electronics, including "electric current control," "DC-DC converters," "electric discharges," and "electric inverters," emphasize the technical focus on energy storage and conversion solutions, such as electric batteries, supercapacitors, and hybrid energy storage systems.

Socioeconomic Impact

Figure 14.4 visually represents the scientific and thematic communities, or clusters, related to the socioeconomic impact of electrification. In this representation, color-coded terms indicate thematic affinities and connections between concepts. The data for constructing the cluster map in VOSviewer come from the following search equation in Scopus: ("Electrification and economic growth" OR "Electrification and quality of life" OR "Energy poverty" OR "Electrification and inequality reduction" OR "Gender and energy access" OR "Electrification and social development") AND PUBYEAR > 2003 AND PUBYEAR < 2024.

The terms situated at the apex of the figure, in blue, pertain to economic matters, including "financial development," "economic growth," and "renewable energy consumption." This indicates a robust correlation between economic advancement and the utilization of renewable energy sources, emphasizing the pivotal role of these energy sources in economic growth and their influence on carbon dioxide emissions and energy intensity. We summarize the most cited articles in this cluster below.

Inglesi-Lotz (2016) studies the impact of renewable energy on economic growth, looking at OECD countries, from 1990 to 2010. The study highlights that promoting renewable energy improves environmental sustainability and boosts economic growth. Based on these findings, the author recommends implementing policies that promote the use of renewable energy. Shahbaz et al. (2017) study the relationship between energy consumption, financial development, and economic growth in India over several decades. They conclude that negative shocks to energy consumption impact economic growth, while positive shocks do not have the same effect. The study also highlights the importance of financial development for improving energy efficiency and supporting economic growth. Zhang et al. (2021) analyze the relationship between public spending, green finance, and economic growth among the Belt

Fig. 14.4 Clusters of keywords related to the socioeconomic impact of electrification

and Road Initiative (BRI) countries, a group of nations participating in China's transnational infrastructure and investment strategy launched in 2013. The study indicates that public spending on green technologies positively impacts economic growth, although the intensity of the positive effect depends on the country's GDP.

The red cluster centers on interdisciplinary themes in renewable energy, energy policy, and greenhouse gases, reflecting a scientific community focused on energy development, policy, and greenhouse gas mitigation. Additionally, the concept of energy justice highlights a growing emphasis on social justice in access to renewable energy and equity in the energy transition.

The following studies stand out in this cluster. Pachauri and Jiang (2008) address the concept of energy transition in households in India and China, highlighting the differences between households in both countries and the influence of policies that promote urbanization, energy prices, and access to energy resources as key factors driving the transition. Newell and Mulvaney (2013) examine the concept of just transition to a low carbon economy, focusing on equity and justice issues for those affected by the fossil fuel economy and for those who do not have secure access to energy. They emphasize the need for policy and institutional frameworks to mediate the transition. Bridge et al. (2013) explore the concept of energy transition from a geographical perspective, pointing out that geographical aspects are crucial for the implementation of effective policies. Qazi et al. (2019) stress the importance of social awareness and public participation in the adoption of renewable energy sources.

The green cluster focuses on issues related to poverty and human well-being, including terms like "poverty," "public health," "indoor air pollution," and "household energy." These topics highlight the connection between energy, public health, and socioeconomic conditions, showing how a lack of access to clean energy harms the health and well-being of vulnerable populations, especially in developing countries and rural areas.

Along these lines, Bouzarovski and Petrova (2015) examine the conceptual difference between "energy poverty" and "fuel poverty" to argue that it is often unproductive to treat the concepts as completely separate phenomena. The authors propose a conceptual framework that overcomes this duality to emphasize that both conditions stem from the same problem: the inability to access adequate domestic energy services, such as lighting, heating, or cooling. Maintaining a rigid distinction between these two terms can hamper analysis and, therefore, effective policymaking by obscuring the sociotechnical and structural factors inherent to domestic energy deprivation. Similarly, Thomson et al. (2017) study the indicators employed by European institutions and national statistical agencies to measure energy poverty, highlighting the complexity and variability of the concept across countries. They propose that measurement should be multifaceted, incorporating factors such as the share of income spent on energy, self-assessments of household energy vulnerability, and housing.

The yellow cluster, comprised of words like "renewable energy sources," "greenhouse gases," "climate change," and "energy justice," centers on the environmental implications of energy consumption and the need for equitable energy solutions. This thematic grouping reflects how renewable energy sources can mitigate greenhouse gas emissions and combat climate change, while also emphasizing the importance of energy justice to ensure that all communities benefit from the transition to sustainable energy systems.

Walker and Day (2012) argue that fuel poverty is a major source of social inequality and distributive injustice, and emphasize the necessity of procedural justice for marginalized groups. The study justifies the formulation of policies that address fuel poverty in the UK by highlighting the importance of engaging disadvantaged groups to recognize their needs. Day et al. (2016) propose a multidimensional energetic poverty definition by including aspects of well-being and the ability to achieve essential functions. This approach, in light of climate change, allows for a greater understanding of how energy services influence social and individual well-being. McCauley and Heffron (2018) promote the integration of distributive, procedural, and normative justice for vulnerable communities during the transition from fossil fuels to sustainable energy sources.

Finally, in purple at the bottom of the figure, are more specific terms related to thermal comfort and energy efficiency in buildings, such as "thermal insulation," "retrofitting," and "thermal comfort." This cluster pertains to the enhancement of housing conditions, which also has socioeconomic implications, particularly regarding energy efficiency and cost savings. In this context, Bird and Hernández (2012) emphasize the potential of addressing energy efficiency in low-income housing through well-designed incentives and financing mechanisms to alleviate energy poverty, benefiting both tenants and landlords and potentially saving billions annually for vulnerable populations. Similarly, Santamouris et al. (2014) illustrate how the economic crisis intensifies the prevalence of inadequate indoor environmental quality in low-income residences. In these settings, suboptimal thermal conditions have a detrimental impact on health and well-being, emphasizing the imperative for enhancing energy efficiency and housing conditions in economically disadvantaged communities. Tirado Herrero et al. (2018) further critique the role of smart home technologies, arguing that while these technologies promise greener and more comfortable homes, they may reinforce unsustainable energy consumption patterns, are often inaccessible to vulnerable populations, and do little to alleviate energy poverty.

Figure 14.5 illustrates the evolution of keywords pertaining to the socioeconomic impact of electrification, created using VOSviewer software.

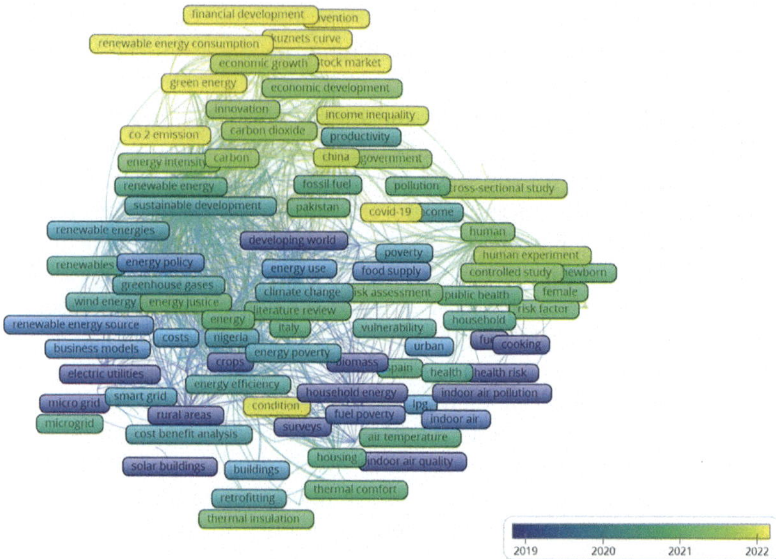

Fig. 14.5 Keywords evolutions pertaining to the socioeconomic impact of electrification

At the apex of the figure in yellow, the more recent keywords like "financial development," "income inequality," and "economic growth" highlight the mounting apprehension about the ramifications of financial advancement and the disparities that emerge during the transition to renewable energy. Furthermore, terms such as "stock market" and "Kuznets curve" are likewise significant, reflecting a focus on the impact of the energy transition on the global economy and financial markets.

Keywords in green and blue at the center of the figure, such as "renewable energy," "sustainable development," and "energy policy," show that these topics have remained relevant in recent years. The focus on renewable energy and sustainable development also reflects the prominence of themes like "greenhouse gases" and "climate change," highlighting the critical role of energy policies in mitigating climate change.

To the right, words like "public health," "poverty," and "indoor air pollution" stand out in green, indicating growing interest in the impact of electrification on public health and living conditions, particularly in vulnerable areas. Additionally, terms related to the pandemic, such as

"COVID-19," also appear, reflecting how researchers and policymakers have considered electrification and renewable energy within the broader context of socioeconomic responses to the global crisis.

Finally, at the bottom of the chart in blue and purple, topics such as "energy efficiency," "retrofitting," and "thermal comfort" remain central to discussions on energy efficiency and improving housing conditions through energy-saving technologies. This emphasizes the need to implement electrification solutions that not only are sustainable but also enhance quality of life.

Future and Trends

Figure 14.6, generated with VOSviewer software, presents a visual representation of the disparate scientific communities that we have grouped into clusters based on their relevance to the fields of electrification and energy transition. In order to construct the cluster map with VOSviewer, we used the following search equation in Scopus to retrieve the relevant data: ("Future of global electrification" OR "Electrification of transport" OR "Electrification and climate change" OR "Emerging trends in energy access" OR "Energy transition and electrification") AND PUBYEAR > 2003 AND PUBYEAR < 2024.

Each cluster is comprised of terms that are interrelated and represent key research areas. The red cluster is primarily concerned with the topic of transportation electrification, encompassing terms such as "vehicles," "electric buses," "commercial fleets," "autonomous vehicles," and "gas emissions." These concepts highlight the imperative of reducing the carbon footprint in the transportation sector through the deployment of electric and autonomous vehicles, in addition to the implementation of sustainable fleets for public and commercial transportation. In this area, Vega-Perkins et al. (2023) demonstrate that over 90% of U.S. households would benefit from reduced emissions and lower transportation energy burdens by adopting electric vehicles (EVs), though low-income households may still face high energy costs. Their study emphasizes the need for policies that promote energy justice and expand access to low-carbon transportation options.

The blue cluster is centered around electric vehicle infrastructure, encompassing terms such as "battery charging," "electric power transmission," and "energy management." This cluster underscores the imperative

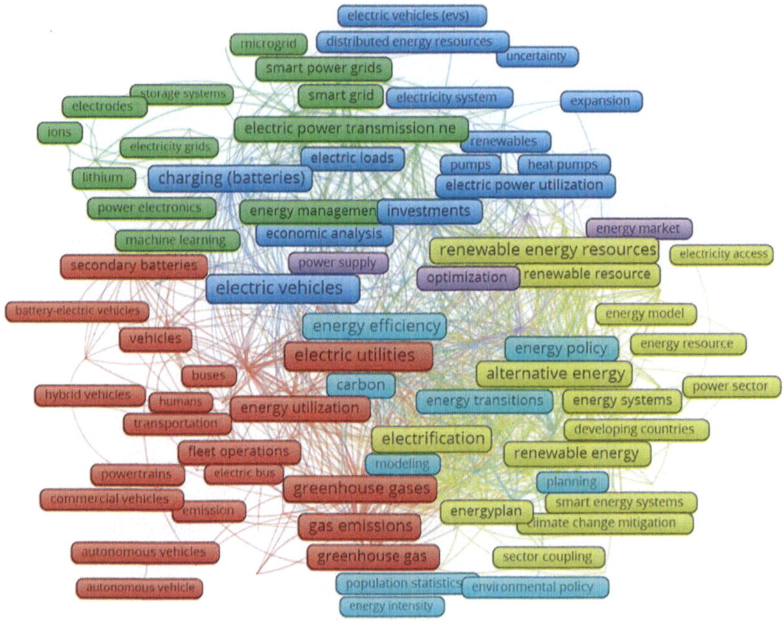

Fig. 14.6 Trends and clusters in electrification research

for the establishment of a resilient infrastructure to facilitate the proliferation of electric vehicles, taking into account the economic and investment requirements necessary for their expansion.

The penetration of EVs in global markets has piqued the interest of researchers, particularly regarding their social, demographic, and economic factors. Some authors have highlighted the importance of sociodemographic characteristics in shaping people's interest in EVs. Chen et al. (2020) found that in Nordic countries, younger men with higher income levels and more children were more likely to purchase EVs due to their environmental values and previous experience with electric vehicles. Similarly, Mohamed et al. (2016) observed that individuals with higher income and education levels were more inclined to adopt EVs, especially when motivated by environmental concerns, greater awareness, and prior experience with EVs.

EV consumers take into account factors like the technical capabilities of vehicles, such as V2G technology and charging stations. Chen et al. (2020) identify charging time and V2G capability as important factors for consumer purchase of electric vehicles. In the same vein, but focusing on electric delivery trucks, Davis and Figliozzi (2013) suggest that reduced operating costs and favorable logistical factors contribute to the competitive advantage of electric delivery trucks, particularly in urban settings.

The green cluster addresses the development of advanced energy technologies, including smart grids, microgrids, energy storage, and lithium-ion batteries. These concepts are fundamental to enhancing the efficiency and resilience of electricity systems, facilitating more effective integration of renewable energies and guaranteeing the stability of energy supply through decentralized solutions. The integration of energy storage systems and their integration into the grid is important to increase the share of renewable energy in the grid. Mcllwaine et al. (2021) highlight the role of energy storage systems in stabilizing power quality and improving services. Although technological and economic factors are important to consider, the need for regulatory frameworks is essential to balance supply and demand in a low-carbon energy system. In this context, Bridge and Faigen (2022) emphasize the growing intersection of battery manufacturing with the automotive and power sectors, particularly in the context of electric vehicles and stationary storage. Their study, which applies a global production network approach, highlights how inter-industry connections, innovation, and state involvement are reshaping the battery production landscape, offering crucial insights for policy and the geopolitical economy of energy transitions. This research is important within the cluster as it deepens the understanding of the evolving energy storage network, which plays a key role in the green energy transformation.

The yellow cluster is concerned with the policy aspects of renewable energy, including terms such as "renewable resources," "electricity access," "energy policy," and "energy transitions." This cluster reflects the necessity for the implementation of policies that facilitate the integration of renewable energies into the global energy system, thereby contributing to climate change mitigation by reducing fossil fuel use. Among the most frequently cited studies in this cluster is that of Batas, Bijelic and Rajakovic (2021) present an initiative to accelerate Serbia's energy transition in alignment with the strategic objectives of the European Union. To this end, the authors propose a transition from small hydroelectric sources to photovoltaic and wind power, aiming to meet the country's renewable energy targets and support its EU membership aspirations. Another important study in this cluster is by Breyer et al. (2022), who demonstrate that a global shift to 100% renewable energy is both feasible and cost-effective, with solar and wind energy at the forefront. They address critical challenges like grid congestion and energy storage, emphasizing the role of policy in advancing a sustainable, net-negative emissions economy. Similarly, Manfren et al. (2020) highlight the crucial role of open energy models and data in supporting transparent, multidisciplinary research to drive effective policy-making. Their work underscores how open resources can enhance science-policy interaction and help resolve technical challenges, such as optimizing energy efficiency and infrastructure operations, that are key to successful energy transitions.

The light blue cluster is oriented towards the advancement of energy efficiency and the promotion of electrification. Its constituent terms include "greenhouse gases," "energy modeling," and "electrification." This cluster highlights the critical need to enhance energy efficiency across all sectors, employing electrification as a strategy to reduce emissions and optimize energy utilization.

The transition to electrification, particularly in the transport sector, is increasingly recognized as a vital strategy for achieving the Sustainable Development Goals (SDGs)—a global framework adopted by the United Nations comprised of 17 interlinked objectives aimed at ending poverty, protecting the planet, and ensuring prosperity for all by 2030—and addressing climate change. Yuan et al. (2021) analyze transport electrification as a significant component of the energy transition, particularly in the context of the Beijing-Tianjin-Hebei region in China. The study employs the EnergyPLAN tool to model various scenarios incorporating EVs along with increased renewable energy capacity. The results indicate that achieving 100% EV penetration could lead to at least an 11% reduction in carbon dioxide emissions by 2050, provided that renewable energy capacity is also increased. Johnsson et al. (2020) further emphasize the critical role of electrification in the context of the construction industry. Their research discusses the need for a robust framework for assessing the impact of sustainability initiatives, specifically in relation to climate change (SDG 13). They identify key strategies for decarbonization, including electrification of construction processes and the introduction of renewable transportation fuels.

The purple cluster addresses the energy market and distributed energy resources, indicating that energy markets are becoming increasingly complex with the integration of distributed energy sources such as rooftop solar panels. The aforementioned clusters collectively illustrate a growing convergence between disparate research domains pertaining to electrification and energy transition. The electrification of transportation, the development of smart infrastructures, the integration of renewable energies, and the formulation of energy policies will serve as the cornerstones for achieving a more sustainable, low-carbon economy in the future. Emerging technologies, such as advanced energy storage and smart grids, will play a pivotal role in this process, while integrated approaches that connect disparate energy sectors, such as electricity, heating, and transportation, will enhance efficiency and curtail overall emissions. In this sense, Tronchin et al. (2018) address the challenges of balancing the supply–demand mismatch in energy systems dominated by renewable sources. They explore how strategies such as demand-side management, storage, and energy-efficient design, combined with scalable modeling techniques, can help create sustainable, cross-sector energy solutions in the built environment. Another important study in this cluster is that of Strielkowski et al. (2021), who examine the role of renewable energy in sustainable development within the electricity sector, highlighting its impact on both suppliers and consumers and the challenges posed by increased electricity demand. The study also addresses issues such as the European gas crisis and identifies key factors for achieving sustainability in the sector.

Figure 14.7 visually represents the temporal evolution of keywords related to future trends in electrification, generated using VOSviewer. Keywords associated with the period before 2018 appear in shades of blue on the far left, including terms like "hybrid vehicles," "emissions," "transportation," and "secondary batteries." This shows that, at the beginning of the analysis, transportation electrification was a central theme, focusing on the transition to hybrid technologies and concerns about pollutant gas emissions. Additionally, terms such as "electric buses," "commercial fleets," and "electric vehicles" are also relatively older, indicating that the electrification of transportation was one of the earliest areas of focus, especially for emissions reduction.

As the analysis progresses, shades of green representing 2019 and 2020 emerge, highlighting keywords such as "energy management," "battery charging infrastructure," "energy efficiency," "renewable energy

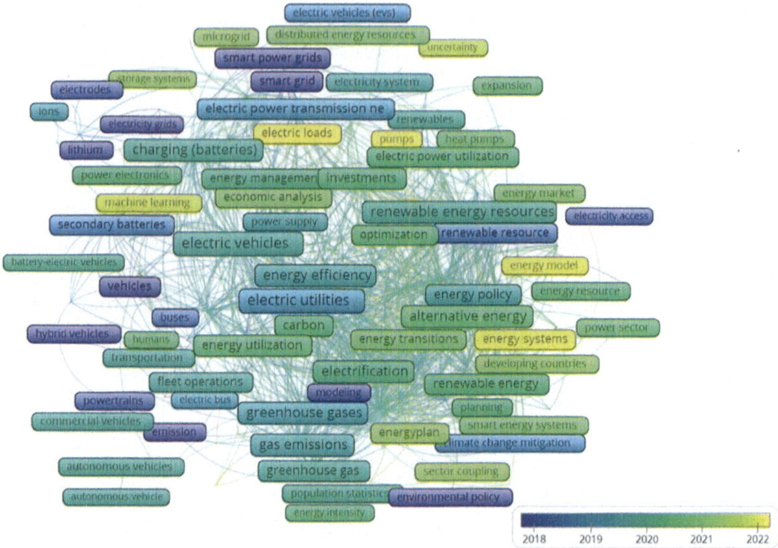

Fig. 14.7 Temporal evolution of keywords associated with future trends in electrification

resources," and "energy policy." This shift reflects a growing focus on the infrastructure supporting electrification, including charging networks and electricity transmission, alongside the need for energy policies that facilitate renewable resource use and promote efficiency.

The most recent keywords, corresponding to 2021 and 2022, are in yellow. In this period, key terms such as "energy systems," "energy transitions," "sector coupling," "climate change mitigation," and "machine learning" dominate. This shift indicates a move towards more integrated and advanced topics, with climate change mitigation becoming a significant priority. The increased focus on optimizing energy systems through machine learning and the growing interest in sector coupling suggest efforts to enhance efficiency and reduce emissions by integrating different energy sectors.

Conclusions

The bibliometric analysis of electrification research conducted from 2003 to 2024 reveals several significant insights into the evolving landscape of this critical field. The notable increase in publications is indicative of not only a growing academic interest but also an urgent response to global energy challenges, including climate change and energy poverty. This surge in research activity underscores the value of interdisciplinary collaboration, as scholars from a range of disciplines—including engineering, environmental science, economics, and social sciences—are increasingly engaged in joint endeavors to address the intricate challenges associated with electrification. It is imperative that this collaborative effort be undertaken to develop integrated solutions that leverage technological advancements while considering their socioeconomic impacts.

The analysis identifies key thematic clusters that highlight areas of concentrated research activity, particularly in renewable energy, smart grids, and the socioeconomic dimensions of electrification. It recommends prioritizing future research at the intersection of these themes, particularly focusing on integrating smart grid technologies with renewable energy sources to improve system efficiency. Additionally, examining the socioeconomic impacts provides valuable insights into the social dimensions of electrification. The focus on themes like energy access, economic growth, and social equity highlights the growing recognition of electrification's role in improving the quality of life and reducing poverty. However, the analysis also reveals a gap in knowledge regarding the effects of electrification on vulnerable populations, suggesting that future studies should prioritize empirical investigations into the social consequences of electrification.

The temporal evolution of keywords shows a clear progression in research focus, moving from initial studies on transportation electrification to contemporary themes addressing climate change mitigation and advanced analytical techniques like machine learning. This shift demonstrates that the field is adapting to emerging global challenges and technological advancements, responding to the changing landscape of research. As the field of electrification evolves, researchers must investigate the role of artificial intelligence and machine learning in optimizing energy systems, enhancing predictive capabilities, and developing more efficient and adaptive energy solutions.

Moreover, the analysis highlights the need for policymakers to create robust policy frameworks that facilitate the integration of renewable energy and smart technologies into existing energy systems. Policymakers must foster an environment that encourages innovation while aligning electrification efforts with broader sustainability goals. This includes creating policies that promote investment in renewable energy infrastructure, provide incentives for research and development in smart technologies, and address the socioeconomic barriers to energy access.

The findings from this bibliometric analysis suggest several potential avenues for future research. There is an urgent need for studies examining the relationship between electrification and environmental sustainability, with a focus on greenhouse gas emissions. Additionally, research should explore the scalability and replicability of successful electrification models across different socioeconomic contexts, especially in rural and underserved areas. Further insight is also necessary into the role of energy storage technologies in enabling the integration of renewable energy to advance the electrification agenda.

References

Akinyele, D. O., & Rayudu, R. K. (2014). Review of energy storage technologies for sustainable power networks. *Sustainable Energy Technologies and Assessments, 8*, 74–91. https://doi.org/10.1016/j.seta.2014.07.004

Bijelic, I. B., & Rajakovic, N. (2021). National energy and climate planning in Serbia: From lagging behind to an ambitious EU candidate? *International Journal of Sustainable Energy Planning and Management, 32*, 47–60. https://doi.org/10.5278/ijsepm.6300

Bird, S., & Hernández, D. (2012). Policy options for the split incentive: Increasing energy efficiency for low-income renters. *Energy Policy, 48*, 506–514. https://doi.org/10.1016/j.enpol.2012.05.053

Bouzarovski, S., & Petrova, S. (2015). A global perspective on domestic energy deprivation: Overcoming the energy poverty–fuel poverty binary. *Energy Research & Social Science, 10*, 31–40. https://doi.org/10.1016/j.erss.2015.06.007

Breyer, C., Khalili, S., Bogdanov, D., Ram, M., Oyewo, A. S., Aghahosseini, A., Gulagi, A., Solomon, A. A., Keiner, D., Lopez, G., Østergaard, P. A., Lund, H., Mathiesen, B. V., Jacobson, M. Z., Victoria, M., Teske, S., Pregger, T., Fthenakis, V., Raugei, M., … Sovacool, B. K. (2022). On the history and future of 100% renewable energy systems research. *IEEE Access, 10*, 78176–78218. https://doi.org/10.1109/ACCESS.2022.3193402

Bridge, G., & Faigen, E. (2022). Towards the lithium-ion battery production network: Thinking beyond mineral supply chains. *Energy Research & Social Science, 89*, Article 102659. https://doi.org/10.1016/j.erss.2022.102659

Bridge, G., Bouzarovski, S., Bradshaw, M., & Eyre, N. (2013). Geographies of energy transition: Space, place and the low-carbon economy. *Energy Policy, 53*, 331–340. https://doi.org/10.1016/j.enpol.2012.10.066

Brown, T., Schlachtberger, D., Kies, A., Schramm, S., & Greiner, M. (2018). Synergies of sector coupling and transmission reinforcement in a cost-optimised, highly renewable European energy system. *Energy, 160*, 720–739. https://doi.org/10.1016/j.energy.2018.06.222

Castillo, A., & Gayme, D. F. (2014). Grid-scale energy storage applications in renewable energy integration: A survey. *Energy Conversion and Management, 87*, 885–894. https://doi.org/10.1016/j.enconman.2014.07.063

Chen, C.-F., Zarazua de Rubens, G., Noel, L., Kester, J., & Sovacool, B. K. (2020). Assessing the socio-demographic, technical, economic and behavioral factors of Nordic electric vehicle adoption and the influence of vehicle-to-grid preferences. *Renewable and Sustainable Energy Reviews, 121*, Article 109692. https://doi.org/10.1016/j.rser.2019.109692

Cheng, F., Liang, J., Tao, Z., & Chen, J. (2011). Functional materials for rechargeable batteries. *Advanced Materials, 23*(15), 1695–1715. https://doi.org/10.1002/adma.201003587

Child, M., Kemfert, C., Bogdanov, D., & Breyer, C. (2019). Flexible electricity generation, grid exchange and storage for the transition to a 100% renewable energy system in Europe. *Renewable Energy, 139*, 80–101. https://doi.org/10.1016/j.renene.2019.02.077

Cook, T. R., Dogutan, D. K., Reece, S. Y., Surendranath, Y., Teets, T. S., & Nocera, D. G. (2010). Solar energy supply and storage for the legacy and nonlegacy worlds. *Chemical Reviews, 110*(11), 6474–6502. https://doi.org/10.1021/cr100246c

Davis, B. A., & Figliozzi, M. A. (2013). A methodology to evaluate the competitiveness of electric delivery trucks. *Transportation Research Part E, 49*, 8–23. https://doi.org/10.1016/j.tre.2012.07.003

Day, R., Walker, G., & Simcock, N. (2016). Conceptualising energy use and energy poverty using a capabilities framework. *Energy Policy, 93*, 255–264. https://doi.org/10.1016/j.enpol.2016.03.019

Gallo, A. B., Simões-Moreira, J. R., Costa, H. K. M., Santos, M. M., & Moutinho dos Santos, E. (2016). Energy storage in the energy transition context: A technology review. *Renewable and Sustainable Energy Reviews, 65*, 800–822. https://doi.org/10.1016/j.rser.2016.07.028

IEA. (2021). *Net zero by 2050*. International Energy Agency (IEA). https://www.iea.org/reports/net-zero-by-2050

IEA. (2024). *World energy outlook 2024*. International Energy Agency (IEA). https://iea.blob.core.windows.net/assets/140a0470-5b90-4922-a0e9-838b3ac6918c/WorldEnergyOutlook2024.pdf

Inglesi-Lotz, R. (2016). The impact of renewable energy consumption on economic growth: A panel data application. *Energy Economics, 53*, 58–63. https://doi.org/10.1016/j.eneco.2015.01.003

Johnsson, F., Karlsson, I., Rootzén, J., Ahlbäck, A., & Gustavsson, M. (2020). The framing of a sustainable development goals assessment in decarbonizing the construction industry—Avoiding "greenwashing." *Renewable and Sustainable Energy Reviews, 131*, Article 110029. https://doi.org/10.1016/j.rser.2020.110029

Li, Z., Khajepour, A., & Song, J. (2019). A comprehensive review of the key technologies for pure electric vehicles. *Energy, 182*, 824–839. https://doi.org/10.1016/j.energy.2019.06.077

Lopes, J. A. P., Moreira, C. L., & Madureira, A. G. (2006). Defining control strategies for microgrids islanded operation. *IEEE Transactions on Power Systems, 21*(2), 916–922. https://doi.org/10.1109/TPWRS.2006.873018

Lund, P. D., Lindgren, J., Mikkola, J., & Salpakari, J. (2015). Review of energy system flexibility measures to enable high levels of variable renewable electricity. *Renewable and Sustainable Energy Reviews, 45*, 785–807. https://doi.org/10.1016/j.rser.2015.01.057

Luo, X., Wang, J., Dooner, M., & Clarke, J. (2015). Overview of current development in electrical energy storage technologies and the application potential in power system operation. *Applied Energy, 137*, 511–536. https://doi.org/10.1016/j.apenergy.2014.09.081

Manfren, M., Nastasi, B., Groppi, D., & Garcia, D. A. (2020). Open data and energy analytics—An analysis of essential information for energy system planning design and operation. *Energy, 213*, 118803. https://doi.org/10.1016/j.energy.2020.118803

Mathiesen, B. V., Lund, H., Connolly, D., Wenzel, H., Østergaard, P. A., Möller, B., Nielsen, S., Ridjan, I., Karnøe, P., Sperling, K., & Hvelplund, F. K. (2015). Smart energy systems for coherent 100% renewable energy and transport solutions. *Applied Energy, 145*, 139–154. https://doi.org/10.1016/j.apenergy.2015.01.075

McCauley, D., & Heffron, R. (2018). Just transition: Integrating climate, energy and environmental justice. *Energy Policy, 119*, 1–7. https://doi.org/10.1016/j.enpol.2018.04.014

McIlwaine, N., Foley, A. M., Morrow, D. J., Al Kez, D., Zhang, C., Lu, X., & Best, R. J. (2021). A state-of-the-art techno-economic review of distributed and embedded energy storage for energy systems. *Energy, 229*, Article 120461. https://doi.org/10.1016/j.energy.2021.120461

Mohamed, M., Higgins, C., Ferguson, M., & Kanaroglou, P. (2016). Identifying and characterizing potential electric vehicle adopters in Canada: A two-stage modelling approach. *Transport Policy, 52*, 100–112. https://doi.org/10.1016/j.tranpol.2016.07.006

Newell, P., & Mulvaney, D. (2013). The political economy of the just transition. *Geographical Journal, 179*(2), 132–140. https://doi.org/10.1111/geoj.12008

Olivares, D. E., Cañizares, C. A., & Kazerani, M. (2014). A centralized energy management system for isolated microgrids. *IEEE Transactions on Smart Grid, 5*(4), 1864–1875. https://doi.org/10.1109/TSG.2013.2294187

Olivares, D. E., Mehrizi-Sani, A., Etemadi, A. H., Cañizares, C. A., Iravani, R., Kazerani, M., Hajimiragha, A. H., Gomis-Bellmunt, O., Saeedifard, M., Palma-Behnke, R., Jiménez-Estévez, G. A., & Hatziargyriou, N. D. (2014). Trends in microgrid control. *IEEE Transactions on Smart Grid, 5*(4), 1905–1919. https://doi.org/10.1109/TSG.2013.2295514

Pachauri, S., & Jiang, L. (2008). The household energy transition in India and China. *Energy Policy, 36*(10), 4022–4035. https://doi.org/10.1016/j.enpol.2008.06.016

Palensky, P., & Dietrich, D. (2011). Demand side management: Demand response, intelligent energy systems, and smart loads. *IEEE Transactions on Industrial Informatics, 7*(3), 381–388. https://doi.org/10.1109/TII.2011.2158841

Parisio, A., Rikos, E., & Glielmo, L. (2014). A model predictive control approach to microgrid operation optimization. *IEEE Transactions on Control Systems Technology, 22*(5), 1813–1827. https://doi.org/10.1109/TCST.2013.2295737

Pogaku, N., Prodanović, M., & Green, T. C. (2007). Modeling, analysis and testing of autonomous operation of an inverter-based microgrid. *IEEE Transactions on Power Electronics, 22*(2), 613–625. https://doi.org/10.1109/TPEL.2006.890003

Qazi, A., Hussain, F., Rahim, N. A., Hardaker, G., Alghazzawi, D., Shaban, K., & Haruna, K. (2019). Towards sustainable energy: A systematic review of renewable energy sources, technologies, and public opinions. *IEEE Access, 7*, 63837–63854. https://doi.org/10.1109/ACCESS.2019.2906402

Ram, M., Child, M., Aghahosseini, A., Bogdanov, D., Lohrmann, A., & Breyer, C. (2018). A comparative analysis of electricity generation costs from renewable, fossil fuel and nuclear sources in G20 countries for the period 2015–2030. *Journal of Cleaner Production, 199*, 687–704. https://doi.org/10.1016/j.jclepro.2018.07.159

Santamouris, M., Alevizos, S. M., Aslanoglou, L., Mantzios, D., Milonas, P., Sarelli, I., Karatasou, S., Cartalis, K., & Paravantis, J. A. (2014). Freezing the poor—Indoor environmental quality in low and very low income households during the winter period in Athens. *Energy and Buildings, 70*, 61–70. https://doi.org/10.1016/j.enbuild.2013.11.074

Schmidt, O., Hawkes, A., Gambhir, A., & Staffell, I. (2017). The future cost of electrical energy storage based on experience rates. *Nature Energy, 2*(8), 1–8. https://doi.org/10.1038/nenergy.2017.110

Serale, G., Fiorentini, M., Capozzoli, A., Bernardini, D., & Bemporad, A. (2018). Model predictive control (MPC) for enhancing building and HVAC system energy efficiency: Problem formulation, applications and opportunities. *Energies, 11*(3), 631. https://doi.org/10.3390/en11030631

Shahbaz, M., Hoang, T. H. V., Mahalik, M. K., & Roubaud, D. (2017). Energy consumption, financial development and economic growth in India: New evidence from a nonlinear and asymmetric analysis. *Energy Economics, 63*, 199–212. https://doi.org/10.1016/j.eneco.2017.01.023

Sinsel, S. R., Riemke, R. L., & Hoffmann, V. H. (2020). Challenges and solution technologies for the integration of variable renewable energy sources—A review. *Renewable Energy, 145*, 2271–2285. https://doi.org/10.1016/j.renene.2019.06.147

Strielkowski, W., Civín, L., Tarkhanova, E., Tvaronavičienė, M., & Petrenko, Y. (2021). Renewable energy in the sustainable development of electrical power sector: A review. *Energies, 14*(24), 8240. https://doi.org/10.3390/en14248240

Tan, K. M., Ramachandaramurthy, V. K., & Yong, J. Y. (2016). Integration of electric vehicles in smart grid: A review on vehicle to grid technologies and optimization techniques. *Renewable and Sustainable Energy Reviews, 53*, 720–732. https://doi.org/10.1016/j.rser.2015.09.012

Thomson, H., Bouzarovski, S., & Snell, C. (2017). Rethinking the measurement of energy poverty in Europe: A critical analysis of indicators and data. *Indoor and Built Environment, 26*(7), 879–901. https://doi.org/10.1177/1420326X1769926

Tirado Herrero, S., Nicholls, L., & Strengers, Y. (2018). Smart home technologies in everyday life: Do they address key energy challenges in households? *Current Opinion in Environmental Sustainability, 31*, 65–70. https://doi.org/10.1016/j.cosust.2017.12.001

Tronchin, L., Manfren, M., & Nastasi, B. (2018). Energy efficiency, demand side management and energy storage technologies—A critical analysis of possible paths of integration in the built environment. *Renewable and Sustainable Energy Reviews, 95*, 341–353. https://doi.org/10.1016/j.rser.2018.06.060

Vega-Perkins, J., Newell, J. P., & Keoleian, G. (2023). Mapping electric vehicle impacts: Greenhouse gas emissions, fuel costs, and energy justice in the United States. *Environmental Research Letters, 18*(1), Article 014027. https://doi.org/10.1088/1748-9326/aca4e6

Walker, G., & Day, R. (2012). Fuel poverty as injustice: Integrating distribution, recognition and procedure in the struggle for affordable warmth. *Energy Policy, 49*, 69–75. https://doi.org/10.1016/j.enpol.2012.01.044

Wang, H., Lei, Z., Zhang, X., Zhou, B., & Peng, J. (2019). A review of deep learning for renewable energy forecasting. *Energy Conversion and Management, 198*, Article 111799. https://doi.org/10.1016/j.enconman.2019.111799

Wang, W., Luo, Q., Li, B., Wei, X., Li, L., & Yang, Z. (2012). Recent progress in redox flow battery research and development. *Advanced Functional Materials, 23*(8), 970–986. https://doi.org/10.1002/adfm.201200694

World Health Organization. (2024). Tracking SDG7: The energy progress report 2024. https://www.who.int/publications/m/item/tracking-sdg7--the-energy-progress-report-2024

Xu, S.-C., He, Z.-X., & Long, R.-Y. (2014). Factors that influence carbon emissions due to energy consumption in China: Decomposition analysis using LMDI. *Applied Energy, 127*, 182–193. https://doi.org/10.1016/j.apenergy.2014.03.093

Yuan, M., Thellufsen, J. Z., Lund, H., & Liang, Y. (2021). The electrification of transportation in energy transition. *Energy, 236*, Article 121564. https://doi.org/10.1016/j.energy.2021.121564

Zakeri, B., & Syri, S. (2015). Electrical energy storage systems: A comparative life cycle cost analysis. *Renewable and Sustainable Energy Reviews, 42*, 569–596. https://doi.org/10.1016/j.rser.2014.10.011

Zhang, D., Mohsin, M., Rasheed, A. K., Chang, Y., & Taghizadeh-Hesary, F. (2021). Public spending and green economic growth in BRI region: Mediating role of green finance. *Energy Policy, 153*, Article 112256. https://doi.org/10.1016/j.enpol.2021.112256

Zhang, N., Cheng, F., Liu, Y., Zhao, Q., Lei, K., Chen, C., Liu, X., & Chen, J. (2016). Cation-deficient spinel $ZnMn_2O_4$ cathode in $Zn(CF_3SO_3)_2$ electrolyte for rechargeable aqueous Zn-ion battery. *Journal of the American Chemical Society, 138*(39), 12894–12901. https://doi.org/10.1021/jacs.6b05958

Zhong, Q.-C., & Weiss, G. (2011). Synchronverters: Inverters that mimic synchronous generators. *IEEE Transactions on Industrial Electronics, 58*(4), 1259–1267. https://doi.org/10.1109/TIE.2010.2048839

CHAPTER 15

Powering Up Iceland: How Iceland Transformed Itself into a Renewable Energy Powerhouse

Jordan Mitchell, Jonas Hlynur Hallgrímsson, and Thorstur Olaf Sigurjónsson

Introduction

Iceland is heralded by the United Nations as a world leader in renewable energy given the island nation derives nearly 100% of its electricity and 85% of total primary energy from renewable sources (Government

J. Mitchell (✉) · T. O. Sigurjónsson
University of Iceland, Reykjavík, Iceland
e-mail: jom27@hi.is

T. O. Sigurjónsson
e-mail: olaf@hi.is

J. H. Hallgrímsson
Landsvirkjun, Reykjavík, Iceland
e-mail: Jonas.Hlynur.Hallgrimsson@landsvirkjun.is

T. Walker et al. (eds.), *Wiring the Future*, Palgrave Studies in Emerging Risk Management and Sustainable Finance,
https://doi.org/10.1007/978-3-032-14717-2_15

of Iceland, 2024; UNEP, 2020). How is this possible? One explanation is the country's unique geographic characteristics and position on two shifting tectonic plates that provide abundant hydro and geothermal power. Another explanation is the steadfast long-term dedication by governments, businesses, and citizens to experiment, innovate, and strive for self-sufficiency. After all, the country was not always the global beacon for renewable energy; until the 1970s, the island nation relied heavily on imported coal and oil. For several decades, its capital city, Reykjavik, was covered in a continuous cloud of "black smoke" (Worland, 2017).

Today, Iceland's primary energy mix is comprised of geothermal power (approximately 65% of all energy consumed), hydroelectricity (20%), and non-renewable sources such as oil and gas, mostly for transportation (15%) (Government of Iceland, n.d.). Homes and water are heated with geothermal energy and return water from space heaters via pipes under Reykjavik's pedestrian areas to heat sidewalks, eliminating the need for snow removal in the winter (Pomrenke, 2023). Electric vehicles (EVs) are becoming more widespread, accounting for approximately 11% of the country's fleet (Statice, 2024). A significant portion of Iceland's electricity is "exported"[1] through three large-scale aluminum smelters, industrial facilities, and data centers, all of which take advantage of the country's low-cost and stable electricity supply.

Despite Iceland's renewable-powered economy and low carbon dioxide (CO_2) emissions, challenges remain. These include the strain on infrastructure due to a booming tourism industry, environmental concerns over new developments, the permitting process for new electricity supply, and transitioning the transportation sector to renewable sources. Innovations such as Carbfix and Orca, with novel technologies to capture and store CO_2 permanently in basalt rock, demonstrate Iceland's commitment to move increasingly close to net zero.

What can the rest of the world learn from the Icelandic experience? Iceland has demonstrated foresight and governmental commitment to a green future throughout the long term. The country aims to create an

[1] Since electricity is challenging to export from an island nation due to the required cables and infrastructure to carry the electricity to other countries, Iceland effectively "exports" electricity by producing energy-intensive goods such as aluminum and services such as data center processing.

environment where scientists, entrepreneurs, and policymakers collaborate to ensure a stable supply of renewable energy, low CO_2 emissions, and responsible development for future generations.

We structure the rest of the chapter in three main parts: past, present, and future. The past details Iceland's early development in electricity and renewables divided into four main periods from 1899 to 2025. The present covers statistics as of 2025 in addition to current energy debates within the country. Finally, the future lays out Iceland's plan for net zero by 2050 and other potential future avenues for development.

The Past

Today, Iceland has a population of 388,790 and enjoys the thirteenth highest GDP per capita in the world at $75,658 USD (Hagstofa, 2024; World Economics, 2023). However, 50 years ago, the United Nations (UN) considered the country a developing nation due to a heavy reliance on imports (Logadottir, 2015). While geothermal energy in the form of naturally occurring hot springs figures into the Viking Sagas dating back over 1,200 years, initiatives to tap into the country's rich sources of geothermal power as part of the primary energy mix started to take form in the early 1900s.

Early Transitions: 1899–1944

Electricity power in Iceland dated back to 1899. The country began importing coal and oil in the early 1900s, as it was more efficient than the predominant energy sources at the time: peat and dung. The first motorized fishing vessel started operating in 1902, and the first car arrived on the island in 1904 (Government of Iceland, n.d.). Modern "renewable" energy can be traced back to 1904 when the first hydropower turbine began operating at Hafnarfjordur near Reykjavik; the initial idea was to develop small-scale, decentralized power throughout the country. In the early 1900s, the Icelandic government was strictly against foreign ownership and did not want to sell rights to the waterfalls. This policy was in stark contrast to Norway's that allowed foreign ownership while stipulating a period in which the power stations would be returned to the government (Askja Energy, n.d.). In 1908, a farmer at Reykir, Stefan B. Jonsson, used geothermal steam to heat his house and in 1911, another farmer, Erlendur Gunnarsson, based in Sturlureykir, invented a way to separate steam from a hot water spring for cooking and heating (Fox Hodgson, 2024; Thorarinsdottir, 2006). These early innovations captured the imagination of Icelanders for years to come, shaping ideas on how the country's geothermal resources might be harnessed.

In 1921, the City of Reykjavik invested in a small-scale hydroelectricity generation facility using water from the Ellidaar river to provide electricity to local homes (Reykjavik Energy, n.d.). To support the project, the city created the public utility Reykjavik Electrical Utility (Rafmagnsveita Reykjavikur), which eventually became Reykjavik Energy (Orkuveita Reykjavikur) when it later merged with the city's water utility (Mitchell et al., 2022). Electrification in the country's second largest city, Akureyri, began in 1922 when the municipal government financed and built a hydroelectricity plant utilizing the water from the Glera river (Nordurorka, n.d.). With the two hydroelectricity plants becoming operational and burgeoning governmental interest in developing future projects, the national government enacted the Inlands Water Act in 1923, largely still in force today, to stipulate how natural water resources were to be protected and developed (Thorarinsdottir, 2006). The next major notable project was the two-turbine, 8.8 megawatts (MW) Ljosafoss hydropower station in 1937 (with an additional 5.8 MW turbine added in 1944) (Landsvirkjun, n.d.).

In addition to growing hydroelectric power, geothermal energy started becoming more prominent on a larger scale for space heating when a three-kilometer pipe from the Laugardalur hot springs was built in 1930 to heat homes and buildings in Reykjavik (Ragnarsson et al., 2020). During World War II, Iceland became a strategic base for the Allies, and investment in district heating abounded. One such example was the expansion of a new 18-kilometer pipe carrying hot water from Reykir to more homes in the capital (Ragnarsson et al., 2020).

The first wave of transition laid the foundation for what was to come—the eventual construction of industrial-scale hydro and geothermal capabilities. Given instability engendered by the two world wars and a changing political landscape, Iceland was focused on securing its own energy sources and avoiding foreign ownership (Davidsdottir, 2023). Despite several early innovations, by the end of World War II and Iceland's formation as a republic in 1944, approximately 80% of primary energy was still from imported fossil fuels (Ragnarsson et al., 2020).

Greater Electrification: 1945–1965

After World War II, electrification started spreading throughout Iceland with the installation of small-scale hydro power turbines—eventually, the individual plants became connected to regional grids, creating a centralized electricity market. The country enacted the Electricity Act in 1946, enabling the State Electricity Authority (Rafmagnsveitur rikisins) in 1947 to build, operate, and distribute electricity (Thorarinsdottir, 2006). Another small-scale hydroelectricity power plant (10 MW) was built in 1950 near Reykjavik (Askja Energy, n.d.) and the country began changing in other ways. The first IBM punch card computer surfaced within Iceland in 1952, marking an important step in embracing computers and technology within the energy sector, and, throughout the 1950s, several power-intensive plants opened, such as the State Fertilizer Plant (Aburdarverksmidjan) in 1953 and the State Cement Plant (Sementsverksmidja rikisins) in 1958 (Reykjavik Energy, n.d.). While the country continued to embrace hydroelectricity and geothermal power, the Icelandic Nuclear Physics Committee formed in 1956 (in subsequent years, no major nuclear power projects were pursued).

The 20-year period from the mid-1940s to the mid-1960s was characterized by the continued development of the nation and efforts to electrify the countryside. The continued aim to harness domestic resources for the common good and avoid foreign ownership remained mainstays of the energy strategy (Davidsdottir, 2023). And, while hydro and geothermal had grown substantially, black clouds of smoke from imported coal continued to hang over the nation's capital. Approximately 65% of the primary energy sources came from non-renewables by 1965 (Hagstofa Islands, n.d.). Given the country's geographical isolation from other energy-producing nations (i.e., 800 kilometers from Scotland and 950 kilometers from Norway), the government understood the vital importance of reducing the country's reliance on foreign energy sources by continuing to invest in the development of domestic energy.

Expansion of Renewables: 1965–1999

In 1965, the government established Landsvirkjun (National Power Company) to build the Burfell Hydropower station. To finance the massive 210 MW plant, Landsvirkjun received a loan from the World Bank for developing countries. The government updated the Energy

Act in 1967, covering all types of energy and establishing the National Energy Authority (Orkustofnun). Another important milestone occurred when Alusuisse inaugurated the country's first aluminum smelting plant in Straumsvik—Icelandic Aluminum Co. Ltd. (ISAL)—powered by hydroelectricity from the new Burfell plant (Askja Energy, n.d.). However, not all Icelanders were in agreement with the direction the country was taking. In 1970, 100 farmers used explosives to blow up a hydroelectric dam due to concerns over the impact to nature and farming, widely considered to be the first incident of environmental activism in the country (Jonsson Ulfhildarson, 2013).

In 1970, the National Energy Authority created a special committee to assess heating in reaction to debates about varied heating costs throughout the country—for example, residents in Reykjavik paid about half the price of more remote areas (Melsted, 2021). The disparate prices were due in part to the developments of the previous decade, when smaller municipalities had built small-scale geothermal district heating utilities.

Two major external events spurred on further development: the significant devaluation of the Icelandic krona and the international energy crisis of the early 1970s (Davidsdottir, 2022). In October 1973, the Organization of the Petroleum Exporting Countries (OPEC) issued an oil embargo—despite half the country relying on imports, "Icelanders remained calm" since the country did not rely on the Middle East for its oil but rather the Soviet Union, with whom they had an agreement dating back to the 1950s (Melsted, 2021, p. 536). In 1974, the then-Minister of Energy, Magnus Kjartansson, presented a plan to the national parliament proposing the elimination of oil and a strident focus on developing domestic energy including geothermal (Melsted, 2021).

Other challenges emerged in the second half of the 1970s as the country looked to build the 60 MW Krafla Geothermal electricity generation plant. Throughout the construction period, the Krafla volcano erupted, inflating and deflating the ground around the facility and affecting the quality of the geothermal fluid (Nielsen et al., 2000). The plant opened in 1978, operating only one of the two 30 MW turbines, eventually adding the other in the late 1990s (Landsvirkjun, n.d.). As the first large-scale geothermal electricity plant in the country, engineers described the Krafla facility as "one of the most intriguing systems" due to its "complex geological conditions and variable production characteristics [at] three exploited [geothermal] sub-fields" (Langella et al.,

2017, p. 91). Another example of harnessing geothermal energy was through the establishment of Hitaveita Sudurnesja (now HS Orka), first as a prototype to supply hot water in 1976 and later as an electricity producer beginning in 1978 with capacity expansions throughout the subsequent decades (HS Orka, n.d.). A year later, in 1979, the Norwegian company Elkem, in partnership with the Icelandic government, inaugurated a 60,000 ton per year ferrosilicon production plant in Akranes (Elkem, 2022).

By 1980, district heating throughout Iceland was generated either by geothermal power or by electricity in the "cold areas" where there was insufficient geothermal activity (Melsted, 2021). As such, district heating had been essentially "decarbonized" (Melsted, 2021). Observers commented that the main impetus for harnessing natural resources for district heating was motivated more by financial reasons and independence than the environmental benefits of decarbonization. As a chief political advisor stated:

> When the oil crisis receded in the 1980s all the interest other countries had shown in renewables disappeared and they fell back on their oil-reliant ways. But we continued to make progress in renewable energy development and now Icelanders are going to reap the benefits. (Aldred, 2008)

Hydroelectricity capacity continued to expand with the 210 MW Hrauneyjafoss Power Station in the southwest in 1981, followed by the underground glacier melt 165 MW Blanda Power Station in the northwest in 1991 (Landsvirkjun, n.d.). Located 230 meters underground, the reservoir leading to the Blanda plant was formed by dams from separate rivers. It was funneled through a 1.3-kilometer canal before dropping vertically to power the turbines in the underground station (IHA, 2017). The Blanda project involved a large-scale revegetation program to compensate for the loss of high-quality sheep grazing areas when the reservoir was created (Landsvirkjun, n.d.).

Throughout the latter half of the 1980s, the Icelandic government entered into negotiations with the Atlantal Group (a consortium of initially four, then three multinationals) to build the country's second aluminum smelter (Hayter, 1997; Skulason, 1994). However, by the early 1990s, the deal fell through and the Icelandic government formed an agreement with another group, Columbia Ventures Corp (CVC), to develop the Nordural facility with Landsvirkjun supplying 900 GWh per year (Walawalker, 1998). Opening in 1998 at Hvalfjordur near Reykjavík, the country's second aluminum smelter had a capacity of 60,000 tons per year. It was built in the record time of 15 months with the majority of the equipment purchased from a shuttered plant in Germany (Walawalker, 1998). Meanwhile, Alusuisse expanded the ISAL plant from 100,000 tons to 162,000 tons in 1997 (Burgert, 1997).

The other major development in the 1990s was the growth of the Blue Lagoon, eventually becoming the country's largest tourist attraction. The site was the result of discharged water from HS Orka's Svartsengi geothermal plant that formed a pool on top of a lava field (Blue Lagoon, n.d.). Through later experimentation, scientists discovered that the mix of the water, silica, algae, and minerals contained healing properties. The country's health officials later recognized the Blue Lagoon as an official dermatology treatment center (Gudmundsottir et al., 2010). The company launched skin care products in 1995 and a modern spa facility in 1999. The Blue Lagoon organization believed they were an early example of a circular economy operation due to its "geothermal ecocycle" (Blue Lagoon, n.d.). Today, the site is revered as it effectively balances ecology, social well-being, and economic development, given its success as a celebrated tourist attraction.

The Master Plan and the Electricity Act: 1999–Present

While conversations about balancing development with environmental protection had generated discussions for decades, the government inaugurated a national plan in the early 2000s (Ramma, n.d). These efforts eventually culminated in the "Master Plan for Nature Protection and Energy Utilization," a tool whereby energy and environmental experts classified power plant operations into three main categories: energy utilization, on-hold, or protection (Ramma, n.d.). The plan's first "experimental" phase unfolded between 1999 and 2003 that involved the evaluation of 43 potential sites (Steingrimsson et al., 2008). The work was organized by a 16-person Steering Committee, which in turn engaged more than 50 experts organized in four main working groups: Nature, Environment, and Cultural Heritage; Recreation, Fishing, Hunting, and Agriculture; Social and Economic Impact; and Identification of Potential Power Projects (Steingrimsson et al., 2008).

Iceland enacted the Electricity Act in 2003, which unbundled the electricity market. The Electricity Act established a legal framework to liberalize Iceland's electricity market, aligning it with European Union (EU) directives (Althingi, 2003). It aimed to enhance competition, improve efficiency, and promote transparency within the sector. The Act separated electricity generation, transmission, and distribution into distinct operations to prevent monopolies and ensure fair access to the grid. Landsnet, a regulated entity, began operating in 2005 to manage the electricity transmission system. Additionally, the Act mandated non-discriminatory grid access, enabling consumers to choose electricity suppliers freely.

Meanwhile, another innovation was taking hold within Iceland—the Icelandic New Energy Company—a joint venture between Vistorka, Shell Hydrogen, Norsk Hydro, and DaimlerChrysler to launch the Ecological City Transport System (ECTOS) project (Bansal et al., 2004). ECTOS aimed to produce green hydrogen, operate a hydrogen fueling station, and power three city buses, with the potential to expand the infrastructure to additional transport and fishing sectors (Arnason & Sigfusson, 2000). The project operated between 2001 and 2005 and evolved into the Sustainable Marine and Road Transport Hydrogen in Iceland (SMART-H2) project to test a fleet of 35 hydrogen-powered cars. Experimentation even extended into the marine sector when a whale-watching ship was the first to be retrofitted with a hydrogen fuel cell (Government of Iceland, 2024).

Electricity capacity continued to expand, and Landsvirkjun began work on the Karahnjukar Dam and Fljotsdalsstod Power Station (690 MW) to sell to the international company Alcoa to support the country's third aluminum smelter that opened in 2007 (Askja Energy, n.d.). The plan for the new hydropower station was met with public opposition over environmental concerns (MBL, 2006).

Then, in 2008, Iceland experienced a major financial crisis with the Icelandic krona losing 50% of its value, the country's stock market plunging 90%, and the financial system effectively collapsing (Sigurjonsson et al., 2018). As a result, political turmoil ensued. Just as the crisis was deepening, Reykjavik Energy, which was facing major scrutiny for its dire financial situation and several questionable expansion activities, was completing the country's largest geothermal plant, the Hellisheidi Power Station (300 MW of electricity and 200 MWth of hot water) (ON Power, n.d.). Despite the near bankruptcy of Reykjavik Energy, the plant was completed in 2010.

As the country began its economic recovery, the Icelandic government launched a series of energetic tourist campaigns, and the tourism sector flourished, putting more strain on the country's energy resources. In 2010, renowned Harvard professor Michael Porter, in coordination with local consultants, laid the foundation for what would become the Icelandic Geothermal Cluster to increase Iceland's overall competitiveness and aid in its resuscitation from the economic crisis by offering infrastructure, drilling, construction, technical consulting, education, and financing of geothermal projects abroad (Porter, 2010).

Another way out of the economic crisis was to use the country's vast electricity sources to power computational servers, marked by the opening of the Verne Global Data Center in Reykjanesbaer in 2010 (Landsvirkjun, n.d.). The overarching plan for energy became a law as the "Master Plan Act" in 2011 and the government continued to position itself as a leader in electricity to manufacturing industries and data centers for its access to internationally competitive prices and stable energy. In 2014, atNorth began operating a data center and while bitcoin mining flourished in the 2010s, by the early 2020s the major growth area was artificial intelligence (AI), as evidenced by Borealis starting operations in 2021 (Landsvirkjun, n.d.). Landsvirkjun and Reykjavik Energy tapped into the nascent sustainable finance market to fund expansionary renewable projects by issuing green bonds in 2018 and 2019, respectively. See Fig. 15.1 for historical electricity generation.

Just as the country had earned the reputation as a "renewable" nation, specific innovations in carbon capture such as Carbfix and Orca garnered worldwide press. They put Iceland at the forefront of greentech innovation. For example, Carbfix, wholly-owned by Reykjavik Energy, started in 2007 as an international research project and became its own legal entity in 2020. Its technology involved storing CO_2 gas from the Hellisheidi Power Station permanently in basalt rock below the earth's surface (Jonsdottir et al., 2023).

Undoubtedly, the country has made a major transition from reliance on coal and oil in the 1940s to 85% of primary energy drawn from renewables by 2025. See Fig. 15.2 for changes in energy sources over time. Figure 15.3 shows total energy use.

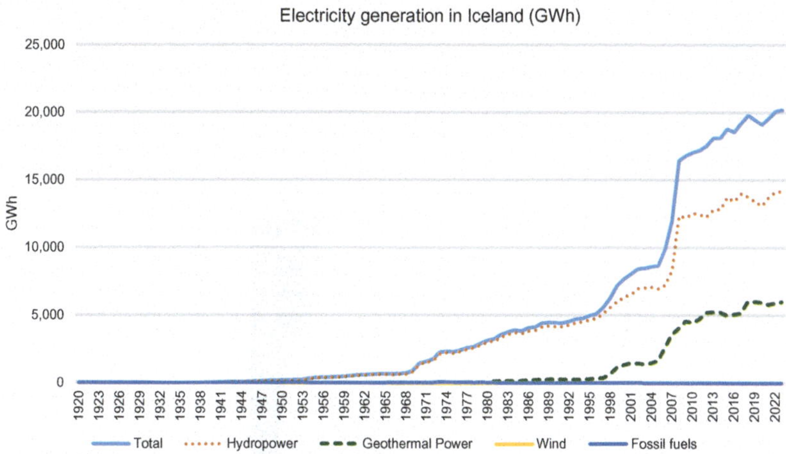

Fig. 15.1 Historical electricity generation in Iceland (GWh) (*Source* Orkustofnun, National Energy Authority [2024b] and Hagstofa Islands [2024])

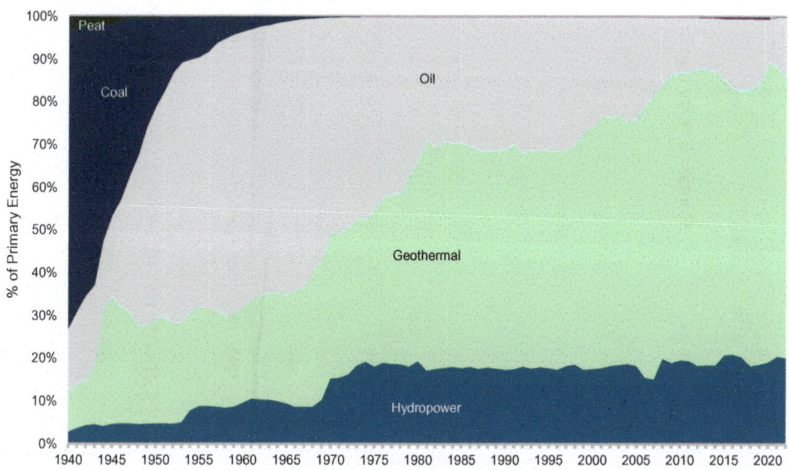

Fig. 15.2 Percentage primary energy in Iceland by energy source (*Source* Orkustofnun, National Energy Authority [2024a])

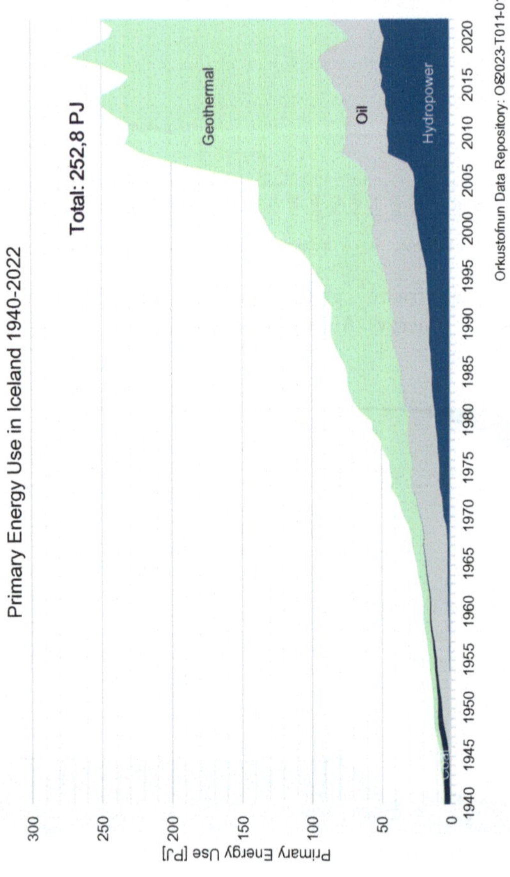

Fig. 15.3 Total primary energy use in Iceland (*Source* Orkustofnun, National Energy Authority [2024a])

The Present

Electricity Industry as of 2025

As of 2025, Iceland produces approximately 20,000 GWh of electricity with three major energy producers: Landsvirkjun, Reykjavik Energy (Orkuveitan), and HS Orka (Orkustofnun, 2024). Iceland's electricity production on a per capita basis is 55,000 kWh, which towers over the EU average of approximately 6,000 kWh per year (Government of Iceland, n.d.).

Landsvirkjun, the National Power Company of Iceland—wholly owned by the national government—maintains a vision of a "sustainable world, powered by renewable energy" and envisions its role "to maximize the value of the renewable energy resources it has been entrusted with, in a sustainable and efficient manner" (Landsvirkjun, n.d.). The company generates approximately 73% of all electricity produced in Iceland, operating a total of 18 power stations: 15 hydropower stations, three geothermal stations, and two wind turbines. Its total installed capacity is 2,162 MW (Orkustofnun, 2024). Landsvirkjun competes internationally by supplying electricity to power-intensive users, such as Alcoa, Nordural, Rio Tinto, Elkem, Verne Global, and many others. Additionally, the company sells electricity to local retailers who serve households and small and medium-sized enterprises (SMEs). Total revenues in 2024 were US$560.9 million and earnings before interest, taxes, depreciation, and amortization (EBITDA) amounted to US$410.3 million (Landsvirkjun, 2025).

The second major energy producer, Reykjavik Energy, is owned by the City of Reykjavik (93.539%), the City of Akranes (5.528%), and the City of Borgarbyggd (0.933%) (Reykjavik Energy, 2024). The company's vision is to "enable a sustainable future" and its stated mission is: "We support growing communities, households and businesses through innovation in energy supply, utility operations and carbon fixation" (Reykjavik Energy, n.d.). The company provides electricity, hot water, cold water, fiber network services, and waste treatment in the southwest of Iceland. Reykjavik Energy operates a number of subsidiaries such as Veitur, Orka natturunnar, Ljosleidarinn, and Carbfix. Veitur is the largest utility in Iceland as it provides electricity distribution, hot and cold water supply, and wastewater treatment (Veitur, n.d.). Orka natturunnar operates two geothermal power plants—Hellisheidi and Nesjavellir—along with a hydroelectric power plant, Andakilsarvirkjun. Total installed capacity is 431 MW, and the company produced 17% of the nation's electricity in 2023. Orka natturunnar sells hot water from its geothermal stations, delivered by Veitur's infrastructure to end-users. The company's consolidated revenues were ISK 66,782 million (US$484.2 million) and EBITDA totaled ISK 39,062 million (US$283.2 million) in 2024 (Reykjavik Energy, 2025).

HS Orka is the third largest energy company in Iceland, and is privately owned by Jardvarmi, consisting of 14 Icelandic pension funds, and Ancala Partners LLP, a UK-based infrastructure firm. Its mission is "to harness renewable energy resources to produce heat and electricity, ensuring sustainable and responsible use of natural resources while providing reliable and efficient services to our community" (HS Orka, n.d.). From its original establishment in 1974, the company has evolved and split into two, HS Orka and HS Veitur, in 2008. HS Veitur provides electricity distribution and hot and cold water services. HS Orka owns and operates two geothermal power plants and two hydroelectric power plants. The total installed capacity of the two plants is 187 MW and the company produced around 7% of the country's electricity in 2023.

Landsnet—Iceland's Transmission System Operator (TSO)—is responsible for managing and operating the high-voltage electricity grid. Although Landsnet was originally a subsidiary of Landsvirkjun, Landsvirkjun sold 93% of its shares to the Icelandic government in 2022, with the remaining being held by Reykjavik Energy (Landsnet, 2022). The company has six main pledges to society: "the secure supply of electricity, a high-quality service and secure transmission system for the future, operations in harmony with society and the environment, efficient operations, informed debate, and targeted management and organization" (Landsnet, n.d.). Landsnet oversees more than 3,000 km of transmission lines and 70 substations, and earned revenues of US$188.95 million and EBITDA of US$70.5 million in 2024 (Landsnet, 2024).

The National Energy Authority (Orkustofnun) is the country's energy regulator, and operates under the Ministry of the Environment, Energy, and Climate. Its mission is "to build knowledge in its areas of operation, such as energy production, utilization, and climate issues, practice efficient and transparent governance, as well as independent and rigorous supervision. The National Energy Authority emphasizes being visionary, reliable, and efficient" (Orkustofnun, n.d.). The regulator has formed international agreements with more than 15 nations for cooperating on projects involving geothermal, district heating, and renewable infrastructure (Orkustofnun, n.d.).

Electricity Usage and Prices

The largest electricity consumers are aluminum smelters (approximately 70%), other power-intensive users such as ferrosilicon manufacturing and data centers (10%), and individual households and SMEs (20%). See Fig. 15.4 for a breakdown of consumption patterns over time.

The cost of electricity has been relatively low in Iceland compared to other nearby countries, given efforts to harness hydro and geothermal energy efficiently. For example, as of 2024, the cost of electricity was about €0.10 per kWh in Iceland, versus an average of approximately €0.25 per kWh in other Nordic countries (Eurostat, 2024). See Fig. 15.5 for the evolution of electricity costs.

Over the last decade, the cost of electricity in Iceland has been relatively flat. It is also noteworthy that Iceland's electricity cost did not increase in 2022 when other countries witnessed dramatic spikes. The main reason for the price increase throughout Europe was the increased

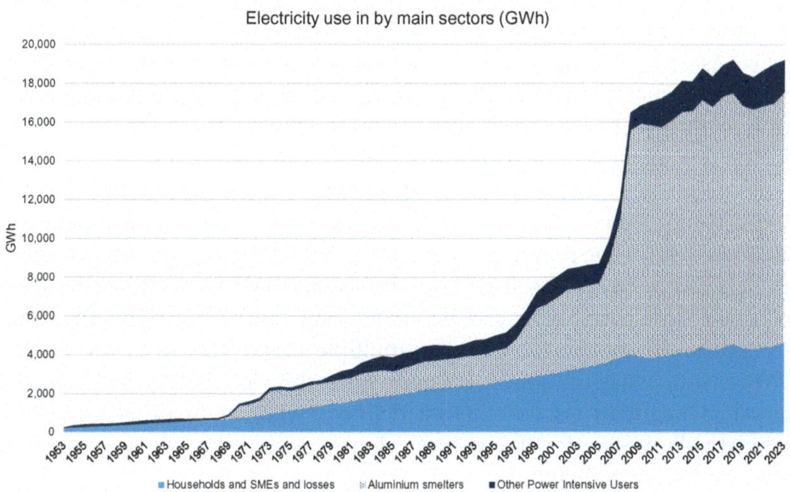

Fig. 15.4 Electricity usage by sector (*Source* Orkustofnun, National Energy Authority [2024a])

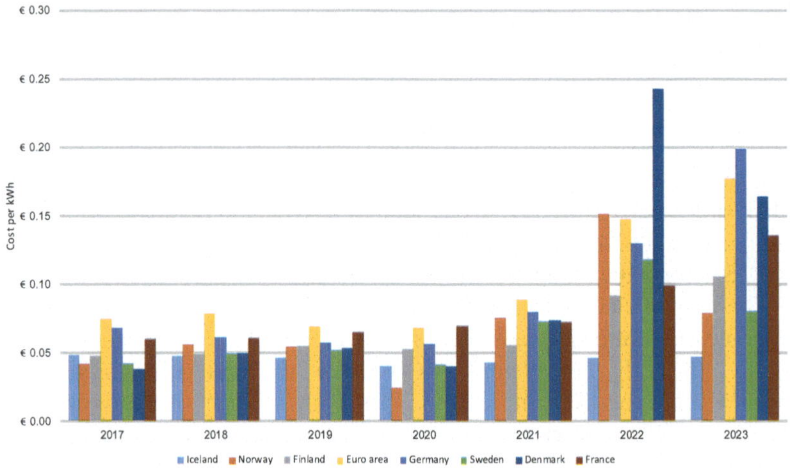

Fig. 15.5 Cost of electricity and supply for households in Iceland and other neighboring countries (Euro/kWh) (*Source* Eurostat [2024])

cost of gas due to sanctions on and lower supply from Russia due to the Ukraine war. Throughout Europe, the marginal producer (i.e., the producer with the highest marginal cost) determines the price of electricity. As such, electricity producers using gas as their main energy source experienced increased costs. Similar to the oil crisis in the 1970s, this situation reaffirmed Iceland's self-sufficiency strategy, as its energy sources for electricity generation relied on natural and local renewables.

Contracts for electricity sales in Iceland are nearly always negotiated through Power Purchase Agreements (PPAs). The PPAs stipulate a negotiated price for power-intensive users and electricity distributors and retailers (who, in turn, sell to households and SMEs). Contract lengths for power-intensive users can be quite long, and often span decades. Contracts with retailers can be for multiple years. The structure of the PPAs is somewhat unique in Iceland, since the electricity price is determined in negotiations well before delivery, whereas in other countries there are allowances for fluctuations. As such, the PPAs allow for greater buyer and seller stability.

Current Debates

Energy matters have been at the forefront of discussion in Iceland for decades. Generally, the debates have revolved around the economic impacts, use of natural resources, and stakeholder management, especially regarding people and businesses in close proximity to the energy infrastructure.

Iceland's long-term energy transition goal is to achieve net zero greenhouse gas (GHG) emissions by 2050 and be fossil fuel-free, with all energy demand fulfilled by renewable resources. Iceland's international climate commitments state a 41% reduction in emissions before 2030 versus 2005 levels. See Fig. 15.6 for GHG emissions over time.

While Iceland has proven its skill at making energy transitions in the past, the next wave involves transitioning away from oil and gas in land transportation, aviation, and marine vessels. The estimated electricity demand gap to complete the transition is around 16,000 GWh. The largest share is in aviation, projected at 8,800 GWh in 2050, followed by marine vessels at 2,900 GWh. A 16,000 GWh increase by 2050 would need to occur in addition to the current 20,000 GWh production in Iceland.

As part of the transition, the current debate centers on whether Iceland can maintain a balance between supply and demand while securing energy self-sufficiency as laid out in the Icelandic Energy Policy for 2050. In

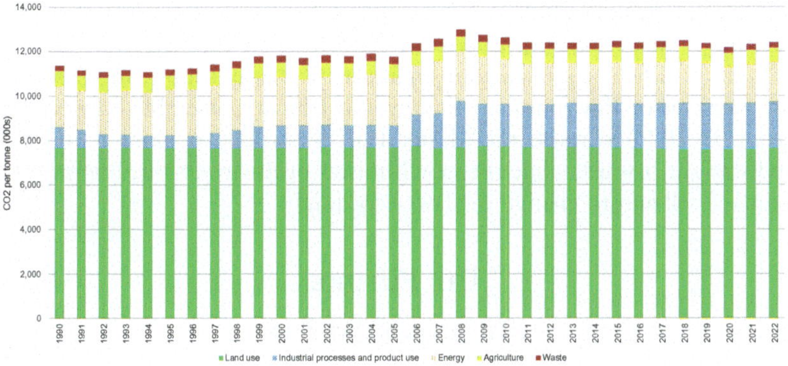

Fig. 15.6 CO_2 emissions in Iceland, per ton (in thousands) (*Source* Umhverfisstofnun [Environmental Agency] [2022])

more recent years, the energy supply has not kept up with the increased growth in demand from numerous potential buyers, such as data centers, e-fuel projects, and food producers. For example, in one report, Iceland's energy regulator found that electricity demand would continue to outpace supply until 2027 (Orkustofnun, National Energy Authority, 2023). Some people in the country believe the situation is drastic and are calling for primary electricity consumption to be curtailed.

Many questions remain on whether energy security can be guaranteed given population and economic growth, a larger fleet of EVs, and a booming tourism sector. This has led to discussions on electricity security for vulnerable customer groups. While electricity networks aim to deliver electricity to customers, no entity has the legal role to make electricity available to meet the expected demand.

Critics within the energy sector have pointed to perceived flaws within the Master Plan itself. Initially, the objective of the Master Plan was to balance diverse perspectives on Iceland's energy resources by conducting comprehensive assessments of energy projects to meet societal energy demands and economic growth while preserving the environment. The Federation of Icelandic Energy and Utility companies believes that the objectives of the Master Plan have not been achieved and suggests that it be repealed. Other critics argue that the plan does not build consensus—whether that be for the development of future power plants or the permanent protection of undeveloped areas from energy projects—in the way that was originally intended (Benediktsson & Waage, 2020; Stjórnarráð Íslands, 2024). Many parties in the country feel that the licensing and permitting process has become too cumbersome without providing sufficient guidance to build a new energy supply. In response, the Minister of Environment, Energy and Climate, assembled a working group in 2024 to revise the Master Plan by making it more actionable and straightforward, while the new Minister of Environment, Energy and Climate will aim to change the law accordingly.

The Future

The Vision of Net Zero

In 2020, the government of Iceland created a "Sustainable Energy Future" vision based on the country as "a land of clean and secure energy" (Government of Iceland, 2020). The vision balances five main elements: energy security, energy transition, energy efficiency and economization, society/economy, and the environment (see Fig. 15.7).

The government further articulated its vision in a comparison chart showing the current state of 2020 reality side-by-side with the key aims for 2050, such as no fossil fuels, diverse energy sources, and a smart and flexible power system, among other elements.

How can Iceland achieve its goal of being net zero by 2050? On the one hand, the goal aligns with Iceland's history of renewable energy innovation and environmental betterment. However, despite generating nearly 100% of its electricity from renewable sources, the country still needs to address emissions outside of electricity production, such as transportation, industry, and even international energy exports (Ministry for the Environment and Natural Resources, 2021). Achieving carbon neutrality will require technological advancements and collaborative efforts among many stakeholders, government agencies, private enterprises, and local communities.

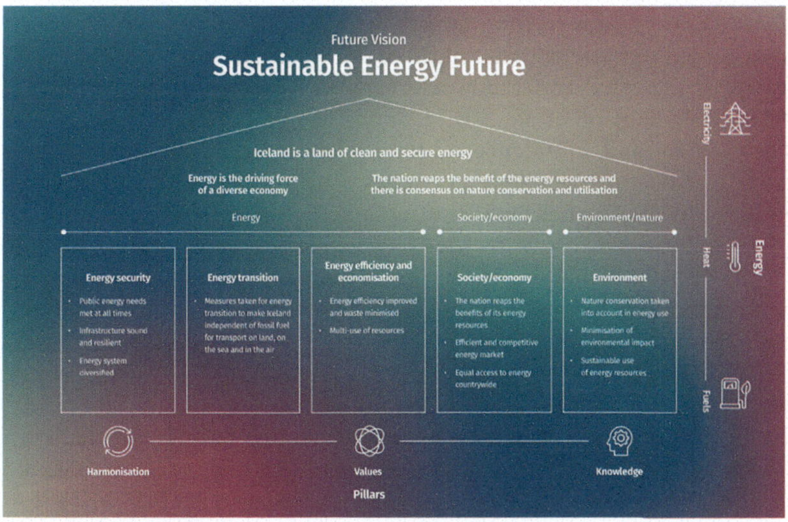

Fig. 15.7 Government of Iceland's future vision (*Source* Government of Iceland, Ministry of Industry and Innovation [2020])

Opportunities and Challenges in Reaching Net Zero

To deliver on the vision of a net zero economy, Iceland will first need to address the immediate shortfall in electricity supply. This challenge is perhaps further compounded by the fact that Iceland competes internationally to attract power-intensive industries seeking stable and competitively priced energy sources. At the same time, power-intensive industries often amplify the carbon footprint, given that producers such as aluminum smelters need to use fossil fuel-based transportation to import raw materials to Iceland. As such, achieving net zero requires balancing economic growth with domestic emission reductions while contributing to global decarbonization efforts through energy exports.

To increase the energy supply, the country can continue to build upon past successes by expanding its hydroelectric and geothermal power. While the potential remains vast, overutilization risks environmental degradation and inefficiency. For example, in contemplating more hydropower plants, energy companies and regulators need to consider the potential disruption to ecosystems and the broader environmental impact. Climate change adds further uncertainty, particularly for hydropower, as glaciers continue to melt, causing risks for hydroelectric power plants to deal with increased water flow (Roberts & Conway, 2018; Sveinsson, 2016; Thorsteinsson et al., 2013). Similarly, geothermal energy projects can result in land subsidence, water depletion, and pollution. Furthermore, hydroelectric development poses a disruption to Iceland's iconic landscapes. This is particularly impactful given the country's eco-tourism appeal and the fact that many tourists seek untouched wilderness, despite debates about whether tourists are negatively affected by hydroelectricity plants in remote areas (Burns & Haraldsdóttir, 2019; Saethorsdottir & Hall, 2018).

Wind power presents another viable option available to increase the renewable electricity supply. Iceland is often mentioned for its unharnessed wind potential, and formal scientific studies have found that even moderately sized wind farms, such as one with 15 modern turbines would produce more power than two smaller hydro plants while being less disruptive to the natural environment (Nawri et al., 2014). Additionally, wind energy strategically located in the highlands, coastal areas, and close to existing hydro and geothermal locations could provide greater grid stability (Magnusdottir, 2024).

Since transportation is one of the last sectors to transition to renewables, many citizens are embracing EVs. While Iceland enjoyed the third-highest percentage rate of new EVs sold in 2023 at 60% (behind Norway at 91% and Sweden at 61%), there is recognition that more than the current ~400 charging stations are necessary to support booming demand (European Environmental Agency, 2024).

In addition to EVs, green hydrogen—produced through electrolysis with renewable electricity—and e-fuels such as e-methanol, e-ammonia, e-diesel, and e-kerosene offer a transformative opportunity to decarbonize transportation while creating export potential. While Iceland has had some experience with test sites dating back to the late 1990s to develop transportation powered by hydrogen fuel cells, past projects had not moved beyond pilots due to the required infrastructure development and research investments. Furthermore, embracing hydrogen and e-fuels would likely require international partnerships to develop vehicles, vessels, and planes. The government forecasts the demand for hydrogen to grow to 364 kilotonnes (kt) by 2040 with 80% attributed to aviation, 17% to maritime applications and fishing, and the remaining 3% to road transportation (Government of Iceland, 2024).

To further optimize the grid and deal with increased demand, new technologies are also at the forefront of the discourse within the country. Smart grids, AI for energy management, and energy storage systems optimize resource use and stabilize renewable energy outputs (International Energy Agency, 2022). In tandem with technological efforts to improve efficiency, many in the country are looking at ways to improve the building stock by upgrading insulation, ensuring building materials meet certain specifications, and preparing for emerging legislation on mandatory life cycle assessments for new buildings (Nordic Sustainable Construction, 2024). Interestingly, Iceland is exempt from the EU directive on building efficiency due to the proliferation of renewables and a relatively new and uniform housing stock (Smari Petersen & Ragnarsson, 2024).

There are several innovative greentech companies operating within the energy sector in Iceland. As previously mentioned, Carbfix has a solution for permanently storing CO_2 within basalt rock safely. One of the company's research projects is Seastone, where it utilizes seawater to dissolve CO_2, thereby eliminating the need to use fresh water (Carbfix, n.d.; Jonsdottir et al., 2023). The other prominent greentech company is Orca, owned by Switzerland's Climeworks, which holds the claim of being the "world's first and largest direct air capture and storage plant, making carbon dioxide removal on large-scale a reality" (Climeworks, n.d.). Although Carbfix and Orca have garnered considerable press coverage for their novel approaches to carbon capture, there are several other startups gaining traction such as: Carbon Recycling International (CRI), which transforms CO_2 to methanol; Dynamic Technology Equipment (DTE), which uses sensors to help aluminum companies maximize efficiency; Laki Power, which develops monitoring technology for electricity grid operators and distributors; and Transition Labs, which works with the UK-based Space Solar to convert sunlight from space into electricity (Djurickovic, 2024; McDonald, 2024).

Energy producers and greentech innovators actively pursue opportunities to export energy and know-how. One of the much-publicized projects is IceLink, which proposes to connect Iceland to the UK via an underwater cable stretching over 1,000 kilometers and transmitting more than 5,000 GWh per year (Sasaki & Nakayama, 2016). While the project has been discussed by the government since the 1980s with Landsvirkjun being involved as the electricity supplier, the project has not gone through a complete feasibility stage.

Aside from efforts to look at the direct export of electricity, the country is a leader in "exporting" its knowledge of geothermal energy, hydro power, and carbon capture. The country's cluster approach to developing a complete set of competencies in geothermal energy has morphed into the Iceland Renewable Energy Cluster, holding "Collaboration—Knowledge—Progress" as key values while serving as the "unifying platform for the entire energy industry in Iceland, bringing together public and private entities and institutions across the full value chain" (IREC, n.d.).

While great promise still exists for domestic and international energy projects, challenges remain for managing stakeholders. Policymakers must ensure that clean energy remains accessible and affordable for all Icelanders, particularly as large-scale infrastructure projects and innovations require substantial financial investments. One study found that "Icelandic decision-makers responsible for energy development face a complex challenge when it comes to designing and implementing a policy that is approved across all stakeholder groups" (Gudlaugsson et al., 2020, p. 168). In short, public participation and consensus-building are vital to overcoming resistance to projects that might disrupt natural landscapes or traditional ways of life. Furthermore, the government must ensure equitable energy access throughout the entire population.

Conclusion

What can the rest of the world learn from Iceland's energy story? Perhaps most salient is how the country has leveraged its unique natural resources to demonstrate a long-term commitment to creating a nation powered by renewables. Of course, this did not happen overnight or with any specific top-down directive. Rather, developments have arisen out of necessity, as witnessed as early as the 1900s when farmers modestly tried to heat their homes and access hot water. Over time, citizens have shared ideas and inspired one another to innovate, even if efforts did not always bear fruit in the short term. Aware that reliance on energy from foreign countries would be costly, the government supported the development of early innovations and invested in novel renewable technology.

As the country continues to scale its renewable energy infrastructure, the government aims to create a regulatory framework that fosters responsible growth while maintaining key values of self-sufficiency and energy security, which are critical for the island nation. While many of these developments ignite debate—from hydroelectric dams to the construction of major aluminum smelters—Icelandic society strives to be transparent, comprehensive, and compassionate toward altering viewpoints. Support for entrepreneurs and partnerships across a wide range of business sectors, universities, and all levels of government has allowed for "moonshot" ideas to be explored (Michelson, 2022). Evidence abounds, ranging from hydrogen fuel cell pilots in the early 2000s to novel carbon capture methods and solar space research and development (R&D) today.

Perhaps many of these collaborations are possible due to the government's ownership stake of major energy companies that pursue a balance between economic benefit and environmental protection. As such, many energy companies have been able to adopt sustainable business models that embed sustainability goals within their strategies, prioritize investments in further renewables expansion, disclose information clearly and transparently, collaborate through partnerships, invest R&D to become more sustainable, establish robust Environmental, Social and Governance (ESG) reporting methods, and support the organization through strong governance (Jonsdottir et al., 2021; Mitchell et al., 2024). Furthermore, the National Energy Authority's database of renewable practices helps increase knowledge and inspire greater collaboration. This helps to reinforce the Icelandic Renewable Energy Cluster, as more individuals and enterprises share their knowledge internationally through educational or commercial exchanges. Iceland's international collaboration and ability to showcase home-grown innovations inspire confidence and attract new talent.

Iceland's energy strategy must remain flexible and forward-looking, capable of adapting to new technologies, shifting global energy dynamics, and evolving environmental standards. Through strategic diversification, technological advancement, and international cooperation, Iceland can secure its energy future while reinforcing its reputation as a trailblazer in renewable energy innovation. By addressing these challenges holistically and inclusively, Iceland can meet its net zero goals and inspire global progress toward a sustainable energy future.

References

Aldred, J. (2008, April 22). Iceland's energy answer comes naturally. *The Guardian*. https://www.theguardian.com/environment/2008/apr/22/renewableenergy.alternativeenergy

Althingi. (2003, March 27). *Raforkulög 2003 nr. 65 27. Mars*. https://www.althingi.is/lagas/nuna/2003065.html

Arnason, B., & Sigfusson, T. I. (2000). Iceland—A future hydrogen economy. *International Journal of Hydrogen Energy, 25*(5), 389–394. https://doi.org/10.1016/S0360-3199(99)00077-4

Askja Energy. (n.d.). *The hydro and geothermal history*. https://askjaenergy.com/iceland-renewable-energy-sources/hydro-and-geothermal-history/

Bansal, P., Mark, K., & Mitchell, J. (2004). *Aiming toward a hydrogen economy: Icelandic New Energy Ltd* (Íslensk Ný Orka). *Ivey Business School*. 9B05M001.

Benediktsson, K., & Waage, E. R. H. (2020, November). Power without politics? Nature, landscape and renewable energy in Iceland. *Annales de geographie, 736*(6), 107–130. https://doi.org/10.3917/ag.736.0107

Benediktsson, K. (2021). Conflicting imaginaries in the energy transition? *Nature and renewable energy in Iceland. Moravian Geographical Reports, 29*(2), 88–100. https://doi.org/10.2478/mgr-2021-0008

Blue Lagoon. (n.d.). *Our story*. https://www.bluelagoon.com/about/our-story

Burns, G. L., & Haraldsdóttir, L. (2019). Hydropower and tourism in Iceland: Visitor and operator perspectives on preferred use of natural areas. *Journal of Outdoor Recreation and Tourism, 25*, 91–101. https://doi.org/10.1016/j.jort.2018.09.003

Burgert, P. (1997, July 8). Alusuisse expands capacity in Iceland (aluminum smelting, Alusuisse-Lonza Holdings AG). *American Metal Market*. Accessed via Factiva.

Carbfix. (n.d.). *Seastone*. https://www.carbfix.com/operations/researchprojects/seastone

Climeworks. (n.d.). *Orca: The first large-scale plant*. https://climeworks.com/plant-orca

Davidsdottir, B. (2022). Towards an Icelandic sustainable energy system. Successful public policy in the Nordic countries: Cases, lessons, challenges. *Successful Public Policy in the Nordic Countries*, 351–369. https://doi.org/10.1093/oso/9780192856296.003.0017

Davidsdottir, B. (2023, October). *Energy development in Iceland*. Presentation to the University of Iceland, Reykjavik, Iceland.

Djurickovic, T. (2024, August 22). *Iceland: Tech innovations from the land of fire and ice*. TechEu. https://tech.eu/2024/08/22/iceland-tech-innovations-from-the-land-of-fire-and-ice/

Elkem. (2022, Nov 24). *Elkem around the world: Akranes, Iceland*. https://magazine.elkem.com/elkem-around-the-world/akranes-iceland/

European Environmental Agency. (2024, October 31). *New registrations of electric vehicles in Europe*. https://www.eea.europa.eu/en/analysis/indicators/new-registrations-of-electric-vehicles

Eurostat. (2024). *Electricity prices components for household consumers—Annual data (from 2007 onwards)*. https://ec.europa.eu/eurostat/databrowser/view/nrg_pc_204_c__custom_13868732/default/table?lang=en

Fox Hodgson, S. (2024). *The Vikings & Geothermal Iceland. Geothermal Rising*. Geothermal Rising. https://geothermal.org/our-impact/stories/vikings-geothermal-iceland

Government of Iceland. (2020). *A Sustainable Energy Future An Energy Policy to the year 2050*. https://www.stjornarradid.is/lisalib/getfile.aspx?itemid=e36477fd-3bc1-11eb-8129-005056bc8c60

Government of Iceland. (2024, April). *Hydrogen and E-fuels roadmap for Iceland*. https://www.government.is/library/Files/240102_URN_RoadmapForIceland_V6.pdf

Government of Iceland. (n.d.). *Legislation*. https://www.government.is/topics/business-and-industry/energy/legislation/

Government of Iceland. (n.d.). *Energy*. https://www.government.is/topics/business-and-industry/energy/

Government of Iceland. (n.d.). *History of fisheries*. https://www.government.is/topics/business-and-industry/fisheries-in-iceland/history-of-fisheries/

Gudlaugsson, B., Fazeli, R., Gunnarsdottir, I., Davidsdottir, B., & Stefansson, G. (2020). Classification of stakeholders of sustainable energy development in Iceland: Utilizing a power-interest matrix and fuzzy logic theory. *Energy for Sustainable Development*, *57*, 168–188. https://doi.org/10.1016/j.esd.2020.06.006

Gudmundsottir, M., Brynjolfsdottir, A., & Albertsson, A. (2010, April). The history of the blue lagoon in Svartsengi. In *Proceedings of the World Geothermal Congress*. https://cohenovate.com/wp-content/uploads/2017/07/the-history-of-blue-lagoon-in-svartsengi.pdf

Gunnarsdottir, I., Davidsdottir, B., Worrell, E., & Sigurgeirsdottir, S. J. E. P. (2022). Indicators for sustainable energy development: An Icelandic case study. *Energy Policy, 164*, 112926. https://doi.org/10.1016/j.enpol.2022.112926

Hagstofa Islands. (2024). *Mannfjöldi eftir sveitarfélögum, kyni, ríkisfangi og ársfjórðungum 2011–2024*. https://px.hagstofa.is/pxis/pxweb/is/Ibuar/Ibuar__mannfjoldi__1_yfirlit__arsfjordungstolur/MAN10001.px/table/tableViewLayout2/

Hayter, R. (1997). *The dynamics of industrial location: The factory, the firm and the production system*. Wiley.

HS Orka. (n.d.). *A cornerstone for half a century*. https://www.hsorka.is/en/about-hs-orka/the-business/a-cornerstone-for-half-a-century/

HS Orka. (n.d.). *Values, mission, and strategic goals*. https://www.hsorka.is/en/about-hs-orka/policies/values-mission-and-strategic-goals/

International Energy Agency. (2022). *Iceland*. https://www.iea.org/countries/iceland/electricity

International Hydropower Association (IHA). (2017, September 7). *Better hydro: Reducing sedimentation and erosion at Blanda, Iceland*. https://www.hydropower.org/blog/better-hydro-reducing-sedimentation-and-erosion-at-blanda-iceland

IREC: Iceland Renewable Energy Cluster. (n.d.). *Collaboration—Knowledge—Progress*. https://energycluster.is/

Jonsdottir, G. E., Sigurjonsson, T. O., Rahnema, A. A., & Mitchell, J. (2021). Applying responsible ownership to advance SDGs and the ESG framework, resulting in the issuance of green bonds. *Sustainability, 13*(3), 7331. https://doi.org/10.3390/su13137331

Jonsdottir, G. E., Mitchell, J., Sigurjonsson, T. O., & Alavi, A. R. (2023). Responsible ownership fostering green tech innovation. *Sustainable Energy Technologies and Assessments*, *55*, 102923. https://doi.org/10.1016/j.seta.2022.102923

Jonsson Ulfhildarson, S. P. (2013, May 24). Preserving the Laxá explosion—Blowing up dams and democracy restrictions. *The Reykjavík Grapevine*. https://www.savingiceland.org/2013/05/preserving-the-laxa-explosion-blowing-up-dams-and-democracy-restrictions/

Kristjansdottir, H. (2015). *Sustainable energy resources and economics in Iceland and Greenland*. Springer International Publishing.

Landsnet. (2022). *Financial Statements 2022*. https://www.landsnet.is/library?itemid=d169db37-fbf9-448d-bb4d-b303ad73d154

Landsnet. (2023). *Financial Statements 2023*. https://www.landsnet.is/library?itemid=13a2a64a-96a9-478c-bc67-f01822ab5b5f

Landsnet. (2024). *Financial Statements 2024*. https://www.landsnet.is/library?itemid=94e7d4aa-baf8-4896-b717-475d41739d7f

Landsnet. (n.d.). *Policy pledges and values*. https://www.landsnet.is/english/about-us/policy-pledges-and-values/

Landsvirkjun. (2025). *Financial Statements 2024*. https://www.landsvirkjun.com/financial-statements

Landsvirkjun. (n.d.). *Blanda Power Station*. https://www.landsvirkjun.com/powerstations/blondustod

Landsvirkjun. (n.d.). *Ljósafoss Power Station*. https://www.landsvirkjun.com/powerstations/ljosafoss

Landsvirkjun. (n.d.). *Krafla Geothermal Station*. https://www.landsvirkjun.com/powerstations/krafla

Landsvirkjun. (n.d.). *Our customers*. https://www.landsvirkjun.com/customers

Landsvirkjun. (n.d.). *Our strategy*. https://www.landsvirkjun.com/strategy

Landsvirkjun. (n.d.). *Power stations*. https://www.landsvirkjun.com/powerstations

Langella, G., Paoletti, V., DiPippo, R., Amoresano, A., Steinunnardottir, K., & Milano, M. (2017). Krafla geothermal system, northeastern Iceland: Performance assessment of alternative plant configurations. *Geothermics, 69*, 74–92. https://doi.org/10.1016/j.geothermics.2017.04.001

Logadottir, H. H. (2015, December). Iceland's sustainable energy story: A model for the world? United Nations Chronicle. *Sustainable Energy, LII*(3). https://www.un.org/en/chronicle/article/icelands-sustainable-energy-story-model-world

Magnusdottir, M. (2024). *Integrating wind and geothermal energy to strengthen the renewable energy portfolio in Iceland* [Master's thesis, Reykjavik University]. Skemman. https://skemman.is/handle/1946/47685

MBL. (2006, May 12). Ríflega 300 manns mótmæla Kárahnjúkavirkjun á Austurvelli. https://www.mbl.is/frettir/innlent/2006/05/12/riflega_300_manns_motmaela_karahnjukavirkjun_a_aust/

McDonald, B. (2024, November 5). *Iceland looking to add space solar power to its sources of renewable energy by 2030*. CBC. https://www.cbc.ca/radio/quirks/iceland-space-solar-1.7370888

Melsted, O. (2021). Eliminating fossil fuels: Iceland's transition from coal and oil to geothermal district heating, 1930–1980. *History and Technology, 37*(4), 527–547. https://doi.org/10.1080/07341512.2022.2033386

Michelson, J. (2022, June 28). *Renewable energy strategies we can learn from Iceland*. Forbes. https://www.forbes.com/sites/joanmichelson2/2022/06/28/iceland-runs-on-85-renewable-energy-7-strategies-we-can-learn-from-them/

Ministry for the Environment and Natural Resources. (2021). *A sustainable energy future: An energy policy to the year 2050*. https://www.stjornarradid.is/library/01--Frettatengt---myndir-og-skrar/ANR/Orkustefna/201127%20Atvinnuvegaraduneytid%20Orkustefna%20A4%20EN%20V4.pdf

Mitchell, J., Sigurjonsson, O., & Alavi, A. R. (2022). *ESG and Reykjavik energy's first green bond*. IESE Business School. https://www.iesepublishing.com/esg-and-reykjavik-energys-first-green-bond-english.html

Mitchell, J., Sigurjonsson, T. O., Kavadis, N., & Wendt, S. (2024). Green bonds and sustainable business models in Nordic energy companies. *Current Research in Environmental Sustainability, 7*, 100240. https://doi.org/10.1016/j.crsust.2023.100240

Nawri, N., Petersen, G. N., Bjornsson, H., Hahmann, A. N., Jonasson, K., Hasager, C. B., & Clausen, N. E. (2014). The wind energy potential of Iceland. *Renewable Energy, 69*, 290–299. https://doi.org/10.1016/j.renene.2014.03.040

Nielsen, G., Maack, R., Gudmundsson, A., & Gunnarsson, G. I. (2000). Completion of Krafla geothermal power plant. In *Proceedings World Geothermal Congress* (pp. 3259–3264).

Nordic Sustainable Construction. (2024, April 17). *Iceland introduces new LCA requirements*. https://www.nordicsustainableconstruction.com/news/2024/april/iceland-lca-requirements

Nordurorka hf. (n.d.). *Norðurorka hf. (Nordurorka Ltd.)*. https://www.no.is/static/files/UM_NO/nordurorka_introduction_2013.pdf

ON Power. (n.d.). *Power Plants*. https://www.on.is/en/about-us/power-plants/

Orkustofnun, National Energy Authority. (n.d.). *History of the National Energy Authority*. https://orkustofnun.is/en/History-of-NEA

Orkustofnun, National Energy Authority. (n.d.). *About the National Energy Authority (Orkustofnun)*. https://orkustofnun.is/en/about_us/national_energy_authority

Orkustofnun, National Energy Authority. (2023). *Electricity*. https://orkustofnun.is/en/information/numerical_data/electricity

Orkustofnun, National Energy Authority. (2024a). *Electricity*. https://orkustofnun.is/en/information/numerical_data/electricity

Orkustofnun, National Energy Authority. (2024b). *Electricity*. https://orkustofnun.is/en/information/numerical_data/electricity

Pomrenke, E. (2023, March 15). Does Reykjavik have heated sidewalks? *Iceland Review*. https://www.icelandreview.com/ask-ir/does-reykjavik-have-heated-sidewalks/

Porter, M. (2010, November 1). *The Icelandic geothermal cluster: Enhancing competitiveness and creating a new engine of Icelandic growth*. Presentation, Reykjavik, Iceland. https://www.hbs.edu/ris/Publication%20Files/2010-1101_Iceland_Geothermal_MEP_d4c507e5-76c5-47f7-9ba2-dcfd7c133091.pdf

Ragnarsson, Á., Steingrímsson, B., & Thorhallsson, S. (2020). Geothermal development in Iceland 2015–2019. In *Proceedings World Geothermal Congress* (Vol. 1, p. 2021).

Ramma. (n.d.). *The master plan*. https://www.ramma.is/english/general-information/

Reykjavik Energy. (2024). *Financial Statements*. https://orkuveitan.is/en/finance/published-financial-material/annual-reports/

Reykjavik Energy. (2025), *Financial Statements*. https://orkuveitan.is/en/finance/published-financial-material/annual-reports/

Reykjavik Energy. (n.d.), *Elliðaárdalur*. https://orkuveitan.is/en/environment/outdoor-areas/ellidaardalur/

Reykjavik Energy. (n.d.). *History of Reykjavik Energy*. https://orkuveitan.is/en/about-or/operations/history-of-reykjavik-energy/

Reykjavik Energy. (n.d.). *About Reykjavik Energy*. https://orkuveitan.is/en/about-or/operations/reykjavik-energy/

Roberts, D., & Conway, M. (2018). Iceland's renewable energy sources & climate change. *Embry-Riddle Aeronautical University*. https://commons.erau.edu/student-works/72

Saethorsdottir, A. D., & Hall, C. M. (2018). Floating away: The impact of hydroelectric power stations on tourists' experience in Iceland. *Sustainability, 10*(7), 2315. https://doi.org/10.3390/su10072315

Sasaki, D., & Nakayama, M. (2016). Effectiveness of public financial support in an electricity transmission project between Iceland and the UK. *Business and Management Studies*, *2*(2), 14–20. https://doi.org/10.11114/bms.v2i2.1518

Shafiei, E., Davidsdottir, B., Fazeli, R., Leaver, J., Stefansson, H., & Asgeirsson, E. I. (2018). Macroeconomic effects of fiscal incentives to promote electric vehicles in Iceland: Implications for government and consumer costs. *Energy Policy, 114*, 431–443. https://doi.org/10.1016/j.enpol.2017.12.034

Sigurjonsson, T. O., Schwarzkopf, D. L., & Bryant, M. (2018). *The return of trust? Institutions and the public after the Icelandic financial crisis*. Emerald Publishing Limited.

Skulason, J. B. (1994). *Foreign investment and the bargaining process: Case study of the aluminum industry in Iceland* [Master's thesis, Simon Fraser University]. Simon Fraser University: https://summit.sfu.ca/_flysystem/fedora/sfu_migrate/7954/b16935901.pdf

Smari Petersen, V., & Ragnarsson, K. H. (2024). Obligations for owners to climate-proof buildings in Iceland. *European Property Law Journal, 13*(1), 104–126. https://doi.org/10.1515/eplj-2024-0006

Statice. (2024, April 17). *The fraction of electric- and plug-in hybrids increased by 1% in 2023.* https://www.statice.is/publications/news-archive/environment/car-ownership-and-use-by-industry-updated-to-2023/

Steingrimsson, B., Bjornsson, S., & Adalsteinsson, H. (2008, November 24). Master plan for geothermal and hydropower development in Iceland. In *African Rift Geothermal Conference*, Entebbe, Uganda. https://www.bgr.de/geotherm/argeoc2/docs/sessions/s4_Steingrimsson_Master_Plan.pdf

Stjórnarráð Íslands (The Government Council of Iceland). (2024, November). *Endurskoðun á lögum um rammaáætlun.* https://www.stjornarradid.is/library/02-Rit--skyrslur-og-skrar/URN/Ramma_a%c3%a6tlun_skyrsla_nov24_NET_NY.pdf

Sveinsson, O. (2016). Energy in Iceland: Adaptation to climate change. *UNU-FLORES Policy Briefs.* United Nations University Institute for Integrated Management of Material Fluxes and of Resources (UNU-FLORES), Dresden. https://collections.unu.edu/eserv/UNU:5756/DNC2015_PolicyBrief_No4.pdf

Thorarinsdottir, R. (2006). *Energy in Iceland. Historical perspective, present status, future outlook.* Orkustofnun (National Energy Authority).

Thorsteinsson, T., Johannesson, T., & Snorrason, A. (2013). Glaciers and ice caps: Vulnerable water resources in a warming climate. *Current Opinion in Environmental Sustainability, 5*(6), 590–598. https://doi.org/10.1016/j.cosust.2013.11.003

Umhverfisstofnun (Environmental Agency). (2022). *Losun gróðurhúsalofttegunda á Íslandi.* https://ust.is/loft/losun-grodurhusalofttegunda/

United Nations Environment Programme (UNEP). (2020, June 24). *Iceland, a world leader in clean energy, supports Africa's push for geothermal power.* UNEP. https://www.unep.org/news-and-stories/story/iceland-world-leader-clean-energy-supports-africas-push-geothermal-power

Veitur. (n.d.). *Operations.* https://www.veitur.is/en/operations

Walawalker, R. (1998, September 1). New smelter warms up in Iceland. *Metal Bulletin Monthly.*

Worland, J. (2017, July 25). *A solution to our clean energy problem may lie right beneath our fee*t. Time. https://www.yahoo.com/news/solution-clean-energy-problem-may-130054336.html

World Economics. (2023). *Iceland's GDP PPP per capita (based on 2023 data and 2017 constant prices).* https://www.worldeconomics.com/GDP-Per-Capita/Iceland.aspx

CHAPTER 16

Water to Watts: Meghalaya's Electrification Strategy of Turning Rainfall into Power

Sarmistha Sarma and Akanksha Jaiswal

Introduction

Meghalaya is a state in the northeastern region of India, with an abundance of greenery and rainwater and scenic beauty. The etymological significance of the name Meghalaya is "Abode of Clouds," a fitting name because this state is one of the wettest places on Earth. Meghalaya is spread across an area of 22,429 square kilometers and is bordered by Assam in the north, and Bangladesh on both the east and the south. The capital city, Shillong, is often called the "Scotland of the East" because of its hilly terrain and humid climate.

As per the 2011 Census, around 2.96 million persons reside in Meghalaya, most of whom live in rural areas. The economy is predominantly

S. Sarma (✉) · A. Jaiswal
Asian Business School, Noida, India
e-mail: sarmistha.sarma@abs.edu.in

A. Jaiswal
e-mail: akankshajaiswal56@gmail.com

T. Walker et al. (eds.), *Wiring the Future*, Palgrave Studies in Emerging Risk Management and Sustainable Finance,
https://doi.org/10.1007/978-3-032-14717-2_16

agrarian, with agriculture, horticulture, and livestock serving as the principal means of livelihood. The state is rich in minerals such as coal, limestone, and uranium, which largely constitute the base of its industries. The state possesses a vast network of rivers, streams, and waterfalls, while its abundant steep hills and deep gorges form a multitude of catchment areas and natural reservoirs. Apart from the big rivers pouring down into the lowlands—Umiam, Myntdu, Simsang, and Kynshi—a host of other rivers can be exploited for hydroelectricity.

The annual heavy rainfall, approximately 12,000 millimeters in regions such as Mawsynram and Cherrapunji, ensures a consistent and reliable availability of water.

The main research objectives of this paper are to investigate the status of the hydro power sector, assess the untapped potential of water resources, and analyze the socio-economic consequences that hydroelectric projects have on local communities. The major contributions of this chapter lie in raising issues and pointing out opportunities for the renewable-energy scene of Meghalaya and proposing strategic recommendations for sustainable development based on hydro power.

Objectives

Developing hydro power strategies is a crucial task for any nation wishing to improve its energy resources. Indeed, for a country like India, coming up with an effective plan can save costs and help ensure that the country remains self-sufficient. As such, investigating the policies and strategies adopted by the Government of Meghalaya for developing hydro power projects allows us to assess these development strategies. In addition, an evaluation is done on how effective the Build, Own, Operate & Transfer (BOOT) model has been in encouraging private investments and revenue generation.

1. Alignment with Sustainable Development Goals:
   Meghalaya's hydro power projects contribute to key SDGs, particularly SDG 30, by promoting sustainable development, preserving the environment, and ensuring efficient natural resource management.
2. Economic and Environmental Impacts to Evaluate:
   These projects offer economic benefits such as job creation, revenue generation, and local development, while also supporting climate action through clean energy production from water resources.

3. Investigating Financial Frameworks and Incentives:
   The study examines the government's financial mechanisms—like upfront premiums and royalty structures—that support project viability and attract private investment in the hydro power sector.
4. Exploring Innovative Approaches:
   Innovations like Pumped Storage Hydro Projects (PSHPs) are explored for their ability to meet peak energy demands while integrating into existing infrastructure with minimal environmental impact.
5. Understanding Social Implications:
   Hydro electrification improves quality of life through reliable energy access and fosters inclusive development by involving communities and safeguarding local water rights.
6. Anticipation of a Replicable Model:
   Meghalaya's strategic approach offers a replicable model for other regions with similar resources, providing valuable insights for policymakers aiming to achieve sustainable and secure energy solutions.

Literature Review

Meghalaya, a north-eastern state in India, has a vast potential for hydro power generation, owing to abnormally high rainfall and resurgence (Das, 2013; Handique & Dutta, 2014). However, Meghalaya is currently facing water management issues with a low Composite Water Index score (Jain & Makkar, 2019). To face these challenges, the state deployed innovative technologies such as rainwater harvesting in high-rise buildings to produce electricity (Martin & Shrivastava, 2013). The hydro power potential of the region was estimated at 63,257 MW—but only a small fraction (3.02%) has so far been used (Das, 2013). Several technical obstacles, political resistance, land acquisition issues, environmental concerns, and regulatory intricacies restrict the hydro potential of the region (Handique & Dutta, 2014). This potential could possibly transform the North-East into the "Powerhouse of India" and will surely lead to infrastructural development in rural communities and improvements in the general quality of life (Das, 2013; Handique & Dutta, 2014).

Hydro Power Opportunities and Challenges in Northeast Region

The Northeast contains abundant opportunities for hydroelectricity, but also many restrictions. For instance, Meghalaya had many opportunities to use hydro power generation in the past, but government agents/authorities have not utilized them yet. Studies indicate that water resources can serve as a critical source of energy, addressing not only India's future energy demands, but also those of neighboring countries such as Bangladesh.

However, large dams constructed in seismically weak areas such as Arunachal Pradesh or Sikkim raise alarm bells in terms of Disaster Risk Reduction (DRR). To this point, careful and thorough Environmental Impact Assessments (EIAs) are necessary. Less disruptive alternatives, such as the construction of small dams, may help reduce flooding risks and would allow for the irrigation of the area while still providing the benefits of the production of energy through electricity generation (Das, 2013).

There are two factors that hinder the hydro power potential of the region: a lengthy approval processes and administrative hurdles as well as a lack of thorough investigations on specific projects. The broader political landscape is further complicated by inter-state disputes and the transboundary nature of rivers, frequently resulting in conflicts with neighboring regions and impeding hydro power developments (Handique & Dutta, 2014).

Meghalaya's Method to Manage Water Resources

Water resource management in Meghalaya presents both challenges and opportunities. Significant groundwater exists, but pollution from industries and coal mining poses important threats. Moreover, rooftop rainwater harvesting in Meghalaya and possibilities for micro-hydro power generation highlight the creative strategies for water and energy management adopted in the region (Martin & Shrivastava, 2013).

Rainwater collecting structures, for example Jalkunds (water ponds), were specifically designed to alleviate water shortages during the dry season, particularly for mountainous farmers. They are economically viable and have been shown to promote agricultural activity during difficult or low-yielding agricultural periods. The successful implementation of these technologies in Meghalaya suggests that there is a potential to scale them up across the region to enhance water and energy security (Saha et al., 2007).

Methane Emission and Hydroelectric Potential in Response to Climate Change in Meghalaya

Another major aspect that influences the availability of water and energy in Meghalaya is how climate change disrupts the hydrological cycle. The study of long-term meteorological data has shown variations in rainfall patterns as well as temperature alterations that may have an impact on the availability of water for hydro power generation. The rise of extreme weather conditions like heavy rainfall and variations in temperatures confirms the necessity of adaptable techniques for managing water resources and generating power in this region (Das et al., 2012). Research also suggests that if better methods are utilized to harvest rainwater and use it for agricultural production, it could help the local economy. In Meghalaya's mid-hills, for instance, high-value crops are grown during winter months to capture rainwater, thereby showing evidence of economic viability. This practice not only increases agricultural productivity, but also promotes the sustainability of farming practices within the region (Singh et al., 2006).

Decreased Precipitation and Its Impact on Hydro Power Generation

Hydro power, a significant source of energy, predominantly relies on perennial rainfall. As a result, climate change is considered a threat to hydro power plants since it has led to reduced water availability in the river basins of Northeastern India. Research indicates that States like Tripura experience frequent shut down of hydroelectric plants whenever there is a water scarcity, which results into low electricity generation. One possible alternative would be to combine smaller hydro power projects with solar photovoltaic (PV) powered pumped storage units. This hybrid technology couples solar energy and water pumps for irrigation purposes and may help to sustain annual performances amid declining rainfall (Bhattacharjee et al., 2019).

Innovations in Rainwater Harvesting for Power Production

The practice of harnessing rainwater for the generation of energy is a new strategy to tackle power shortages mostly experienced in places that face seasonal blackouts. One study proposes capturing the kinetic energy of rainwater through organized disposal pipes, small generating turbines, and piezoelectric generators. This system was designed to supply energy to buildings located in areas affected by high temperature season power cuts and shows both the pros and cons of this technology. The research stresses how careful pipe design can be used to maximize power production, addressing also the practicality and efficiency aspects of the suggested system (Tinaikar, 2013).

Rainfall Patterns in Meghalaya: Changes and Implications

Meghalaya is one of the most well-known areas for rainfalls and is one of the rainiest regions in India. It has been observed that Meghalaya experienced a decline in low- and medium-intensity rainfall values while high-intensity values increased. Local experts are particularly concerned about such changes because it may lead to limited water supplies during dry seasons, which could be detrimental for both agriculture and the population at large. To identify these changes more accurately, the study relied on gridded precipitation data from Umiam and Umtru watersheds along with novel trend analyzing methods. Yet, they present potential consequences for the future assessments of water availability, thus warranting a revision of the existing watershed management measures (Marak et al., 2020).

Machine Learning in Forecasting Rain for Management of Hydro Power

The emergence of machine learning has enhanced the accuracy of rainfall projection, which is essential to improve hydro power plant operations. In a rural micro hydel plant located in Kerala, an energy management system was developed utilizing XGBoost regression models to predict rainfall, water flow rates, and electricity generation. Several regression-based machine learning models were tested during the research, and it was found that XGBoost had better accuracy. This better forecasting ability improves real-time power consumption efficiency and asset management, therefore proving that machine learning can aid in promoting sustainable energy usage practices in rural areas (Valsan et al., 2022).

Geomorphological Responses to Extreme Rainfall in Eastern India

In Eastern India, including parts of Meghalaya, extreme rainfall occurrences result in significant geomorphological alterations. The response to such occurrences depends on the annual rainfall totals and the frequency of the heavy rainfalls. Research on the major areas like the Darjeeling Himalayas, the fringe of Bhutan Himalayas, and the Cherrapunji region shows that although they are all vulnerable to change, they have different responses to extreme weather events. For instance, the deforestation in Cherrapunji led to the loss of fertile soils, causing it to become a "sterile" system built on fast overland flow. This change highlights the intricate relationship between human activities, land-use, and geomorphic processes in such places (Soja & Starkel, 2007).

Rainwater Harvesting for Maintaining Sustainable Water Management

In Northeastern India, an area characterized by its abundant though unevenly distributed water supply resources, rainwater collection offers an alternative to store water supply. This not only supports increasing water demand, but also avoids runoffs, reduces flooding risks, curtails soil erosion, and enhances groundwater quality as well as agricultural yield. Additionally, the area's special geography and temperature are well suited for rainwater collection, transforming small-scale farming into a more viable profession because of the constant availability of water (Samuel & Satapathy, 2008).

Assisted Forest Restoration and Soil Hydrological Function

The presently conducted research in Meghalaya related to the restoration of degraded forest shows positive outcomes regarding soil hydrological functioning. Nonetheless, these projects still face multiple challenges. It was also established through research in Khasi Hills that while ANR can increase the hydraulic conductivity of soils and their preferred flow paths, its effectiveness on such parameters may not reach that of undisturbed sacred forests. This investigation focused on topsoil Ksat under different vegetation types and found that sacred trees had the highest Ksat values, followed by ANR zones, and the lowest were in the degraded grasslands. Therefore, it can be concluded that although ANR is able to enhance some aspects of hydrology in soils, there is a need for more efforts to restore its natural capacity for water storage and infiltration completely (Ghimire et al., 2021).

The Synoptic Scale Disturbances Impact the Rainfall of Meghalaya

In Meghalaya, particularly in Cherrapunji and Mawsynram, rainfall patterns, both within days and months, are considerably affected by synoptic scale disturbances and orographic interactions. Regions famed for their high rainfall within the country tend to have wide divergencies from mean rainfall, with annual and monthly peaks overly exceeding average values. It is necessary to understand these variations for water supply management as well as to lessen the negative effects encountered during extreme weather events in one of the rainiest areas on earth (Murata et al., 2007).

Methodology

Justification for the Qualitative Method

The qualitative method was chosen to capture the intricate and context-specific aspects of Meghalaya's electrification strategy, enabling the collection of rich, detailed data through interviews, focus groups, and case studies—insights often missed by quantitative approaches. This method effectively addressed the region's unique social, economic, and environmental conditions, providing a deeper understanding of the strategy's implementation. Its inherent flexibility also allowed researchers to adapt their approach in response to emerging challenges and evolving circumstances throughout the project.

Data Collection Method

The data collection process combined multiple qualitative techniques to gain a comprehensive understanding of Meghalaya's electrification strategy. Document analysis of government reports, policy papers, project plans, and legal frameworks helped uncover core and ancillary issues affecting implementation. Semi-structured interviews with key stakeholders—including officials, engineers, community leaders, and residents—provided in-depth insights into planning, execution, and impact. Focus group discussions with local communities highlighted collective perspectives and social dynamics underlying project challenges and success. Field observations offered real-time evidence of infrastructure development and community interaction with power sources, while case studies of specific projects illustrated best practices and served as valuable models for future initiatives.

Here are the details for the data sources:

Policy Papers

The *Meghalaya Hydro Power Policy 2023*, published by the Department of Power, Government of Meghalaya in January 2023, outlines the state's strategic approach to utilizing its abundant rainfall for hydro power generation. The policy emphasizes regulatory frameworks, sustainability measures, and incentives to attract both private and public investment. Key sections include an introduction and background, defined policy objectives and targets, investment incentives, environmental and social safeguards, and mechanisms for implementation and monitoring, all aimed at promoting sustainable and inclusive hydro power development in the state.

Project Reports

The *Feasibility Study on Small and Micro Hydel Projects in Meghalaya*, published by NEEPCO in March 2023, provides a detailed analysis of the technical, environmental, and financial viability of small and micro hydro power sources across various districts in the state, including site assessments, potential energy output, and recommendations for implementation. Complementing this, the *Community-Based Hydro Power Projects: Case Studies from Meghalaya* report by MEDA, released in May 2023, showcases successful community-driven hydro initiatives, such as the Umiam River and Myntdu-Leshka projects, highlighting local participation, project design, operational challenges, and offering policy recommendations and best practices for future development. Together, these reports offer valuable insights into both the feasibility and community-centric models of hydro power generation in Meghalaya.

Stakeholder Interviews

Two key studies provide in-depth insights into community involvement and the socio-economic impacts of hydro power projects in Meghalaya. The *Community Involvement in Hydro Power Projects* study, conducted by the Centre for Environment Protection and Sustainable Development (CEPSD) from February to April 2023, involved local leaders, engineers, environmentalists, and policymakers to explore community participation, technical and operational aspects, environmental and social impacts, and implementation challenges. Similarly, the *Impact of Hydro Power Projects on Local Livelihoods* study by IIT Guwahati (June–August 2023) engaged farmers, fishermen, developers, officials, and NGOs to examine economic benefits, employment, effects on agriculture and fishing, community concerns, and future policy support. These sources support the project *Water to Watts: Meghalaya's Electrification Strategy of Turning Rainfall into Power* by highlighting critical stakeholder perspectives and socio-environmental dimensions of hydro power development.

Analytical Framework Implementation

The project adopted a comprehensive, multi-stage analytical framework to assess hydro power potential in Meghalaya. It began with problem identification through literature review, energy needs analysis, and stakeholder engagement. Data collection involved reviewing key policy documents like the Meghalaya Hydro Power Policy 2023, technical feasibility studies, and conducting stakeholder interviews. Technical assessments included GIS mapping, hydrological modeling, and engineering evaluations to determine viable project sites and infrastructure needs. Economic and financial analyses assessed project viability through cost-benefit evaluations, sensitivity analysis, and ROI modeling. Environmental and social impact assessments were conducted to evaluate ecological and community implications, supported by stakeholder consultations. Policy and regulatory reviews examined relevant state and national frameworks, identifying incentives and potential barriers. Community involvement was ensured through focus group discussions, benefit-sharing models, and grievance redressal systems. Finally, a detailed implementation and monitoring framework with clear timelines, KPIs, and evaluation mechanisms was established. This holistic approach ensures technically feasible, economically viable, and socially inclusive hydro power developments aligned with Meghalaya's electrification and sustainability goals.

Case Studies

1. **Umiam Hydro Power Station Renovation Project** (Tables 16.1, 16.2, 16.3 and 16.4)

This is a report examining Meghalaya and Northeast India's strategies for harnessing hydro power. It considers its vast, unexploited hydro power potential and the related challenges and prospects that come with them. Most of the waterbodies in Meghalaya are highland rivers fed by heavy rains throughout the year. The NER serves as India's primary power supplier to its neighboring countries and the people within, such as

Table 16.1 Project cost and funding of Umiam Hydro Power Station Renovation Project

Category	*Amount (Million Yen)*
Total project cost	2090
Loan amount	1700
Disbursed amount	1693

Note The data related to cost and funding are from the Umiam Hydro Power Station Renovation Project

Table 16.2 Energy output improvement of Umiam Hydro Power Station Renovation Project

Fiscal year	*Target production (GWh)*	*Actual production (GWh)*	*Achievement (%)*
Pre-renovation Avg (98-01)	143	110.1	77.0
FY 2003	143	116.4	81.4
FY 2004	143	139.3	97.2
FY 2005	143	98.9	69.2

Note The data related to energy output improvement are from the Umiam Hydro Power Station Renovation Project

Table 16.3 Efficiency and economic outcome of Umiam Hydro Power Station Renovation Project

Indicator	*Pre-renovation (%)*	*Post-renovation (%)*
Financial Internal Rate of Return (FIRR)	6.97	16
Economic Internal Rate of Return (EIRR)	19.12	17

Note The data related to efficiency and economic outcome are from the Umiam Hydro Power Station Renovation Project

Table 16.4 Socio-economic impact of Umiam Hydro Power Station Renovation Project

Category	*Impact (%)*
Umiam Contribution to Meghalaya's Generation	20
Industrial Electricity Consumption Increase (1998–2004)	638

Note The data related to socio-economic impact are from the Umiam Hydro Power Station Renovation Project

Bangladeshi who suffer from severe shortages of electricity. Consequently, the document suggests that using this energy would make the northeast region of India "the powerhouse" of the entire country. Nevertheless, the successful completion of hydro power projects relies on solving socio-environmental challenges related with it as well as meeting technical requirements, including site selection criteria like seismicity.

In Meghalaya, India, the Umiam Stage-I Hydroelectric Project stands out as a significant infrastructure investment. Some of its key features are:

Location: The Umiam river is used in the Sumer Village found in the Ri-Bhoi district.

Dam Specifications: The structure is made up of a concrete gravity dam that is 73 m tall.

Reservoir: The full reservoir level (FRL) is of 981.46 m, while the minimum draws down level (MDDL) is 960.12 m; hence giving it an active storage capacity of about 142.35 million cubic meters (Mm^3).

Spillway: The top level for the spillway, which is ogee type gated, stands at 969.26 meters tall.

Head Race Tunnel: With a diameter of 3.05 m and length of 2058 m, the head race tunnel has been designed for a discharge rate of 28.12 cubic meters per second (cumecs).

Powerhouse: The surface powerhouse contains four units, each one having an installed capacity of nine Mw using vertical Francis turbines.

Penstock: Two steel lined penstock pipes are used, both having diameters measuring 1.98 m and lengths of around 618.70 m.

Switchyard: The outdoor switchyard consists of four generator bays plus seven-line bays.

All these factors indicate how much more renewable energy potential and strategic importance to regional power supply makes this project incomparable to others in its class.

Development Challenges

- Seismic Vulnerability: In seismic locations where massive dams are being built, there are certain obstructions that might harmfully affect them. Hence it becomes necessary for transdisciplinary groups to carry out detailed Environmental Impact Assessments (EIA) for social-environmental assessment prior to starting any project.
- Land Acquisition and Local Participation: Most land in Meghalaya is owned privately, making land acquisition for hydro power projects difficult. Therefore, it is important to incorporate local communities into the planning and decision-making processes for project acceptance and success.
- Regulatory and Infrastructure Issues: The hydro power sector within the NER faces several regulatory challenges such as lengthy approval processes and insufficient infrastructure, which affects power evacuation. As a result, project implementation may be delayed while expenses rise.
- Water Wars: Many rivers in the region are transboundary so there are water-sharing conflicts, especially with Bangladesh. Therefore, collaborative frameworks like the Joint River Commission need to be established to solve the problems as well as for mutual benefits.

Development Strategies

- Concentrate on small-scale projects: Instead of building enormous dams, we advocate for smaller constructions like cheek dams and micro-hydro systems. These initiatives provide local energy solutions, assist with rural electrification, and improve irrigation in agriculture.
- Integration with regional power grids: If a South Asia regional power grid system were established, hydro power resources could be shared, allowing electricity generated in Meghalaya to be efficiently transmitted to other parts of India, including power-deficient areas in the North and the West.
- Community participation and awareness: Ongoing community awareness activities are required to educate residents on the benefits of hydro power projects. Involving locals in project development increases acceptance while meeting community needs.
- Innovative Water Management: The publication underlines the need of effective water management methods, namely rainwater collecting systems such as Jalkund, which may store water for agricultural use and increase water availability during dry seasons.
- Climate Change Adaptation: As rainfall patterns change because of climate change, adaptable hydro power-generating systems must be developed. This involves combining solar photovoltaic systems with hydroelectric facilities to maintain a continuous power supply even during seasons of low rainfall.

Conclusion

Hydro power development plans in Meghalaya offer a huge potential to use the region's natural resources to generate sustainable electricity. Meghalaya can realize its hydro power potential by resolving problems such as seismic risk, land acquisition, regulatory concerns, and community participation. Small-scale initiatives, increased regional collaboration, and the adoption of innovative water management methods will be critical for the effective implementation of these policies, ultimately contributing to India and its neighboring nations' energy security (Tables 16.5, 16.6, 16.7 and 16.8).

2. Mawphu Hydroelectric Project (2 x 42.5 mw) Meghalaya[1]

In the document, we present the Terms of Reference for a Comprehensive Environmental Impact Assessment (EIA) for the Mawphu Hydroelectric Project located in Meghalaya, India, with a target generation capacity of 85 MW of hydro power. It helps in identifying the environmental aspects and consequences of the project over its entire life cycle from its construction to the operation phase.

Table 16.5 Project overview of Mawphu Hydro Electric Project

Location	*East Khasi Hills, Meghalaya*
Capacity	2*42.5 MW (Total 85 MW)
Category	Category "A" Project (Requires Environmental Clearance from the Ministry of Environment, Forest and Climate Change)
Project status	Under Environmental Impact Assessment (EIA) process
Total cost	Not Mentioned
Local area development plan budget	0.5% of the project cost

Note The data related to project overview are from the Mawphu Hydro Electric Project

Table 16.6 Energy output of Mawphu Hydro Electric Project

Parameter	*Details*
Installed capacity	85 MW
Number of units	2
Capacity per unit	42.5 MW
Design discharge	Not specified
Minimum assured release	To be determined through scientific study

Note The data related to energy output are from the Mawphu Hydro Electric Project

[1] Refer to Fig. 16.1 for visual representation (Author's Own Input).

Table 16.7 Environmental management from Mawphu Hydro Electric Project

Category	*Key provisions*
Compensatory afforestation	Forest Protection and Restoration Plans
Wildlife conservation	Management plans for habitat protection
Fisheries conservation	Fish ladder installation and aquatic habitat monitoring
Catchment area treatment	Erosion control measures
Public health management	Water supply, sanitation, and disease control plans
Disaster management	Dam break analysis and flood modeling
Solid waste management	Waste disposal planning for labor camps

Note The data related to environment management are from the Mawphu Hydro Electric Project

Table 16.8 Socio-economic impact of the Mawphu Hydro Electric Project

Impact	*Description*
Employment generation	During construction and operation phases
Land acquisition	Impact on agricultural and forest land
Community development	Local Area Development Plan (0.5% of project cost)
Rehabilitation and resettlement	Plan for Project Affected Families (PAFs)
Tribal area development	Special provisions for indigenous communities

Note The data related to socio-economic impact are from the Mawphu Hydro Electric Project

Objectives of EIA Study

The key objectives include:

- The Evaluation of Environmental Impacts: To assess physical, biological, and socio-cultural effects caused by the project.
- Baseline Data Collection: This involves the collection of data about water quality, air quality, noise levels, forestry activities, wildlife population sizes, and aquatic ecosystems.
- Environmental Management Plan (EMP): This entails coming up with strategies that would help mitigate negative impacts that were identified during evaluation.

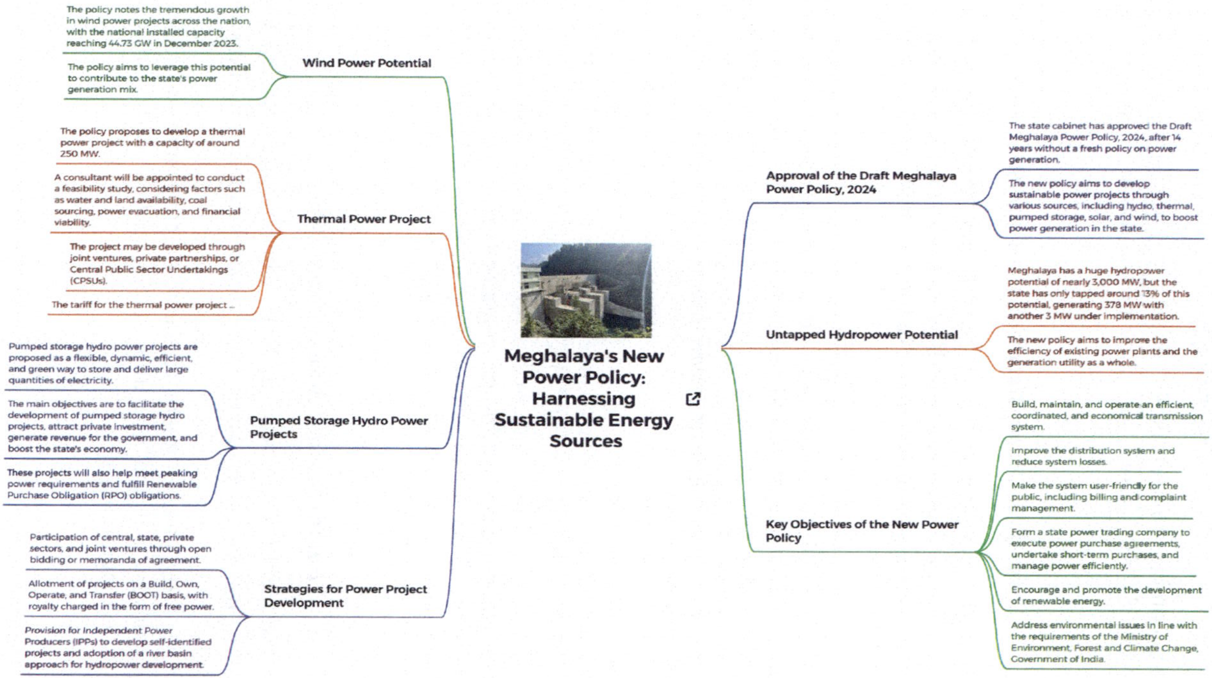

Fig. 16.1 Comprehensive environmental impact assessment (EIA) study for Mawphu hydroelectric project, Meghalaya

Methodology

The EIA shall conform to MOEF&CC criteria that necessitate a comprehensive survey of three seasons, including pre-monsoon, monsoon, and winter.

Some of the Major Methodologies Include

Scoping: through a scoping matrix, EIA identifies possible impacts in terms of environmental components such as land, water resources, aquatic and terrestrial ecology, socio-economic factors, air and noise pollution, and public health.

Baseline Studies: Field data collection will be done to establish current environmental conditions including ambient air quality, river water quality, soil characteristics, and biodiversity indices.

Impact Prediction

Air Quality: increased emissions from construction activities may affect local air quality.

Water Resources: there may be changes in water quality and hydrological patterns due to construction and operational activities.

Biodiversity: habitat loss could occur with this project, affecting both terrestrial and aquatic species including migratory fish.

Socio-Economic Effects: labor influx may result in cultural conflicts and could put pressure on local infrastructures while at the same time creating job opportunities.

Environmental Management Plan

To address the identified impacts, an EMP will be conceived and will include: Mitigation measures: Steps to decrease harmful influences on surroundings.

Monitoring Programs: Ongoing appraisal of environmental factors throughout construction and operation stages, making certain that it complies with ecological norms.

Conclusion

To make sure that environmental issues become part of project planning and implementation phases, there is an EIA for Mawphu Hydroelectric Project. This project aims at reconciling energy generation with environment-friendly practices by identifying possible impacts and finding ways of managing them that meet legal frameworks and community requirements.

3. New Umtru Hydel Electric Project[2]

The document enfolds the development strategies for the New Umtru Hydroelectric Project (NUHEP). The document primarily focuses on its capital cost approval, operational framework, and regulatory compliance (Tables 16.9, 16.10, 16.11, 16.12, 16.13, 16.14 and 16.15).

Table 16.9 Project overview of the New Umtru Hydel Electric Project

Location	*Ri-Bhoi district, Meghalaya*
River	Umtru River (tributary of the Brahmaputra River)
Installed capacity	40 MW (2 × 20 MW)
Type	Run-of-River with Pondage
Date of Commercial Operation (COD)	1st July 2017

Note The data related to project overview are from the New Umtru Hydel Electric Project

Table 16.10 Project cost of the New Umtru Hydel Electric Project

Description	*Cost (INR crore)*	*Remarks*
Approved project cost (2007)	226.40	Initial sanctioned cost
Final audited capital cost	584.00	As of March 2019
Actual capital cost (claimed)	607.25	Additional costs for deferred works and liabilities
Interest during construction	173.29	Due to project delays

Note The data related to project cost are from the New Umtru Hydel Electric Project

[2] Refer to Fig. 16.2 for visual representation (Author's Own Input).

Table 16.11 Energy output improvements from the New Umtru Hydel Electric Project

Year	*Gross generation (MU)*	*Net generation (MU)*	*Auxiliary consumption (%)*	*Transformation loss (%)*
FY 2017–2018	173.61	171.01	1.0	0.5
FY 2018–2019	179.82	177.12	1.0	0.5
FY 2019–2020	219.00	215.72	1.0	0.5
FY 2020–2021	219.00	215.72	1.0	0.5

Note The data related to energy output improvements are from the New Umtru Hydel Electric Project

Table 16.12 Funding pattern of the New Umtru Hydel Electric Project

Capital funding	*Amount (INR Cr.)*	*Percentage*
Loan	418.65	69%
Equity	188.60	31%
Total capital cost	607.25	100%

Note The data related to funding pattern are from the New Umtru Hydel Electric Project

Table 16.13 Efficiency and performance of the New Umtru Hydel Electric Project

Normative Annual Plant Availability Factor (NAPAF)	62.60%
Design energy	235 million Units (MU) per year
Overload capacity	10%

Note The data related to efficiency and performance are from the New Umtru Hydel Electric Project

Table 16.14 Socio-economic impact of the New Umtru Hydel Electric Project

Impact	*Description*
Employment generation	During construction and operation phases
Land acquisition	Acquisition of land for the reservoir and infrastructure
Local area development	0.5% of the project cost allocated for community development

Note The data related to socio-economic impact are from the New Umtru Hydel Electric Project

Table 16.15 Environmental clearance of the New Umtru Hydel Electric Project

Approved by the Ministry of Environment & Forests (MoEF) in	2006
Clearance valid for	5 years

Note The data related to environmental clearance are from the New Umtru Hydel Electric Project

Key Features

Infrastructure

The project entails a new dam, spillway, and powerhouse aimed at optimizing water flow as well as energy output. Some of its key components include:

Dehal dam: It has six radial gates for flood management with a live storage capacity of 2.38 million cubic meters.

Powerhouse: It contains two 20 MW generating units with continuous overload capability ensuring that there is always some reliable supply during peak demand.

Head Race Tunnel (HRT) and Pressure Shaft: These structures facilitate transport of water from the dam to the turbines.

Compliance with Environmental Regulations and Statutes

The project has been given permission by the Ministry of Environment & Forests, which guarantees that it will meet national environmental standards. The Government of Meghalaya has also granted administrative approval, which means that the ₹629 Crores developmental project can commence.

Financial Aspects

Capital Cost and Tariff:

This document gives an outline of the process of approval for the capital cost as well as Annual Revenue Requirement (ARR) for the fiscal years 2017–2018 through 2020–2021. The amount spent on it was analyzed according to Meghalaya State Electricity Regulatory Commission (MSERC) regulations to ensure regulatory compliance when it comes to spending.

Additionally, Capitalization:

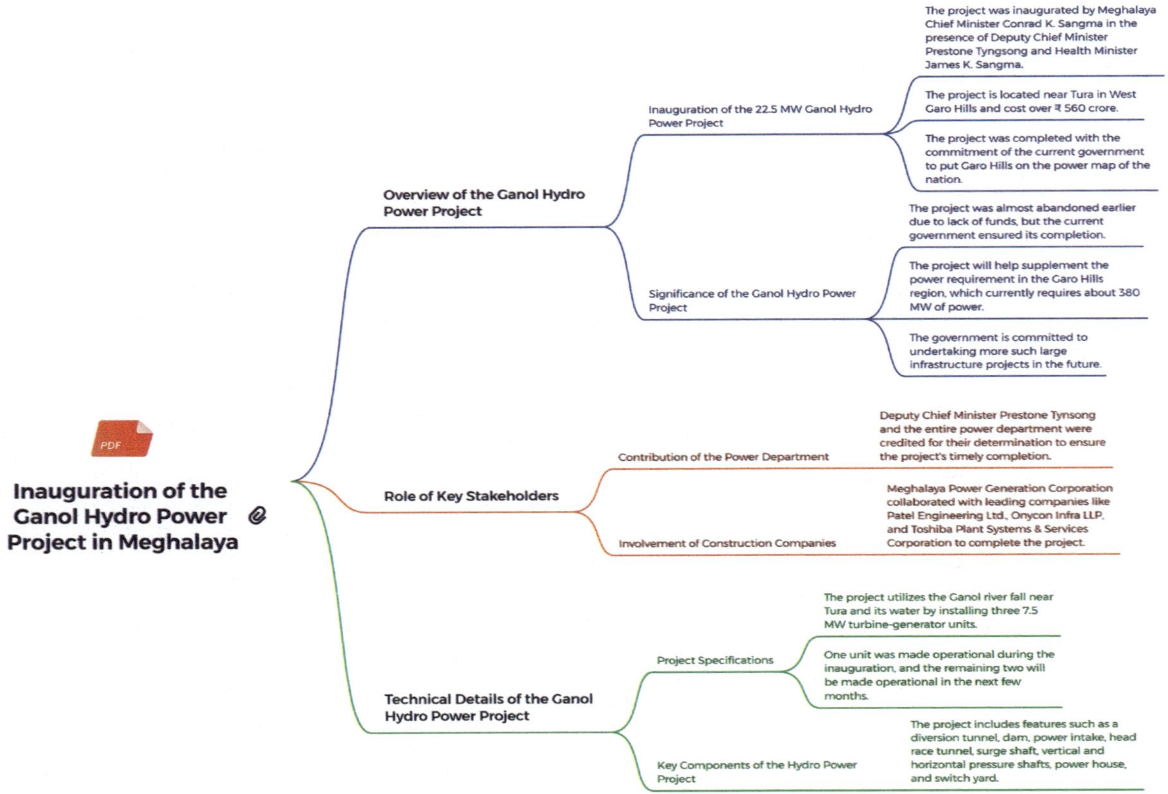

Fig. 16.2 Development of the new Umtru hydro electric project (NUHEP) in Meghalaya

There are certain provisions in regulations that allow additional capitalization that accommodates costs related to un-discharged liabilities and deferred works; this is very crucial for the efficient operation of the project. As such, any changes in costs can be made because of the changing financial requirements of the project.

Conclusion
The New Umtru Hydroelectric Project is a strategic move for the development of energy in Meghalaya with an eye on increasing hydro power generation without neglecting environmental concerns. In addition to optimizing currently available resources, this program intends to offer sustainable energy options to the region that are in line with India's larger ambitions on energy safety and economic growth.

Hydro Power Development Strategies

1. **Meghalaya Power Policy 2024**[3]
   The Meghalaya Power Policy 2024 outlines a comprehensive framework to advance hydro power development in the state through structured governance, regulatory compliance, and sustainable practices. A State-Level Committee, chaired by a Power Department Minister, is tasked with project identification, land facilitation, feasibility review, and recommending projects to a High-Power Committee for final approval. The policy details project allocation methods, including bidding and negotiations, and assigns approval authority to the High-Power Committee. It also defines financial aspects such as upfront premiums, processing fees, and free power obligations. Emphasizing sustainable development, the policy mandates adherence to environmental guidelines and the modernization of existing plants. It requires detailed project reports and feasibility studies before implementation, while aligning itself with legal frameworks like the Electricity Act of 2003. Overall, the policy aims to ensure systematic, legally compliant, and ecologically responsible hydro power growth in Meghalaya.

[3] Refer to Fig. 16.3 for visual representation (Author's Own Input).

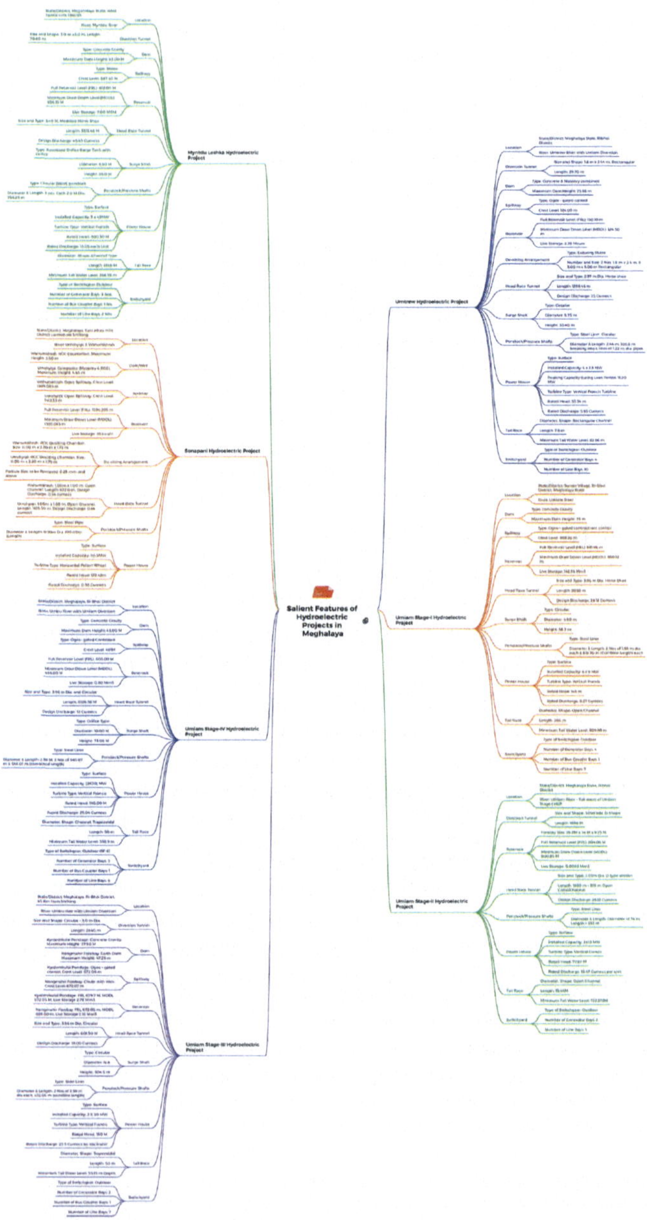

Fig. 16.3 Meghalaya's new power policy: harnessing sustainable energy sources

2. **Inauguration of Ganol Hydro Power Project byGovernment of Meghalaya**[4]
   Meghalaya, in Northeast India, holds significant untapped hydroelectric potential that could meet both local and regional energy demands, including those of neighboring countries like Bangladesh. However, development faces challenges such as seismic risks requiring thorough Environmental Impact Assessments, difficulties in land acquisition due to private ownership, and delays from regulatory hurdles and inadequate power evacuation infrastructures. Solutions include focusing on small-scale hydro projects like micro-hydel systems to support local electrification, promoting regional cooperation through a South Asian Regional Power Grid for efficient energy sharing, and adopting innovative water management techniques like Jalkund to improve water storage and agricultural resilience.
3. **Salient Features of Hydroelectric Projects**[5]
   Hydro power projects in Meghalaya, India, follow structured strategies and technical specifications to ensure efficient and sustainable development. Key components include detailed planning of infrastructure such as diversion tunnels, dams, spillways, reservoirs, head race tunnels, surge shafts, penstocks, and surface-type powerhouses. Projects like Umtru feature mixed concrete-masonry dams with 25.98 m height and 2.38 MCM storage, while Umiam Stage-I has a 73 m high concrete gravity dam. Tunnel and penstock designs vary, such as the 2.97 m diameter horse-shoe tunnel at Umtru and dual 1.98 m diameter penstocks at Umiam. Powerhouses host vertical Francis turbines with capacities ranging from 4 × 2.8 MW (Umtru) to 3 × 42 MW (Myntdu Leshka). Environmental measures include de-siltation systems like scouring sluices, while switchyards are equipped with line bays, generator bays, and bus coupler bays, as seen in Umiam Stage-I. These elements collectively contribute to sustainable hydro power development through robust technical design and environmental safeguards.
4.

[4] Refer to Fig. 16.4 for visual representation (Author's Own Input).

[5] Refer to Fig. 16.5 for visual representation (Author's Own Input).

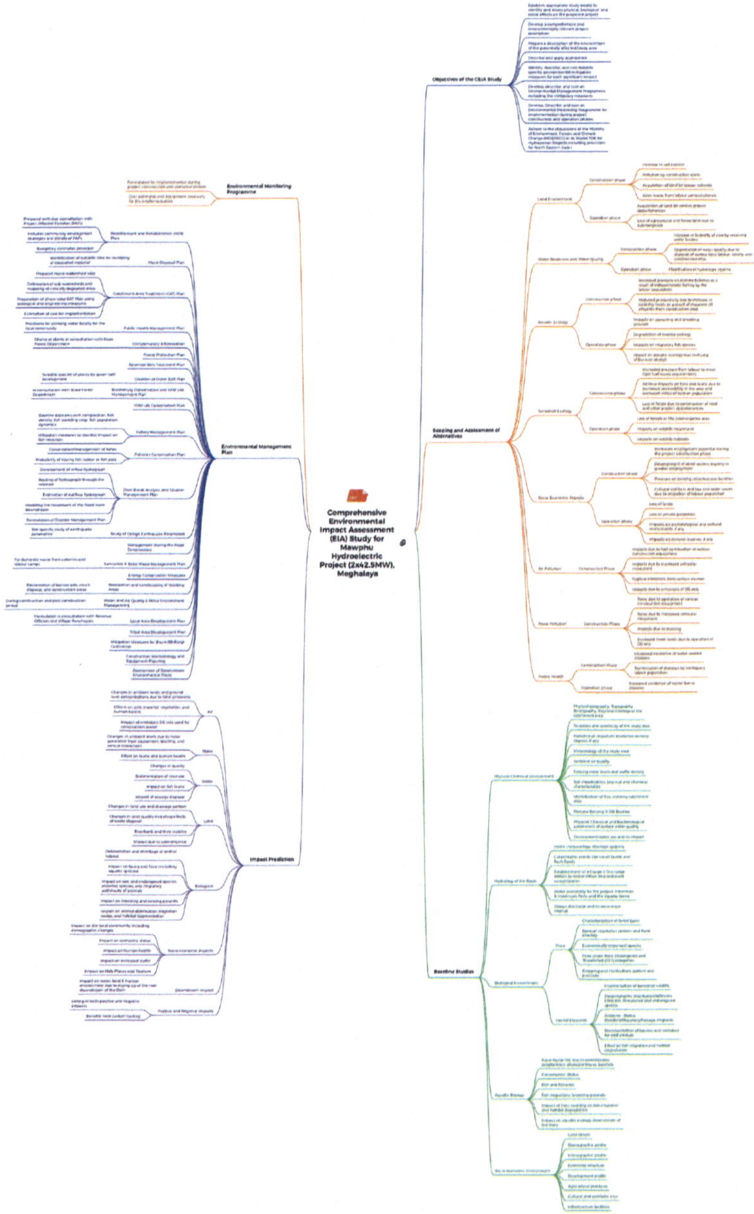

Fig. 16.4 Inauguration of the Ganol hydro power project in Meghalaya

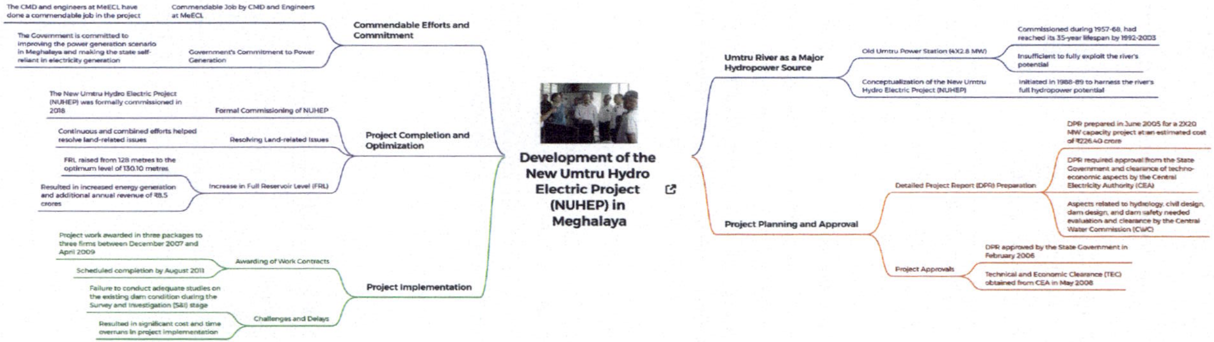

Fig. 16.5 Salient features of hydroelectric projects in Meghalaya

Meghalaya Power Distribution Sector Improvement Project Sector Assess

This paper provides a summary of strategies for hydro power development in Meghalaya, India, and highlights the fact that there is more than enough hydro power potential in this federal state, amounting to about 3000 MW with only about 11.8% utilized so far. To meet a peak demand of around 400 MW, the state depends upon its eight hydro power stations and central stations. The Meghalaya Power Policy proposes the establishment of 24 new power plants having an overall generation capacity of 1440 MW; however, progress has been rather slow as only fifteen projects are now under implementation. The state intends to use build–own–operate–transfer models to increase hydro power production and thereby enhance revenues as well as promote renewable sources of energy in remote areas through private involvement. Also, the transmission infrastructure is being renovated for supporting new hydro power projects that have an estimated 99% availability in their networks.

Apart from this, it was pointed out in the paper that there should be an improvement on the efficiency of electricity distribution since currently the losses account for over 30%. In response to this situation, government measures have been initiated to reduce these losses and achieve long-term financial sustainability. The strategies aim at maximizing hydro power resources, upgrading physical structures, and promoting fair access to energy throughout Meghalaya's varied landscape (Figs. 16.1 and 16.2).

Alignment with SDG 30

Water electrification in Meghalaya is crucial for various reasons, mainly for the sustainable use of natural resources to attain the SDG 30. By using most of Meghalaya's abundant water resources to generate hydro power, this initiative strives to help achieve several targets given in the Sustainable Development Goals, including:

SDG 7: Affordable and Clean Energy—Hydro power projects are undertaken with a view to ensure access to affordable, reliable, and sustainable energy for all. This reduces the dependence on fossil fuels, mitigating environmental pollution and greenhouse gas emissions.

SDG 8: Decent Work and Economic Growth—Hydro power plants create employment opportunities both in their construction and operational phases while stimulating local economies at their places of operations. In return, these revenues can be reinvested in other developmental activities.

SDG 9: Industry, innovation and infrastructure—Industry has become so much more productive. The industry can generate innovations with the help of hydroelectric infrastructure while sustaining it.

SDG 11: For Sustainable Cities and Communities—In both urban centers and rural areas, reliable electricity means better health care services, education systems such as schools, and hospitals among other social amenities that improve standards of living generally.

SDG 13: Climate Action—Through the exploitation of renewable water resources, Meghalaya contributes to mitigating climate change. Hydro power is a pure source of energy that reduces carbon footprints.

The successful execution of hydro power projects in Meghalaya brings along with it various economic benefits. They provide employment, attract private investment, and generate huge revenue from royalties and sales from electricity production. Such hydroelectric infrastructure can increase the stability of the power supply in the state, supporting industries and improvement in the quality of life.

Hydro power is an environmentally friendly and renewable energy source. It takes advantage of the natural flow of water, which makes it sustainable. Additionally, hydro power projects have comparatively lesser negative impacts on the environment than their fossil fuel counterparts, resulting in decreased air pollution and preservation of wildlife habitats.

It is very important to research water electrification in Meghalaya as an example of how areas with huge potential in natural resources are striving to make use of the electrification for sustainable development. Taking a cue from SDG 30, Meghalaya's disposition toward the development of hydro power has become a prime example for other areas that face challenges in reconciling economic growth and environmental concerns. While meeting the energy requirements, harnessing Meghalaya's natural water resources has added another feather in the cap toward the global efforts for sustainable development.

Suggestions for Future Research

The exploration into the development of hydro power in Meghalaya must focus on a few specific areas. Initially, it is important to conduct a more comprehensive study with regards to possible seismic threats brought about by the construction of big dams in this region. For instance, alternative types of less disruptive energy generation like small-scale hydro projects may serve as an appropriate solution that is safer and will not affect local communities negatively. Furthermore, there is a need for research on community participation models in such ventures on hydro power. In essence, local communities need to know how they can become integral stakeholders in future projects' planning, implementation, and benefit-sharing processes.

Also, the potential incorporation of hydro power with other renewable energy sources, chiefly solar photovoltaic (PV) systems, is another area for future research. The negative effects caused by fluctuating water availability resulting from climate change could be reduced through hybrid technologies that incorporate solar energy and hydro power. Besides, it is also important to focus on the socio-economic effects of hydro power schemes, including local livelihoods, employment creation, and economic growth. The aim of this research should be to develop methods that increase benefits while reducing adverse effects on communities at large. Lastly but not least, regulation and policies governing hydro power in this part of the world need to be studied. Discovering possible advancements in such frameworks would help make it easier to approve projects, thus improving the overall effectiveness of hydro power generation.

Findings and Conclusion

Hydro power development, if executed well, could go a long way in closing the energy gap in Meghalaya. However, they come with several disadvantages. Seismic hazards and problems regarding land acquisition are some of the regulatory impediments against infrastructural development that the hydro power industry in Meghalaya must contend with. However, small-scale energy generation, renewable energy integration, and improved community participation could at least help mitigate some of these challenges. Combined with technical support, environmental sustainability, and socio-economic viability shall enhance successful hydro power growth in Meghalaya. Growth in the future should concentrate on small-scale projects that have a less adverse impact on the environment and can win the communities approval. One such instrument that would strengthen regional cooperation arrangements and enhance cross-border power sharing for energy security in this region would be the South Asia Regional Power Grid. Systems of rainwater harvesting should also be investigated for the design of hydro power projects to alleviate the situation of water scarcity during dry periods. Relatively easier project execution and approval shall have to entail improving regulations and their physical applications, which, therefore, should command the utmost priority among concerned governments. Following these recommendations, Hydro power development can help Meghalaya maximize its hydro power potential for energy security and mitigate the global climate action agenda.

Undertaking hydro power development pertaining to the state of Meghalaya had a lot of insight with key implications for the power strategy of the region. At the face of what is arguably an area most prone to environmental disturbances, it is a crucial finding that the hydro power resources are vast but largely untapped, thereby accounting for an unjustifiably small percentage of its total possible capacity. Provided that the issues of seismic vulnerability, land acquisition, and guidelines are adequately tackled, there is a great opportunity for energy generation and economic development. Furthermore, this study highlights the importance of community involvement in such projects. This kind of engagement foster community support while ensuring the equitable sharing of energy benefits among a family or a community. Furthermore, hydro power development provides common ground for linking solar and hydro power projects for the development of a stronger and more sustainable energy system. Such initiatives are expected to mitigate weather-related risks, especially in terms of water supply variability arising from climate change. The study's implications are unambiguous: Meghalaya must take an all-encompassing approach to technical, environmental, and social issues if it is to reap or harvest the maximum benefits related to hydro power while also exploring innovative solutions and regional cooperation.

Appendix: Questionnaire for Stakeholder Interviews

Section 1: Introduction

1. **Introduction and Consent**
    - Introduce yourself and explain the purpose of the interview.
    - Ensure confidentiality and seek consent to record the interview (if applicable).

Section 2: General Information

2. **Respondent Details**
    - Name:
    - Age:
    - Gender:
    - Occupation:
    - Location:
    - Role in the Community/Organization:

Section 3: Current Energy Situation

3. **CurrentEnergy Access**
    - How reliable is the current electricity supply in your area?
    - What are the primary sources of energy for households and businesses?
    - Are there frequent power outages? If yes, how do they affect daily life and economic activities?

Section 4: Potential of Hydro Power

4. **Awareness and Perception of Hydro Power**
    - Are you aware of the state's plans to develop hydro power projects using rainfall?
    - What is your opinion on using hydro power as a source of energy?

- Do you think hydro power projects can meet the energy needs of your community?

Section 5: Technical Feasibility

5. **Site Identification and Suitability**

- Are there any existing small or micro hydro power projects in your area?
- What are the key geographical and environmental features that you think are suitable for hydro power projects?
- Have there been any assessments or studies conducted to identify potential sites for hydro power?

Section 6: Economic and Financial Considerations

6. **Economic Impact**

- How do you think hydro power projects will impact the local economy?
- Are you aware of any government incentives or financial support for developing hydro power projects?
- Would you or your community be willing to invest in or support hydro power projects financially?

Section 7: Environmental and Social Impact

7. **Environmental Concerns**

- What are the potential environmental impacts of hydro power projects in your area?
- How can these environmental impacts be mitigated?
- Do you think the benefits of hydro power outweigh the environmental concerns?

8. **Social Impact**

- How would hydro power projects affect local communities, including potential displacement or changes in livelihoods?
- What measures can be taken to ensure that local communities benefit from hydro power projects?

- How can conflicts or concerns related to hydro power projects be addressed?

Section 8: Policy and Regulatory Framework

9. **Regulatory Environment**

    - Are you familiar with the current policies and regulations governing hydro power development in Meghalaya?
    - What are the main regulatory challenges faced in developing hydro power projects?
    - What policy changes or support mechanisms would help facilitate the development of hydro power projects?

Section 9: Community Engagement and Participation

10. **Community Involvement**

    - How involved have local communities been in the planning and implementation of hydro power projects?
    - What are the best ways to engage with and involve local communities in hydro power projects?
    - How can benefit-sharing models be designed to ensure that local communities receive direct benefits from hydro power projects?

Section 10: Implementation and Monitoring

11. **Implementation Challenges**

    - What are the main challenges faced during the implementation of hydro power projects?
    - How can these challenges be addressed to ensure successful project completion?

12. **Monitoring and Evaluation**

    - What mechanisms should be put in place to monitor the progress and impact of hydro power projects?
    - How can communities be involved in the monitoring and evaluation process?

Section 11: Conclusion

13. **Final Thoughts**

- Is there anything else you would like to share about the potential for hydro power in Meghalaya?
- Do you have any additional suggestions or recommendations for the study?

Closing

- Thank the respondent for their time and insights.
- Provide contact information for follow-up questions or additional input.

This questionnaire aims to capture a comprehensive range of insights from stakeholders, including their perceptions, concerns, and recommendations related to the development of hydro power projects in Meghalaya.

References

Bhattacharjee, S., & Nayak, P. (2019). PV-pumped energy storage option for convalescing performance of hydroelectric station under declining precipitation trend. *Renewable Energy*. https://doi.org/10.1016/J.RENENE.2018.12.021

Das, A. K., Slong, A., & Chowdhury, P. (2012). Trend analysis of long term weather variables in mid altitude Meghalaya, North-East India.

Das, A. K., Munda, G. C., Thakur, N. S., Yadav, R. K., Ghosh, P. K., Ngachan, S. V., Bujarbaruah, K. M., Lal, B. V., Das, S. K., Mahapatra, B., Islam, M., & Dutta, K. K. (2014). Rainwater harvesting and integrated development of agri-horti-livestock-cumpisciculture in high altitudes for livelihood of Tribal farmers. *The Indian Journal of Agricultural Sciences.*

Das, P. K. (2013). North–East, 'The Power House of India': Prospects and problems. *IOSR Journal of Humanities and Social Science, 18*, 36–48.

Ghimire, C., Zwartendijk, B., Pde, F., & Bruijnzeel, L. (2021). Changes in soil hydraulic conductivity and preferential flow pathways after assisted forest restoration on degraded land in the Khasi Hills (Meghalaya, NE India). https://doi.org/10.5194/egusphere-egu21-6559.

Handique, K., & Dutta, A. (2014). Power and North East: The hydro power scenario of North East.

Jain, N., & Makkar, N. S. (2019). A comparative study on water management practices in India with special reference to Gujarat and Meghalaya. *Ushus—Journal of Business Management*.

Marak, J., Sarma, A., & Bhattacharjya, R. (2020). Innovative trend analysis of spatial and temporal rainfall variations in Umiam and Umtru watersheds in Meghalaya, India. *Theoretical and Applied Climatology, 142*, 1397–1412. https://doi.org/10.1007/s00704-020-03383-1

Martin, S., & Shrivastava, K. K. (2013). Feasibility of rainwater harvesting in high rise building for power generation.

Murata, F., Hayashi, T., Matsumoto, J., & Asada, H. (2007). Rainfall on the Meghalaya plateau in northeastern India—One of the rainiest places in the world. *Natural Hazards, 42*, 391–399. https://doi.org/10.1007/S11069-006-9084-Z

Saha, R., Ghosh, P. K., Mishra, V. K., & Bujarbaruah, K. M. (2007). Low-cost micro-rainwater harvesting technology (Jalkund) for new livelihood of rural hill farmers. *Current Science, 92*, 1258–1265.

Samuel, M., & Satapathy, K. (2008). Concerted rainwater harvesting technologies suitable for hilly agro-ecosystems of Northeast India. *Current Science, 95*, 1130–1132.

Singh, R. D., Lama, T., Saikia, U. S., & Satapathy, K. K. (2006). Economics of rainwater harvesting and recycling for winter vegetable production in Mid Hills of Meghalaya. *Journal of Agricultural Engineering (India)*.

Soja, R., & Starkel, L. (2007). Extreme rainfalls in Eastern Himalaya and southern slope of Meghalaya Plateau and their geomorphologic impacts. *Geomorphology, 84*, 170–180. https://doi.org/10.1016/J.GEOMORPH.2006.01.040

Sovacool, B. K., Dhakal, S., Gippner, O., & Bambawale, M. J. (2011). Halting hydro: A review of the socio-technical barriers to hydro power plants in Nepal. *Energy, 36*(5), 3468–3476. ISSN 0360-5442. https://doi.org/10.1016/j.energy.2011.03.051

Tinaikar, A. (2013). Harvesting Energy from Rainfall. *International Journal of Literature and Arts, 2*, 130. https://doi.org/10.11648/J.IJRSE.20130203.18.

Valsan, V., Sai, A., Devidas, A., & Ramesh, M. (2022). Regression based prediction of rainfall for energy management in a Rural Islanded micro-hydro grid in Kerala. In *2022 IEEE Global Conference on Computing, Power and Communication Technologies (GlobConPT)*, 1–7. https://doi.org/10.1109/GlobConPT57482.2022.9938353.

CHAPTER 17

Consumer Behavior in the Energy Transition: The EKB Decision-Making Approach

Jakub Kubiczek

Introduction

Understanding consumer behavior is foundational in social research and holds significant practical value for various organizations. Consequently, over the years, numerous models of decision-making models in consumer behavior have been proposed, such as the Nicosia model (Nicosia, 1966), the Howard and Sheth model (Howard & Sheth, 1969), and the McCarthy, Perreault, and Quester model (McCarthy et al., 1997). However, the most commonly used and recognizable model remains the Engel, Kollat, and Blackwell (EKB) model (Engel et al., 1968).

The EKB model, despite being developed over 50 years ago, continues to prove exceptionally useful in research on consumer behavior. Its versatility and universality allow it to adapt to various contexts and markets

J. Kubiczek (✉)
University of Economics in Katowice, Katowice, Poland
e-mail: jakub.kubiczek@uekat.pl

T. Walker et al. (eds.), *Wiring the Future*, Palgrave Studies in Emerging Risk Management and Sustainable Finance,
https://doi.org/10.1007/978-3-032-14717-2_17

(Ashman et al., 2015). As a result, the EKB model is widely employed to analyze diverse consumer behaviors. For example, Chae et al. (2006) use it to analyze purchasing decisions in the women's sports fashion market, while Dimitriou and AbouElgheit (2019) examine the consumer behaviors of Generation Z in the context of travel choices. Similarly, Sihi (2018) investigates the application of virtual and augmented reality technologies in high-involvement purchasing decisions, and Kaur et al. (2024) study consumer behaviors in the Metaverse.

A particularly notable application of the EKB model is in the area of green consumer behaviors, which support the pursuit of sustainable development (Kubiczek et al., 2024). Contemporary research demonstrates that sustainability increasingly influences purchasing decisions. This context represents a significant and complex issue requiring an analysis of the decision-making background and factors influencing the adoption of pro-environmental attitudes among consumers. Joshi and Rahman (2017) explore the determinants of purchasing behaviors related to sustainability, as well as perceptions of sustainability knowledge and environmental concern. Paswan et al. (2017) examine attitudes determining eco-friendly consumer behaviors. By placing the consumer at the center of research, it is possible not only to assess their behaviors but also to develop strategies for effectively influencing them, encouraging the incorporation of environmental aspects into their decision-making process (Zhang & Benyoucef, 2016).

The EKB model is also applied in research on energy consumption, including in the context of energy transition, which is a key element of green consumer behaviors. Zhang and Dong (2020) highlight the significance of this model in analyzing decisions related to the choice of energy sources and providers. Gushterov (2018) emphasizes that decisions in this area are a classic example of decision-making issues within the framework of consumer behavior. Understanding the factors influencing energy choices enables a better understanding of overall consumer behaviors.

This chapter seeks to highlight the relevance and applicability of the EKB model in analyzing consumer behavior within the context of the energy transition. The main research question guiding this chapter is: how can the EKB model be used to analyze consumer behaviors in the context of energy transition?

Energy transition can be defined in broad terms as a shift in the sources of energy supply. Yet, today, the concept is predominantly associated with the transition from fossil fuels to renewable energy sources (RES).

This process can be viewed at the macroeconomic level, encompassing the transformation of entire economies toward low-carbon systems, as well as at the microeconomic level, which refers to the decision-making processes of households adopting renewable energy solutions. As such, the macroeconomic level will be the focus of this chapter.

The discussion in the following pages focuses on the individual stages of the EKB model, linking them to the specific characteristics of consumer decisions within the energy transition process. By applying the EKB model to the context of energy transition, this chapter contributes to the literature by bridging consumer behavior theory with the study of sustainable energy choices. Most importantly, the EKB model's universal structure makes it adaptable to various national contexts, allowing country-specific factors to be accounted for in analyzing consumer decision-making in the energy transition.

The EKB Model as a Framework for Evaluating Consumer Behavior

In 1968, Engel, Kollat, and Blackwell proposed a model of consumer decision-making. This conceptual framework has since undergone numerous modifications to align with evolving research contexts (Chae et al., 2006). The EKB model remains one of the foundational theories in the field of consumer behavior (Ashman et al., 2015). Its versatility and adaptability to diverse contexts make it an invaluable research tool, enabling the analysis of complex decision-making processes (Kaur et al., 2024).

The EKB model describes the consumer decision-making process, illustrating how decisions are made among available alternatives. Due to its structure and clarity, this model is regarded as one of the most significant achievements in the study of consumer behavior (Schiffman & Kanuk, 2008). It serves as both a theoretical foundation for analysis and a practical tool supporting researchers and practitioners in understanding the mechanisms behind consumer choices.

Value, as a subjective concept, implies that the same product or service can hold or have a different value for different consumers. Therefore, flexibility in creating a research framework like the EKB model is crucial. Various goods fulfilling the same need may be evaluated differently depending on individual preferences and the consumer's context.

Moreover, even the same individual may make different purchasing decisions depending on the situation, as consumer choices are influenced by numerous factors simultaneously (Oliveira & De Muylder, 2012). The EKB model accommodates these variables, allowing for the analysis of factors affecting consumer decisions at every stage of the decision-making process.

The EKB model emphasizes the complexity of decision-making, presenting the consumer as a rational decision-maker who systematically processes and evaluates information to maximize benefits. The model assumes that consumers proceed through an orderly sequence of stages, including need recognition (also called problem recognition), information search, evaluation of alternatives, purchase decision, and post-purchase evaluation (Kubiczek et al., 2024).

A graphical representation of the EKB model, illustrating the individual stages and their interconnections, is presented in Fig. 17.1. This visualization facilitates our understanding of the flow of the decision-making process and highlights the analytical potential of the model, which can be applied in a variety of research contexts.

The first stage in the decision-making process is the recognition of a need or problem. This is followed by a search for information about solutions, which involves gathering relevant information from various sources. The third stage is the evaluation of alternatives, which is carried out according to the consumer's personal criteria, leading, then, to the establishment of preferences. Once a decision is made, the consumer moves on to the fourth stage, which is the purchase decision. The final step is the post-purchase evaluation. The EKB model assumes that the process begins with a state of unmet needs and desires (Osei & Abenyin, 2016).

Adopting the EKB model provides a systematic and comprehensive approach that facilitates our understanding of how consumers engage (with what), acquire knowledge, and exhibit specific behaviors, even in virtual environments (Kaur et al., 2024). As Sihi (2018) highlights,

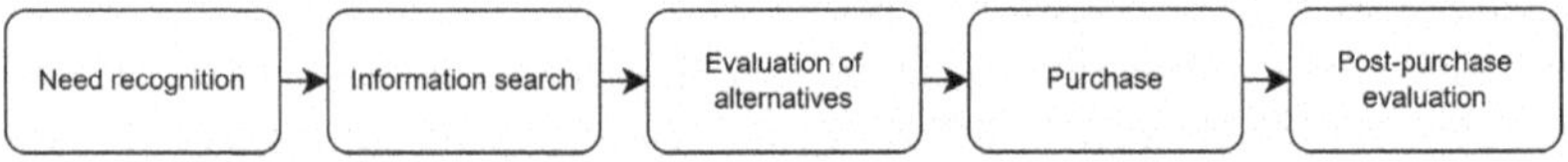

Fig. 17.1 Stages in the EKB model (*Source* Own study based on Darley et al. [2010] and Longart et al. [2016])

this model is particularly useful in addressing contemporary consumer decision-making challenges.

Tuan-Pham and Higgins (2005) made minor modifications to the EKB model, detailing the components of each stage that deserve attention during analyses using this framework. These suggestions, presented in Table 17.1, enable the further development of the model and its application to specific decision-making issues. This flexibility ensures that the EKB model remains a universal research tool adaptable to evolving contexts and research needs.

This detailed breakdown enhances the EKB model by specifying the components involved at each stage, allowing researchers and practitioners to better analyze consumer decision-making processes. These elements provide a nuanced understanding of how needs are recognized, information is gathered, and decisions are made and evaluated, offering insights into both the emotional and cognitive aspects of consumer behavior.

In the considered context, it is worth emphasizing that a significant portion of the cognitive and physical effort associated with purchasing products takes place even before the actual purchasing behavior occurs (Wilkie, 1994). For this reason, it is crucial to understand how consumers approach decision-making (Chen & Dubinsky, 2003). For instance, high-involvement consumers actively gather and analyze information

Table 17.1 Components of the stages in the EKB model

Stage	*Components*
Need recognition	Experience, endogenous activation
Information search	Extent, internal vs. external, content, attribute vs. alternative-based, global vs. local
Consideration set formation	Size, composition, criteria order, construction process
Evaluation of alternatives	Sensitivity to content, strategy, endogenous activation
Purchase (choice)	Rules, status quo/default/deferral, risk-taking context effects/variety-seeking
Post-purchase evaluation	Satisfaction/dissatisfaction intensity, emotional responses, dissonance/regret, process satisfaction

Note 1 In the presented example, the EKB model consists of six stages. However, consideration set formation can be treated as the initial component of the evaluation of alternatives stage
Source Own study based on Tuan-Pham and Higgins (2005)

about products or services of interest to them, while low-involvement consumers do not engage in such activities (Hong, 2015).

The decision regarding energy transition is unique because it requires a high level of consumer involvement, both cognitively and practically, as it often entails evaluating complex technical options (e.g., choosing between different types of photovoltaic panels or heat pumps), comparing financial offers, and navigating regulatory requirements. Moreover, household members must actively coordinate the installation process and assess the potential impact on their daily routines.

In the context of energy transition, consumers must consider both the short-term and long-term effects of their decisions. Short-term consequences include high initial costs—renewable energy technologies are often only partially subsidized, and in some countries financial support may be entirely unavailable. Additionally, implementing such a change often involves time-consuming renovations or requires the construction of new installations, which may pose a barrier for many households.

On the other hand, the long-term benefits for the consumer are significant. By producing renewable energy, consumers can generate electricity without negatively impacting the environment, thereby contributing to global efforts toward sustainable development. Moreover, investing in clean energy can prove financially advantageous for households, as self-generated renewable energy—often cheaper or even free—can eventually offset the initial costs. Consequently, the entire investment has the potential to deliver returns for the consumer both environmentally and economically.

Need Recognition

The consumer decision-making process begins with the recognition of a need, which occurs when a discrepancy between the current state and the desired state is noticed or acknowledged. In the context of energy transition, this need can stem from two primary reasons:

1. **The Need for Energy Consumption as an Essential Good**

   Electricity is indispensable for the functioning of modern households, powering appliances such as refrigerators, induction cooktops, microwaves, radios, televisions, and phones. Rising electricity prices may prompt actions aimed at reducing consumption, such as unplugging chargers when not in use. However, the ability to limit energy consumption is finite—it is not feasible to completely stop using it. Therefore, individuals unwilling to compromise their current level of consumption often seek alternatives that can reduce costs in the long term. Renewable energy sources present such an alternative, offering lower energy bills over time. Moreover, the search for a new energy source may also be driven by dissatisfaction with the current electricity provider, for reasons such as high energy rates or issues with the reliability of energy supply.

2. **The Desire to Reduce Negative Environmental Impact**
   The second reason for the need for energy transition is an intrinsic motivation to minimize one's negative impact on the environment. Consumers with pro-environmental attitudes may feel internal pressure to opt for energy derived from renewable sources. However, cultivating such attitudes often begins with environmental education, typically acquired at a young age, often in schools. For adult consumers who lack developed ecological awareness, changing habits can be more challenging due to a preference for convenient, often less eco-friendly solutions.

Once consumers recognize this need, they enter the stage of actively seeking information, which may come from various sources. This process enables them to identify available alternatives, assess their advantages and disadvantages, and, in accordance with the EKB model, take informed steps toward fulfilling the identified need.

Information Search

The energy transition process is exceptionally complex, consisting of numerous stages and elements that require consumers to make thoughtful decisions at various levels of sophistication. This complexity is especially apparent during the information search stage of the EKB model. Consumers have access to a wide range of information sources, including government programs, energy suppliers, independent experts, online reviews, and recommendations from friends and family. The scope and intensity of their information search vary depending on individual needs, prior knowledge, and personal circumstances.

The energy transition process is demanding not only because of its duration but also due to the broad range of decisions that must be made.

Technical infrastructure and available technologies that can be implemented in households play a significant role. Technical conditions, such as the type of building or existing installations, often necessitate substantial modifications, and not all technologies prove suitable in every case. Moreover, technically optimal solutions often entail the highest investment costs. Consumers frequently require detailed information on integrating new technologies with the existing infrastructure. For example, they may wonder whether photovoltaic panels will be compatible with their current heating system, what changes need to be made to the electrical installation, or whether additional devices, such as energy storage batteries, will be necessary.

The energy transition also necessitates the consideration of legal and regulatory aspects, which can significantly impact the entire process. Consumers often seek information on required permits, technical standards, and procedures related to connecting to the power grid. Regulations concerning tax reliefs and other financial benefits from investing in renewable energy sources are also crucial. For many, finding specialized companies that handle the installation and maintenance of equipment is equally important. In this context, consumers gather offers to obtain information about costs or warranty durations.

From an economic perspective, consumers often look for information on financing options for investments. These may include municipal or government support programs, typically available on official websites or through other information channels. The imitation effect is also a significant factor influencing consumer decisions. Many individuals learn about energy transition opportunities from neighbors or family members who have already implemented such solutions and can share their experiences.

Considering the energy transition process within the EKB model framework requires a comprehensive analysis that considers the diversity of information sought and its sources. The varied levels of consumer knowledge play a key role here. For instance, individuals with technical education, such as environmental engineering, often possess greater knowledge in this area and may not feel the need to explore technological possibilities in detail. This diversity in knowledge and information needs is fundamental to fully understanding consumer behavior in the context of energy transition.

It is also worth noting that the consumer decision-making process can be interrupted at any stage if the acquired information leads to the conclusion that none of the available solutions meet their expectations or financial capabilities. For example, a consumer with sufficient data may conclude that the available technologies are too expensive relative to their budget, resulting in the decision to forego the switch.

Evaluation of Alternatives

Based on the gathered information, consumers compare technological options and the offers of companies providing energy transition services. This process typically begins with an assessment of financial capabilities, narrowing the selection to solutions that fit within available budgetary resources. In this analysis, information about potential funding sources plays a crucial role, including subsidies, tax incentives, municipal or government programs, and leasing options or loans dedicated to renewable energy projects. When full financing is available through support programs, consumers may be more inclined to choose technological solutions with higher initial costs that promise greater efficiency or savings in the long term.

After limiting the options to those meeting the financial criteria, consumers proceed to analyze other aspects. Key criteria include the quality of the technology and its compatibility with the existing infrastructure in the household. It is essential that the chosen solution is tailored to specific conditions, such as the type of building, installation size, or individual energy needs. Consumers may also compare the energy efficiency of devices, their durability, and the availability of technical support.

A critical element of the decision-making process is analyzing reviews of companies offering installation and maintenance services. In this context, consumers often rely on recommendations from friends or family as well as online reviews. The selection criteria include service costs, the quality of execution, warranty duration, availability of technical support, and the professionalism and experience of the contractors. Choosing a reliable provider is crucial, as it affects the system's performance and the user's satisfaction level.

Another important factor influencing consumer decisions is the project implementation time. Many people prefer solutions that can be implemented relatively quickly, especially when changes to the energy infrastructure are urgently needed due to issues such as system failures or rising energy costs.

In the context of the EKB model, understanding how consumers search for, process, and utilize information in the decision-making process regarding energy transition is crucial. This can be analyzed by identifying information sources and evaluating selection criteria.

Purchase

The purchase phase represents the culmination of the consumer decision-making process, where all prior analyses, comparisons, and gathered information translate into a specific decision and its execution. In the context of energy transition, this phase involves selecting the appropriate technology, choosing a service provider, and completing all related formalities. At this stage, the consumer decides to purchase a specific product, such as photovoltaic panels, heat pumps, or energy storage systems, taking into account criteria such as cost, efficiency, durability, and compatibility with the existing infrastructure. The choice of the company responsible for delivering and installing the technology also plays a crucial role.

This stage also requires completing formalities related to financing the investment, which often includes utilizing grants, tax incentives, or preferential loans. The process necessitates that the consumer provides appropriate technical documentation and cost estimates, and meet the requirements specified by supporting institutions. Meeting these formalities can sometimes be time-consuming and challenging for consumers, particularly in the case of more complex projects. Once all the necessary financial procedures are completed, the consumer finalizes the purchase by making a payment for the selected technology or service. This payment could be a one-time transaction or installment-based, depending on the chosen financing method.

The purchase phase also involves planning and agreeing on the installation schedule and any additional modernization actions, such as adapting the electrical installation or thermomodernization of the building. Although the actual installation is part of the post-purchase phase, these details are often discussed and determined during the purchase stage. This is the moment when previous preparations and analyses materialize into concrete action, and the decisions made by the consumer have a direct impact on the outcome of the energy transition process.

This phase is, therefore, an important stage not only from the consumer's perspective but also in the context of research on consumer behavior, as it can be analyzed in terms of the decisions made and how purchases are finalized within the framework of the energy transition.

Post-Purchase Evaluation

The post-purchase evaluation phase involves assessing the consumer's experience after making a purchase and implementing the chosen solution. In the context of energy transition, this is the stage where consumers compare their expectations with the actual performance of the purchased technologies and services. This evaluation is highly significant both for consumer satisfaction and for recommendations to other potential users.

During this phase, consumers analyze the effectiveness of the purchased solution and its alignment with initial expectations, including the manufacturer's promises regarding energy production efficiency. For instance, in the case of photovoltaic panel installations, consumers might verify whether the savings on energy bills meet their forecasts. Similarly, the performance of heat pumps or energy storage systems might be evaluated in terms of efficiency, reliability, and ease of use. If issues arise with the operation of renewable energy infrastructure, an important component of post-purchase evaluation is also assessing satisfaction with after-sales services, such as technical support, response times to complaints, or company assistance.

When the outcomes fall short of expectations, consumers may experience post-purchase dissonance, often accompanied by feelings of dissatisfaction, financial loss, or the need for additional investments to improve the functionality of the purchased solution.

Analyzing this phase within the EKB model framework enables a deeper understanding of how consumers perceive the value of their decisions and what factors influence their satisfaction or dissatisfaction. Energy transition research based on the EKB approach should focus on aspects such as the level of satisfaction, frequency of reported issues, ways of addressing them, and the impact of consumer opinions on the market and company actions.

Summary

This chapter highlights the significant potential of the EKB model as both a theoretical and practical framework for understanding consumer decision-making in the context of the energy transition. By dissecting each stage of the decision-making process, it provides valuable insights into the motivations, barriers, and cognitive mechanisms that shape consumer choices regarding renewable energy adoption.

The EKB model offers a flexible and universal tool for analyzing consumer behavior, with broad applications across various research areas—from everyday shopping decisions to choices involving high emotional and financial engagement. In the context of energy transition, its application requires adaptation to the unique nature of this process, which is exceptionally complex and multifaceted. Energy transition necessitates the consideration of diverse factors such as environmental awareness, economic aspects, and the availability of technology.

By dividing the decision-making process into stages (need recognition, information search, evaluation of alternatives, purchase decision, and post-purchase evaluation), the EKB model enables the analysis of each stage separately. This structure allows a complex process to be broken down into smaller, more manageable components, thereby increasing the precision of analyses.

Applying this model to consumer behavior in the context of energy transition provides a comprehensive view of consumers' motivations, barriers, and decision-making processes. This facilitates a deeper understanding of why consumers engage—or fail to engage—in specific pro-environmental actions, which can, in turn, serve as a basis for developing conclusions and practical recommendations. Consequently, the EKB model functions not only as a research tool fulfilling cognitive objectives but also as a practical instrument supporting the development of policies and initiatives aimed at a sustainable future.

References

Ashman, R., Solomon, M. R., & Wolny, J. (2015). An old model for a new age: Consumer decision making in participatory digital culture. *Journal of Customer Behaviour, 14*(2), 127–146. https://doi.org/10.1362/147539215X14373846805743

Chae, M., Black, C., & Heitmeyer, J. (2006). Pre-purchase and post-purchase satisfaction and fashion involvement of female tennis wear consumers. *International Journal of Consumer Studies, 30*(1), 25–33. https://doi.org/10.1111/j.1470-6431.2005.00434.x

Chen, Z., & Dubinsky, A. J. (2003). A conceptual model of perceived customer value in e-commerce: A preliminary investigation. *Psychology & Marketing, 20*(4), 323–347. https://doi.org/10.1002/mar.10076

Darley, W. K., Blankson, C., & Luethge, D. J. (2010). Toward an integrated framework for online consumer behavior and decision making process: A review. *Psychology & Marketing, 27*(2), 94–116. https://doi.org/10.1002/mar.20322

Dimitriou, C. K., & AbouElgheit, E. (2019). Understanding generation z's travel social decision-making. *Tourism and Hospitality Management, 25*(2), 311–334. https://doi.org/10.20867/thm.25.2.4

Engel, J. F., Kollat, D. T., & Blackwell, R. D. (1968). *Consumer behaviour*. Rinehart & Winston.

Gushterov, N. (2018). Modelling consumer behaviour on the energy market. *Икономика, 21*(8(1-ENG)), 25–43.

Hong, I. B. (2015). Understanding the consumer's online merchant selection process: The roles of product involvement, perceived risk, and trust expectation. *International Journal of Information Management, 35*(3), 322–336. https://doi.org/10.1016/j.ijinfomgt.2015.01.003

Howard, J. A., & Sheth, J. N. (1969). *The theory of buyer behavior*. John Wiley & Sons.

Joshi, Y., & Rahman, Z. (2017). Investigating the determinants of consumers' sustainable purchase behaviour. *Sustainable Production and Consumption, 10*, 110–120. https://doi.org/10.1016/j.spc.2017.02.002

Kaur, J., Mogaji, E., Paliwal, M., Jha, S., Agarwal, S., & Mogaji, S. A. (2024). Consumer behavior in the metaverse. *Journal of Consumer Behaviour, 23*(4), 1720–1738. https://doi.org/10.1002/cb.2298

Kubiczek, J., Hadasik, B., Krawczyńska, D., Przedworska, K., Madarász, E. Z., & Ryczko, A. (2024). Perspective of created value in consumer choice: Comparison of economic and ecological dimensions. *SAGE Open, 14*(1), 21582440241238516. https://doi.org/10.1177/21582440241238516

Longart, P., Wickens, E., & Bakir, A. (2016). Consumer decision process in restaurant selection: An application of the stylized EKB model. *Market-Tržište, 28*(2), 173–190. https://doi.org/10.22598/mt/2016.28.2.173

McCarthy, E. J., Perreault, W. D., & Quester, P. G. (1997). *Basic marketing: A managerial approach* (2nd Australasian ed.). Irwin.

Nicosia, F. M. (1966). *Consumer decision processes: Marketing and advertising implications*. Prentice-Hall.

Oliveira, W. A. D., & De Muylder, C. F. (2012). Value creation from organizational project management: A case study in a government agency. *Journal of Information Systems and Technology Management, 9*(3), 497–514. https://doi.org/10.4301/s1807-77520120003000004

Osei, B. A., & Abenyin, A. N. (2016). Applying the Engell–Kollat–Blackwell model in understanding international tourists' use of social media for travel decision to Ghana. *Information Technology & Tourism, 16*(3), 265–284. https://doi.org/10.1007/s40558-016-0055-2

Paswan, A., Guzmán, F., & Lewin, J. (2017). Attitudinal determinants of environmentally sustainable behavior. *Journal of Consumer Marketing, 34*(5), 414–426. https://doi.org/10.1108/jcm-02-2016-1706

Schiffman, L. G., & Kanuk, L. L. (2008). *Consumer behavior*. Pearson International Edition.

Sihi, D. (2018). Home sweet virtual home: The use of virtual and augmented reality technologies in high involvement purchase decisions. *Journal of Research in Interactive Marketing, 12*(4), 398–417. https://doi.org/10.1108/JRIM-01-2018-0019

Tuan-Pham, M., & Higgins, E. T. (2005). Promotion and prevention in consumer—The state of the art and theoretical propositions. In S. Ratneshwar & D. G. Mick (Eds.), *Inside consumption—Consumer motives, goals and desires* (pp. 8–43). Routledge.

Wilkie, W. L. (1994). *Consumer behavior*. Van Hoffman Press.

Zhang, K. Z. K., & Benyoucef, M. (2016). Consumer behavior in social commerce: A literature review. *Decision Support Systems, 86*, 95–108. https://doi.org/10.1016/j.dss.2016.04.001

Zhang, X., & Dong, F. (2020). Why do consumers make green purchase decisions? Insights from a systematic review. *International Journal of Environmental Research and Public Health, 17*(18), 6607. https://doi.org/10.3390/ijerph17186607

Index

T. Walker et al. (eds.), *Wiring the Future*, Palgrave Studies in Emerging Risk Management and Sustainable Finance,
https://doi.org/10.1007/978-3-032-14717-2

The manufacturer's authorised representative in the EU is Springer Nature Customer Service Centre GmbH, Europaplatz 3, 69115 Heidelberg, Germany. If you have any concerns regarding our products, please contact ProductSafety@springernature.com

Printed and bound by CPI Group (UK) Ltd, Croydon, CR0 4YY
21/09/2026
02219963-0005